IEEE Recommended Practice for Industrial and Commercial Power Systems Analysis

Published by
The Institute of Electrical and Electronics Engineers, Inc.

IEEE

Recognized as an
American National Standard (ANSI)

**IEEE
Std 399-1990**

IEEE Recommended Practice for Industrial and Commercial Power Systems Analysis

Sponsor

**Power Systems Engineering Committee
of the
IEEE Industry Applications Society**

Approved May 31, 1990

IEEE Standards Board

Approved October 12, 1990

American National Standards Institute

Abstract: IEEE Std 399-1990, *IEEE Recommended Practice for Industrial and Commercial Power Systems Analysis* (ANSI), is intended as a practical, general treatise for engineers on power system analysis theory and as a reference work on the analytical techniques that are most commonly applied to the computer-aided analysis of electric power systems in industrial plants and commercial buildings.
Keywords: Industrial and commercial power systems analysis, power system analysis theory, power system studies.

Third Printing
March 1996

ISBN 1-55937-044-0

Library of Congress Catalog Number 90-085501

December 15, 1990   *SH13433*

Foreword

(This Foreword is not a part of IEEE Std 399-1990, IEEE Recommended Practice for Industrial and Commercial Power Systems Analysis.)

This Recommended Practice is intended as a practical, general treatise for engineers on power system analysis theory and as a reference work on the analytical techniques that are most commonly applied to the computer-aided analysis of electric power systems in industrial plants and commercial buildings. The "Brown Book" should be considered a useful supplement to several other texts on the subject of power system analysis which appear in the references and bibliography sections of various individual chapters of this book. The Brown Book is both complementary and supplementary to the rest of the Color Book Series.

Two new and important chapters have been added: Chapter 13, Cable Ampacity Studies, and Chapter 15, Coordination Studies. The Computer Services chapter (Chapter 13 in IEEE Std 399-1980 (ANSI)) has been revamped with significant new material added and has become Chapter 5, Computer Solutions and Systems, in the 1990 edition. All the other chapters in this new 1990 edition have been revised and updated, in some cases quite substantially, to reflect current technology.

To many of the members of the working group who wrote and developed this Recommended Practice, the Brown Book has become a true labor of love. The dedication and support of each individual member is clearly evident in every chapter of the Brown Book. These individuals deserve our many thanks for their excellent contributions.

At the time it recommended these practices, the working group of the Power Systems Analysis Subcommittee of the Power Systems Engineering Committee of the IEEE Industry Applications Society had the following members:

M. Shan Griffith, *Chair and Technical Editor*
Douglas W. Durand, *Assistant Technical Editor*

R. M. Bucci
L. E. Crawford
J. W. Feltes
Damian Gonzalez
Alton Knight
Clifton A. LaPlatney
Richard H. McFadden

Marco W. Migliaro
Philip Nobile
A. D. Patton
Carlos Pinheiro
A. E. Ronat
David Shipp

Farrokh Shokooh
David Smith
R. P. Stratford
George A. Terry
George W. Walsh
Claus Wiig
Glenn E. Word

The following members of the IEEE Power Analysis Subcommitee balloted and approved this Recommended Practice for submission to the IEEE Standards Board:

IEEE Recommended Practice for Industrial and Commercial Power Systems Analysis

Working Group Members and Contributors

M. Shan Griffith, *Working Group Chair*

Chapter 1 — Introduction: George A. Terry, *Chair*; J. W. Feltes, Claus Wiig

Chapter 2 — Applications of Power System Analysis: Damian Gonzalez, *Chair*

Chapter 3 — Analytical Procedures: M. Shan Griffith, *Chair*; Douglas W. Durand, Carlos Pinheiro

Chapter 4 — System Modeling: Marco W. Migliaro, *Chair*; J. W. Feltes, Claus Wiig

Chapter 5 — Computer Solutions and Systems: Glenn W. Word, *Chair*

Chapter 6 — Load Flow Studies: J. W. Feltes, *Chair*

Chapter 7 — Short-Circuit Studies: David H. Smith, *Chair*; Claus Wiig

Chapter 8 — Stability Studies: Richard H. McFadden, *Chair*

Chapter 9 — Motor Starting Studies: M. Shan Griffith, *Chair*; Douglas W. Durand, A. J. Williams

Chapter 10 — Harmonic Analysis Studies: David Shipp, *Chair*; Walter Chumakov, R. P. Stratford, Claus Wiig

Chapter 11 — Switching Transient Studies: Carlos Pinheiro, *Chair*; George W. Walsh

Chapter 12 — Reliability Studies: A. D. Patton and C. R. Heising, *Co-Chairs*

Chapter 13 — Cable Ampacity Studies: Farrokh Shokooh, *Chair*; M. Shan Griffith, Glenn E. Word

Chapter 14 — Ground Mat Studies: L. E. Crawford and M. Shan Griffith, *Co-Chairs*

Chapter 15 — Coordination Studies: A. E. (Elizabeth) Ronat, *Chair*; Alton Knight, J. D. Langhans

Contents

Chapter 1
Introduction

1.1 General Discussion. IEEE Std 399-1990, IEEE Recommended Practice for Industrial and Commercial Power System Analysis, commonly known as the IEEE Brown Book, is published by the Institute of Electrical and Electronics Engineers (IEEE) as a reference source to give plant engineers a better understanding of the purpose for and techniques involved in power system studies. The IEEE Brown Book can also be helpful in system and data acquisition for an engineering consultant to perform necessary engineering studies prior to designing a new system or expanding an existing power system. Such information will help ensure high standards of power system reliability and maximize the utilization of capital investment.

The IEEE Brown Book has been prepared on a voluntary basis by engineers and designers functioning as a Working Group within the IEEE, under the Industrial and Commercial Power Systems Department of the Industry Applications Society. This Recommended Practice is not intended as a replacement for the many excellent texts available in this field. The IEEE Brown Book complements the other IEEE color book standards, and it emphasizes up-to-date techniques in system studies that are most applicable to industrial and commercial power systems.

1.2 History of Power System Studies. The planning, design, and operation of a power system requires continual and comprehensive analyses to evaluate current system performance and to establish the effectiveness of alternative plans for system expansion.

The computational work to determine power flows and voltage levels resulting from a single operating condition for even a small network is all but insurmountable if performed by manual methods. The need for computational aids led to the design of a special purpose analog computer (ac network analyzer) as early as 1929. It provided the ability to determine flows and voltages during normal and emergency conditions and to study the transient behavior of the system resulting from fault conditions and switching operations.

The earliest application of digital computers to power system problems dates back to the late 1940's. Most of the early applications were limited in scope because of the small capacity of the punched card calculators in use during that period. Large-scale digital computers became available in the mid-1950's, and the initial success of load flow programs led to the development of programs for short-circuit and stability calculations.

Today, the digital computer is an indispensable tool in power system planning where it is necessary to predict future growth and simulate day-to-day operations over periods of 20 years or more.

1.3 Applying Power System Analysis Techniques to Industrial and Commercial Power Systems.

As computer technology has advanced, so has the complexity of industrial and commercial power systems. These power systems have grown in recent decades with capacities far exceeding that of a small electric utility system.

Today's intensely competitive business environment forces plant or building management personnel to be very aware of the total owning cost of the power distribution system. Therefore, they demand assurances of maximum return on all capital investments in the power system. The use of digital computers makes it possible to study the performance of proposed and actual systems under many operating conditions. Answers to many questions regarding impact of expansion on the system, short-circuit capacity, stability, load distribution, etc., can be intelligently and economically obtained.

1.4 Purposes of This Recommended Practice

1.4.1 Why a Study? As is stated in Chapter 2, the planning, design, and operation of industrial and/or commercial power systems require several studies to assist in the evaluation of the initial and future system performance, system reliability, safety, and the ability to grow with production and/or operating requirements. The studies most likely to be needed are load flow studies, cable ampacity studies, short-circuit studies, coordination studies, stability studies, and routine motor starting studies. A brief summary of the above studies and a discussion pertaining to data collection and preparation of the data files required to perform the desired study on a digital computer are presented in Chapter 2.

Additional studies relating to switching transients, reliability, grounding, harmonics, and special motor starting considerations may also be required. The engineer in charge of system design must decide which studies are needed to ensure that the system will operate safely, economically, and efficiently over the expected life of the system.

1.4.2 How to Prepare for a Power System Study. For a plant engineer to solve a power system analysis problem, he or she must be thoroughly familiar with the fundamentals of power electrical engineering. He or she can then analyze the problem, prepare the necessary equivalent circuits and obtain appropriate system data before using a computer program to perform repetitive calculations. Failure to use a valid analytical procedure to establish a sound basic approach to the problem could lead to disastrous consequences in both the design and operation of a system.

Furthermore, a basic understanding of power engineering is essential to correctly interpret the results of computer calculations. It is important to emphasize the need for a thorough foundation and base of experience in power system engineering in addition to modern, effective computer software.

Chapter 3 offers an excellent review of the most essential fundamentals in a system study.

To set up a computer program for system analysis, certain basic data must be gathered with accuracy and proper presentation. The extent of system representation, restrictions in terms of nodes (buses) and branches (lines and transformers), balanced three-phase network and single-line diagram, impedance diagram, etc., are all important inputs to a meaningful system study.

Chapter 4 deals with system modeling and data requirements to illustrate how these basic inputs for a study can be properly prepared or organized.

Once the basic preparations are completed, the next step is to look for an actual computer program. The computer industry has evolved rapidly and, today, economical personal computers (PC's) may be purchased and used by consultants, equipment manufacturers, and plant engineers. Programs are available—written for personal computers—which allow in-house calculation of the studies outlined in this standard. These programs are user friendly and provide a "menu" that directs even the inexperienced engineer through the process and ultimately to the completion of most studies.

Chapter 5 discusses the basic computation methods, various types of computer systems and their requirements, and the availability of commercial computing services and their capabilities.

1.4.3 The Most Common System Studies. For the engineer, the following sections address the most common studies for the design or operation of an industrial or commercial power system:

Chapter 6, Load Flow Studies
Chapter 7, Short-Circuit Studies
Chapter 8, Stability Studies
Chapter 9, Motor Starting Studies
Chapter 10, Harmonic Analysis Studies
Chapter 11, Switching Transient Studies
Chapter 12, Reliability Studies
Chapter 13, Cable Ampacity Studies
Chapter 14, Ground Mat Studies
Chapter 15, Coordination Studies

The purpose of each study and what can be achieved by it are briefly explained.

Figure 1 is a typical composite single-line diagram for a large industrial power system that may be used for modeling in Chapters 6 through 15. Bus and line number designations have been shown on the diagram in a manner to be discussed in Chapter 4.

After studying these chapters, an engineer should be better equipped to prepare necessary data and criteria for a specific computer study. The study can be performed in-house or by an outside consultant. There is a growing number of consulting firms that specialize in performing system studies at a reasonable cost.

Studying these chapters will provide the basic understanding of the studies needed to coordinate the data and criteria for specific studies and will also serve as a reference to those analysts for whom studies are a principal activity.

1.5 References. The following standards were used as references in the preparation of this standard:

[1] IEEE Std 141-1986, IEEE Recommended Practice for Electric Power Distribution for Industrial Plants (ANSI).[1]

[2] IEEE Std 142-1982, IEEE Recommended Practice for Grounding of Industrial and Commercial Power Systems (ANSI).

[3] IEEE Std 241-1983, IEEE Recommended Practice for Electric Power Systems in Commercial Buildings (ANSI).

[4] IEEE Std 242-1986, IEEE Recommended Practice for Protection and Coordination of Industrial and Commercial Power Systems (ANSI).

---

[1] IEEE publications are available from the Institute of Electrical and Electronics Engineers, IEEE Service Center, 445 Hoes Lane, P.O. Box 1331, Piscataway, NJ 08855-1331.

Chapter 2
Applications of Power System Analysis

2.1 Introduction. The planning, design, and operation of industrial power systems require engineering studies to evaluate existing and proposed system performance, reliability, safety, and economics. Studies, properly conceived and conducted, are a cost-effective way to prevent surprises and to optimize equipment selection. In the design stage, the studies identify and avoid potential deficiencies in the system before it goes into operation. In existing systems, the studies locate the cause of equipment failure and misoperation and determine corrective measures for improving system performance.

The complexity of modern industrial power systems make studies difficult, tedious, and time consuming to perform manually. The computational tasks associated with power systems studies have been greatly simplified by the use of digital computer programs. Sometimes, economics and study requirements dictate the use of an analog computer — a transient network analyzer (TNA) — which provides a scale model of the power system.

2.1.1 Digital Computer. The digital computer offers engineers a powerful tool to perform efficient system studies. Computers permit optimal designs at minimum costs, regardless of system complexity. Advances in computer technology, like the introduction of the personal computer with its excellent graphics capabilities, have not only reduced the computing costs but also the engineering time needed to use the programs. Study work formerly done by outside consultants can now be perormed in-house. User-friendly programs utilizing interactive menus and online help facilities guide the engineer through the task of using a digital computer program.

2.1.2 Transient Network Analyzer (TNA). The TNA is a useful tool for transient overvoltage studies. The use of microcomputers to control and acquire the data from the TNA allows the incorporation of probability and statistics in switching surge analysis. One of the major advantages of the TNA is that it allows for quick reconfiguration of complex systems with immediate results, avoiding the relatively longer time associated with running digital computer programs for these systems.

2.2 Load Flow Analysis. Load flow studies determine the voltage, current, active, and reactive power and power factor in a power system. Load flow studies are an excellent tool for system planning. A number of operating procedures can be analyzed, including contingency conditions, such as the loss of a generator, a transmission line, a transformer, or a load. These studies will alert the user to conditions that may cause equipment overloads or poor voltage levels. Load flow studies can be used to determine the optimum size and location of capacitors for power factor improvement. Also, they are very useful in determining system voltages under conditions of suddenly applied or disconnected loads. The results of a load flow study are also starting points for stability studies. Digital computers are used extensively in load flow studies due to the complexity of the calculations involved.

2.3 Short-Circuit Analysis. Short-circuit studies are done to determine the magnitude of the currents flowing throughout the power system at various time intervals after a fault occurs. The magnitude of the currents flowing through the power system after a fault vary with time until they reach a steady-state condition. This behavior is due to system characteristics and dynamics. During this time, the protective system is called on to detect, interrupt, and isolate these faults. The duty imposed on this equipment is dependent upon the magnitude of the current, which is dependent on the time from fault inception. This is done for various types of faults (3 phase, phase to phase, double phase to ground, and phase to ground) at different locations throughout the system. The information is used to select fuses, breakers, and switchgear ratings in addition to setting protective relays.

Modeling is done using symmetrical components, a powerful tool for analyzing unsymmetrical faults and mutual coupling. Digital computer programs using symmetrical components are especially helpful for analyzing large and complex systems, such as industrial power system networks.

2.4 Stability Analysis. The ability of a power system, containing two or more synchronous machines, to continue to operate after a change occurs on the system is a measure of its stability. The stability problem takes two forms: steady state and transient. *Steady-state* stability may be defined as the ability of a power system to maintain synchronism between machines within the system following relatively slow load changes. *Transient* stability is the ability of the system to remain in synchronism under transient conditions, i.e., faults, switching operations, etc.

In an industrial power system, stability may involve the power company system and one or more in-plant generators or synchronous motors. Contingencies, such as load rejection, sudden loss of a generator or utility tie, starting of large motors or faults (and their duration), have a direct impact on system stability. Load-shedding schemes and critical fault-clearing times can be determined in order to select the proper settings for protective relays.

These types of studies are probably the single most complex ones done on a power system. A simulation will include synchronous generator models with their controls, i.e., voltage regulators, excitation systems, and governors. Motors are sometimes represented by their dynamic characteristics as are static VAR compensators and protective relays.

2.5 Motor Starting Analysis. The starting current of most ac motors is several times normal full load current. Both synchronous and induction motors can draw five to ten times full load current when starting them across the line. Motor starting torque varies directly as the square of the applied voltage. If the terminal voltage drop is excessive, the motor may not have enough starting torque to accelerate up to running speed. Running motors may stall from excessive voltage drops, or undervoltage relays may operate. In addition, if the motors are started frequently, the voltage dip at the source may cause objectionable flicker in the lighting system.

By using motor starting study techniques, these problems can be predicted before the installation of the motor. If a starting device is needed, its characteristics and ratings can be easily determined. A typical digital computer program will calculate speed, slip, electrical output torque, load current, and terminal voltage data at discrete time intervals from locked rotor to full load speed. Also, voltage at important locations throughout the system during startup can be monitored. The study can help select the best method of starting, the proper motor design, or the required system design for minimizing the impact of motor starting on the entire system.

2.6 Harmonic Analysis. A harmonic-producing load can affect other loads if significant voltage distortion is caused. The voltage distortion caused by the harmonic-producing load is a function of both the system impedance and the amount of harmonic current injected. The mere fact that a given load current is distorted does not always mean there will be undue adverse effects on other power consumers. If the system impedance is low, the voltage distortion is usually negligible in the absence of harmonic resonance. However, if harmonic resonance prevails, intolerable harmonic duties are likely to result.

Some of the primary effects of voltage distortion are:
(1) Control/computer system interference
(2) Heating of rotating machinery
(3) Overheating/failure of capacitor banks

When the harmonic currents are high and travel in a path with significant exposure to parallel communication circuits, the principal effect is telephone interference. This problem depends on the physical path of the circuit as well as the frequency and magnitude of the harmonic currents. Harmonic currents also cause additional line losses and additional stray losses in transformers.

Watthour meter error is often a concern. At harmonic frequencies, the meter may register high or low depending on the harmonics present and the response of the meter to these harmonics. Fortunately, the error is usually small.

Analysis is commonly done to predict distortion levels for addition of a new harmonic-producing load or capacitor bank. The general procedure is to first develop a model that can accurately simulate the harmonic response of the present system and then to add a model of the new addition. Analysis is also commonly done to evaluate alternatives for correcting problems found by measurements.

Only very small circuits can be effectively analyzed without a computer program. Typically, a computer program for harmonic analysis will provide the engineer with the capability to compute the frequency response of the power system and to display it in a number of useful graphical forms. The programs provide the capability to predict the actual distortion based on models of converters, arc furnaces, and other nonlinear loads.

2.7 Switching Transients Analysis. Switching transients severe enough to cause problems in industrial power systems are most often associated with inadequate or malfunctioning breakers or switches and the switching of capacitor banks and other frequently switched loads. The arc furnace system is most frequently studied because of its high frequency of switching and the related use of capacitor banks.

By properly using digital computer programs or the TNA, these problems can be detected early in the design stage. In addition to these types of switching transient problems, digital computer programs and the TNA can be used to analyze other system anomalies, such as lightning arrester operation, ferroresonance, virtual current chopping, and breaker transient recovery voltage.

2.8 Reliability Analysis. When comparing various industrial power system design alternatives, acceptable system performance quality factors (including reliability) and cost are essential in selecting an optimum design. A reliability index is the probability that a device will function without failure over a specified time period. This probability is determined by equipment maintenace requirements and failure rates. Using probability and statistical analyses, the reliability of a power system can be studied in depth with digital computer programs.

Reliability is most often expressed as the frequency of interruptions and expected number of hours of interruptions during one year of system operation. Momentary and sustained system interruptions, component failures, and outage rates are used in some reliability programs to compute overall system reliability indexes at any node in the system, and to investigate sensitivity of these indexes to parameter changes. With these results, economics and reliability can be considered to select the optimum power system design.

2.9 Cable Ampacity Analysis. Cable ampacity studies calculate the current-carrying capacity (ampacity) of power cables in underground or above ground installations. This ampacity is determined by the maximum allowable conductor temperature. In turn, this temperature is dependent on the losses in the cable, both I^2R and dielectric, and thermal coupling between heat-producing components and ambient temperature.

The ampacity calculations are extremely complex. This is due to many considerations. The heat transfer through the cable insulation and sheath, and, in the case of underground installations, heat transfer to duct or soil as well as from duct bank to soil are examples of thermal considerations involved. Other considerations include the effects of losses caused by proximity and skin effects. In addition, the cable-shielding system may introduce additional losses depending on the installation. The analysis involves the application of thermal equivalents of Ohm's and Kirchoff's laws to a thermal circuit.

2.10 Ground Mat Analysis. Under ground-fault conditions, the flow of current will result in voltage gradients within and around the substation, not only between structures and nearby earth, but also along the ground surface. In a properly designed system, this gradient should not exceed the limits that can be tolerated by the human body.

The purpose of a ground mat study is to provide for the safety and well-being of anyone that can be exposed to the potential differences that can exist in a station during a severe fault. The general requirements for industrial power system grounding are similar to those of utility systems under similar service conditions. The differences arise from the specific requirements of the manufacturing or process operations.

Some of the factors that are considered in a ground-mat study are:
(1) Fault-current magnitude and duration
(2) Geometry of the grounding system
(3) Soil resistivity
(4) Probability of contact
(5) Human factors such as:
    (a) Body resistance
    (b) Standard assumptions on physical conditions of the individual

2.11 Protective Device Coordination Analysis. The objective of a protection scheme in a power system is to minimize hazards to personnel and equipment while causing the least disruption of power service. Coordination studies are required to select or verify the clearing characteristics of devices such as fuses, circuit breakers, and relays used in the protection scheme. These studies are also needed to determine the protective device settings that will provide selective fault isolation. In a properly coordinated system, a fault results in interruption of only the minimum amount of equipment necessary to isolate the faulted portion of the system. The power supply to loads in the remainder of the system is maintained. The goal is to achieve an optimum balance between equipment protection and selective fault isolation that is consistent with the operating requirements of the overall power system.

Short-circuit calculations are a prerequisite for a coordination study. Short-circuit results establish minimum and maximum current levels at which coordination must be achieved and aids in setting or selecting the devices for adequate protection. Traditionally, the coordination study has been performed graphically by manually plotting time-current operating characteristics of fuses, circuit breaker trip devices, and relays, along with conductor and transformer damage curves — all in series from the fault location to the source. Log-log scales are used to plot time versus current magnitudes. These "coordination curves" show graphically the quality of protection and coordination possible with the equipment available. They also permit the verification/confirmation of protective device characteristics, settings, and ratings to provide a properly coordinated and protected system.

With the advent of the personal computer, the light-table approach to protective device coordination is being replaced by computer programs. Some of the programs utilize graphics to provide a visual representation of the device coordination as it is developed. In the future, computer programs are expected to use expert systems based on practical coordination algorithms to further assist the protection engineer.

Chapter 3
Analytical Procedures

3.1 Introduction. With the development of the digital computer and advanced computer programming techniques, power system problems of the most complex types can be rigorously analyzed. Previously, solutions were usually only approximate and errors were introduced by many simplifying assumptions necessary to permit classical longhand calculating procedures. For progress to be realized in using the computer for power system analysis work, it has been necessary for the specialist involved in the creation of power system analysis computer programs to understand thoroughly the application of basic analytical solution methods that apply. It is also important for those concerned with assembling and preparing data for input to a power system analysis computer program and those interpreting and applying results generated by such a program to understand the application of analytical solution methods.

This section attempts, first, to identify and document the basic analytical solution methods that are valid for determining the voltage and current relationships which exist during various power system network events and operating conditions. Secondly, these basic analytical solution methods will be demonstrated where not otherwise self-evident. Finally, critical restraints that must be respected to avoid serious error in applying analytical solution methods will be discussed.

Whether a power system analysis problem is to be solved directly or by a computer program, proper application of sound analytical solution methods is essential for three reasons. First, accuracy of the solution to each individual problem being considered will be directly affected. Second, and perhaps the most important because of the significant expense involved, accuracy of the solution determines the validity and effectiveness of any remedial measures suggested. Finally, extension of erroneous results to related problems or to what appears to be a trivial modification of the original problem, possibly in combination with other misapplied or misunderstood techniques, can lead to a compounding of initial error and a progression of incorrect conclusions.

The most common causes of errors in circuit analysis work are:
(1) Failure to use a valid analytical procedure because the analyst is unaware of its existence or applicability
(2) Careless or improper use of *cookbook* methods that have neither a factual basis, nor support in the technical literature, nor a valid place in the electrical engineering discipline

(3) Improper use of a valid solution method due to application beyond limiting boundary restraints or in combination with an inaccurate simplifying assumption

Many situations occur in industrial and commercial power systems that illustrate some or all of these common causes of error, as well as the resulting evils. Any problem investigated as a part of the general types of power system analysis studies covered in other sections of this Recommended Practice and described as follows would qualify.

  (1) Short-circuit studies
  (2) Load analysis studies
  (3) Load flow studies
  (4) Stability studies
  (5) Motor starting studies
  (6) Harmonic studies
  (7) Reliability studies
  (8) Ground mat studies
  (9) Switching transient studies

3.2 The Fundamentals. The following identify the more important analytical solution methods that are either available as or are the basis for valid techniques in solving power system network circuit problems.

  (1) Linearity
  (2) Superposition
  (3) The Thevenin Equivalent Circuit
  (4) The Sinusoidal Forcing Function
  (5) The Phasor Representation
  (6) The Fourier Representation
  (7) The Laplace Transform
  (8) The Single-Phase Equivalent Circuit
  (9) The Symmetrical Component Analysis
(10) The Per Unit Method

Rigorous treatment of these analytical techniques is available in several circuit analysis tests [1][2], [2], [5], [7], [8], [9], and is beyond the scope of this discussion. In the following sections, however, a brief qualitative explanation of each principle is presented, along with a review of major benefits and restraints associated with the use of each principle.

3.2.1 Linearity. Probably the simplest concept of all, linearity is also one of the most important because of its influence on the other principles. Linearity is best understood by examination of Fig 2. The simplified network represented by the single-impedance element Z in Fig 2(a) is linear for the chosen excitation and response functions, if a plot of response magnitude (current) versus source excitation magnitude (voltage) is a straight line. This is the situation shown for case A in

---

[2]Numbers in brackets correspond to those in the References at the end of each chapter.

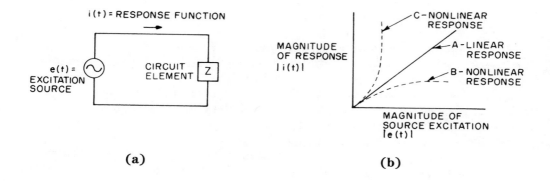

(a)

(b)

**Fig 2
Linearity**

Fig 2(b). When linearity exists, the plot applies either to the steady-state value of the excitation and response functions or to the instantaneous value of the functions at a specific time.

When linear dc circuits are involved, the current doubles if the voltage is doubled. The same holds for linear ac circuits if the frequency of the driving voltage is held constant. In a similar manner, it is possible to predict the response of a *constant impedance* circuit (that is, constant R, L, and C elements) to any magnitude of dc source excitation or *fixed frequency* sinusoidal excitation based on the known response at any other level of excitation. For the chosen excitation function of voltage and the chosen response function of current, both dotted curves B and C are examples of the response characteristic of a nonlinear element.

With the circuit element represented by any of the response curves shown in Fig 2 (including the linear element depicted by curve A), the circuit will, in general, become nonlinear for a different response function, for example, power. If, for example, the element was a constant resistance (which would have a linear voltage-current relationship), the power dissipated would increase by a factor of f if voltage were doubled ($P = I^2R$).

An important limitation of linearity, therefore, is that it applies only to responses that are linear for the circuit conditions described (that is, a constant impedance circuit will yield a current that is linear with voltage). This restraint must be recognized in addition to the previously mentioned limitations of constant source excitation frequency for ac circuits and constant circuit element impedances for ac or dc circuits. Excitation sources, if not independent, must be linearly dependent. This restraint forces a source to behave just as would a linear response (which, by definition, is also linearly dependent).

3.2.2 Superposition. This very powerful principle is a direct consequence of linearity and can be stated as follows:

In any linear network containing several dc or fixed frequency ac excitation sources (voltages), the total response (current) can be calculated by algebraically adding all the individual responses caused by each independent source acting alone, i.e., all other sources inactivated (voltage sources shorted, current sources opened).

An example which illustrates this principle is shown in Fig 3. The equation written is for the sum of the currents from each individual source V_1 and V_2. Although Fig 3 also illustrates a way this principle might actually be used, more often its main application is in support of other calculating methods. The only restraint associated with superposition is that the network should be linear. All limitations associated with linearity apply.

The nonapplicability of superposition is why all but the very simplest nonlinear circuits are almost impossible to analyze using hand calculations. Although most real circuit elements are nonlinear to some extent, they can often be accurately represented by a linear approximation. Solutions to network problems involving such elements can be readily obtained.

Problems involving complex networks having substantially nonlinear elements can practically be solved only through the use of certain simplification procedures, or through adjustment of calculated results to correct for nonlinearity. Both of these approaches can potentially lead to significant inaccuracy. Tiresome iterative calculations performed in an instant by the digital computer make accurate solutions possible when an equation can be written mathematically to describe the nonlinear circuit elements.

3.2.3 The Thevenin Equivalent Circuit. This powerful circuit analysis tool is based on the fact that any active linear network, however complex, can be represented by a single voltage source, V_{OC}, equal to the open-circuit voltage across any two terminals of interest, in series with the equivalent impedance, Z_{EQ}, of the network viewed from the same two terminals with all sources in the network inactivated (voltage sources shorted, current sources opened). Validity of this representation requires only that the network be linear. Existence of linearity is a necessary restraint. Application of the Thevenin equivalent circuit can be appreciated by referring to the simple circuit of Fig 3 and developing the Thevenin equivalent for the network with the switch in the open position as illustrated in Fig 4. After connecting the 6 Ω load to the Thevenin equivalent network by closing the

Fig 3
Superposition

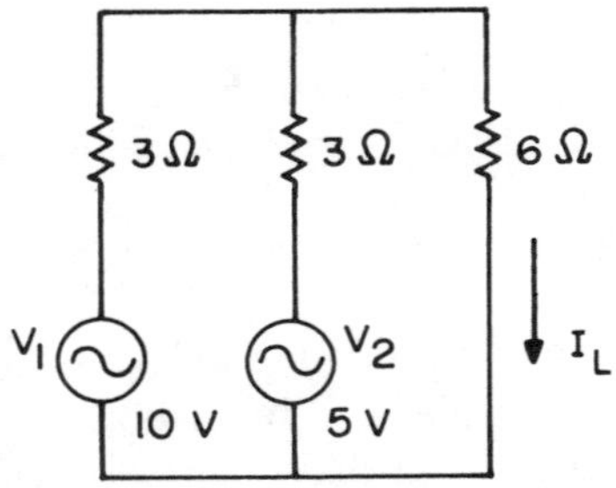

$$I_L = I_{v_1} + I_{v_2}$$

$$= \frac{10}{\left(3+\frac{6\cdot3}{6+3}\right)}\left(\frac{6\cdot3}{6+3}\right)\cdot\frac{1}{6} + \frac{5}{\left(3+\frac{6\cdot3}{6+3}\right)}\left(\frac{6\cdot3}{6+3}\right)\cdot\frac{1}{6}$$

$$= \frac{10}{5}\cdot2\cdot\frac{1}{6} + \frac{5}{5}\cdot2\cdot\frac{1}{6}$$

$$= \frac{2\cdot2}{6} + \frac{2}{6} = \frac{2}{3} + \frac{1}{3} = 1\ A$$

switch, the solution for I_L is the same as before, 1 A. Use of the simple Thevenin equivalent shown for the entire left side of the network makes it easy to examine circuit response as the load impedance value is varied.

The Thevenin equivalent circuit solution method is equally valid for complex impedance circuits. It is the type of representation shown in Fig 4 that is the basis for making per unit short-circuit calculations, although the actual values for the source voltage and branch impedances would be substantially different from those used in this case. (The circuit property of linearity would, incidentally, allow them to be scaled up or down.) The network shown in Fig 4(a), with the 6 Ω resistance shorted and the other resistances visualized as reactances, might well serve as an oversimplified representation of a power system about to experience a bolted fault with the closing of the switch.

The V_1 branch of the circuit would correspond to the utility supply while the V_2 branch might represent a large motor running unloaded, immediately adjacent to the fault bus, and highly idealized so as to have no rotor flux leakage. For such a model, the 5 V source corresponds to the pre-fault, air-gap voltage behind a stator leakage (subtransient) reactance of 3 Ω [3]. In a more realistic situation where rotor leakage is evident, a model that accurately describes the V_2 branch in detail before and after switch closing is much more difficult to develop, because the air-gap voltage decreases (exponentially) with time and varies (linearly) with the

**Fig 4
The Thevenin Equivalent**

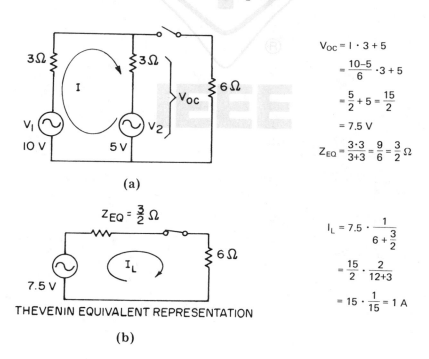

$$V_{OC} = I \cdot 3 + 5$$

$$= \frac{10-5}{6} \cdot 3 + 5$$

$$= \frac{5}{2} + 5 = \frac{15}{2}$$

$$= 7.5 \text{ V}$$

$$Z_{EQ} = \frac{3 \cdot 3}{3+3} = \frac{9}{6} = \frac{3}{2} \Omega$$

(a)

$$I_L = 7.5 \cdot \frac{1}{6 + \frac{3}{2}}$$

$$= \frac{15}{2} \cdot \frac{2}{12+3}$$

$$= 15 \cdot \frac{1}{15} = 1 \text{ A}$$

THEVENIN EQUIVALENT REPRESENTATION

(b)

steady-state rms magnitude of the motor stator current following application of the fault. The problem of accounting for motor internal behavior is avoided altogether by use of a Thevenin equivalent. This permits the V_2 branch to be represented by the apparent motor impedance effective at the time following switch closure. In shunt with the equivalent impedance for the remainder of the network, the Thevenin equivalent impedance, Z_{EQ}, for the motor (at any point in time of interest) is simply connected in series with the pre-fault open-circuit voltage, V_{OC}, to obtain the corresponding current response to switch closing.

The current response obtained in each branch of a network using a Thevenin equivalent circuit solution represents the change of current in that branch. The actual current that flows is the vector sum of currents before and after the particular switching event being considered. See Fig 5.

Fig 5
Current Flow of a Thevenin Equivalent Representation

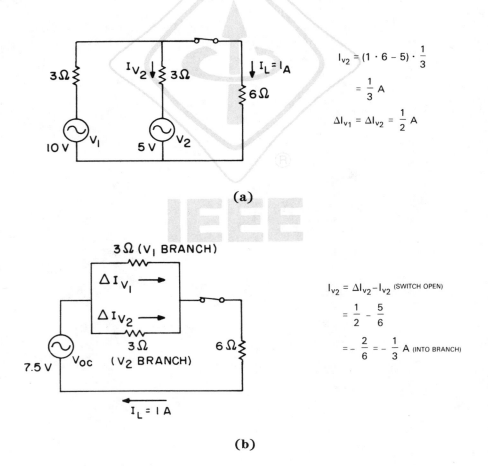

$$I_{v_2} = (1 \cdot 6 - 5) \cdot \frac{1}{3}$$

$$= \frac{1}{3} \text{ A}$$

$$\Delta I_{v_1} = \Delta I_{v_2} = \frac{1}{2} \text{ A}$$

(a)

$$I_{v_2} = \Delta I_{v_2} - I_{v_2} \text{ (SWITCH OPEN)}$$

$$= \frac{1}{2} - \frac{5}{6}$$

$$= -\frac{2}{6} = -\frac{1}{3} \text{ A (INTO BRANCH)}$$

(b)

In Fig 5(a), the current flowing in the V_2 branch circuit is shown to be ⅓ A. A more detailed representation of the Thevenin equivalent circuit previously examined in Fig 4 is shown in Fig 5(b). Here, the solution for the same current I_{V2} is determined by subtracting the current flowing in the V_2 branch prior to closing the switch (⅚ A from inspection of the circuit in Fig 4(a)) from the current $I_{V2} = ½$ A, calculated to be flowing in the Thevenin equivalent for this V_2 branch.

In the branch of the circuit defined by the switch itself, the *change of current* due to closing is normally the response of interest. This means the solution to the Thevenin equivalent is sufficient. The resultant current in the other branches, however, cannot be determined by the solution to the Thevenin equivalent network alone.

In the case where the V_2 branch represents a motor switched onto a bolted fault, the motor contribution is the locked-rotor current minus the pre-fault current as illustrated in Fig 6 and not just the locked-rotor current as it is so often carelessly described. As a rule, this effect is never as significant as the example suggests, even when the motor is loaded prior to the fault; the load current is much smaller than the locked-rotor current and almost 90° out of phase with it.

A Norton equivalent circuit, which can be developed directly from the Thevenin equivalent circuit, consists of a current source of magnitude V_{OC}/Z_{EQ} in parallel with the same equivalent impedance, Z_{EQ}. This representation is not generally as useful in power system analysis work.

3.2.4 The Sinusoidal Forcing Function. It is a most fortunate truth in nature that the excitation sources (driving voltage) for electrical networks, in general, have a sinusoidal character and can be represented by a sine wave plot of the type illustrated in Fig 7. There are two important consequences of this circumstance. First, although the response (current) for a complex R, L, C network represents the solution to at least one second-order differential equation, the result will also be a sinusoid of the same frequency as the excitation and different only in magnitude and phase angle. The relative character of the current with respect to the voltage for simple R, L, and C circuits is also shown in Fig 7.

The second important concept is that when the sine waveshape of current is forced to flow in a general impedance network of R, L, and C elements, the voltage

**Fig 6
Fault Flow**

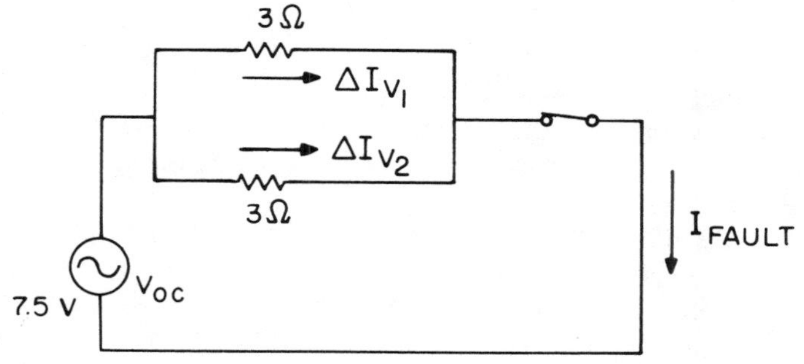

$$I_{FAULT} = \frac{15}{2} \cdot \frac{2}{3} = 5 \text{ A}$$

$$\Delta I_{v1} = \Delta I_{v2} = 2.5 \text{ A} = \frac{5}{2} = \frac{15}{6}$$

$$I_{v2} = \Delta I_{v2} - I_{v2} \text{ (SWITCH OPEN)}$$

$$I_{v2} = \frac{15}{6} - \frac{5}{6} = \frac{10}{6} \text{ A}$$

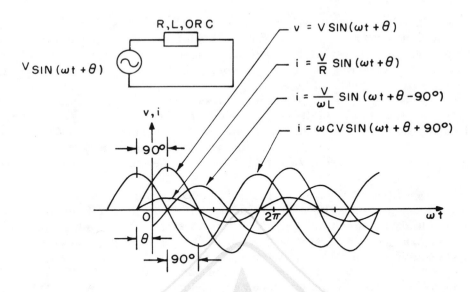

Fig 7
The Sinusoidal Forcing Function

drop across each element will always exhibit a sinusoidal shape of the same frequency as the source. The sinusoidal character of all the circuit responses makes the application of the superposition technique to a network with multiple sources surprisingly manageable. The necessary manipulation of the sinusoidal terms is easily accomplished using the laws of vector algebra, which evolve from the next technique to be reviewed.

The only restraint associated with the use of the sinusoidal forcing function concept is that the circuit must be comprised of linear elements, that is, R, L, and C are constant as current or voltage varies.

3.2.5 Phasor Representation. Phasor representation allows any sinusoidal forcing function to be represented as a phasor in a complex coordinate system as shown in Fig 8. As indicated, the expression for the phasor representation of a sinusoid can assume any of the following shorthand forms:

Exponential: $E\,e^{j\theta}$

Rectangular: $E\cos\theta + jE\sin\theta$

Polar: $E\,\underline{/\theta}$

For most calculations, it is more convenient to work in the *frequency domain* where any angular velocity associated with the phasor is ignored, which is equivalent to assuming the coordinate system rotates at a constant angular velocity of ω.

The impedances of the network can likewise be represented as phasors using the vectorial relationships shown. As illustrated, the circuit responses (current) can be obtained through the simple vector algebraic manipulation of the quantities involved. The need for solving complex differential equations to determine the circuit responses is completely eliminated.

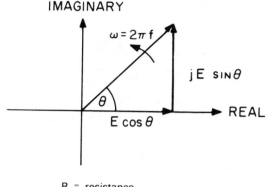

$$Ee^{j\theta} = E \cos\theta + jE \sin\theta = E \;\underline{/\theta}$$

$$I = \frac{E}{Z} = \frac{|E|}{|Z|}\frac{e^{j\theta}}{e^{j\Phi}} = \frac{|E|}{|Z|}e^{j(\theta-\Phi)}$$

$$Z = R + j\,\omega L - j\,\frac{1}{\omega C} = R + jX$$

$$X_L = \omega L$$

$$X_c = \frac{1}{\omega C}$$

$$|Z| = +\sqrt{R^2 + X^2}$$

$$\Phi = \tan^{-1}\frac{X}{R}$$

R = resistance
X_L = inductive ractance
X_C = capacitive reactance
Z = impedance

**Fig 8
The Phasor Representation**

The restraints that apply are:
(1) The sources must all be sinusoidal.
(2) The frequency must remain constant.
(3) The circuit R, L, and C elements must remain constant (that is, linearity must exist).

3.2.6 The Fourier Representation. This powerful tool allows any nonsinusoidal periodic forcing function, of the type plotted in Fig 9, to be represented as the sum of a dc component and a series (infinitely long, if necessary) of ac sinusoidal forcing functions. The ac components have frequencies that are integral harmonics of the periodic function fundamental frequency. The general mathematical form of the *Fourier series* is also shown in Fig 9.

The importance of the Fourier representation is immediately apparent. The response to the original driving function can be determined by first solving for the response to each Fourier series component forcing function and summing all the individual solutions to find the total superposition. Since each component response solution is readily obtained, the most difficult part of the problem becomes the determination of the component forcing function. The individual harmonic voltages can be obtained, occasionally in combination with numerical integration approximating techniques through several well-established mathematical procedures. Detailed discussion of their use is better reserved for the many excellent texts [2], [5] that treat the subject.

There are several abstract mathematical conditions that must be satisfied to use a Fourier representation. The only restraints of practical interest to the power systems analyst are that the original driving function must be periodic (repeating) and the network must remain linear.

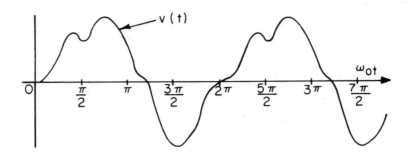

$$V(t) = V_0 + V_1 \cos \omega_0 t + V_2 \cos 2 \omega_0 t + \ldots$$
$$+ V_1' \sin \omega_0 t + V_2' \sin 2 \omega_0 t + \ldots$$

Fig 9
The Fourier Representation

3.2.7 The Laplace Transform. In the solution of circuit transients by classical methods, the models of circuit elements are represented with sets of differential equations. In addition, for a specific problem, a set of initial conditions must be known in order to solve the differential equations for the unknown quantity. An alternative technique for solving a transient problem is by the use of the Laplace transform. The proper use of this technique eliminates the need for the solution of the differential equations and simplifies all mathematical manipulations to elementary algebra.

It is helpful to keep in mind that the concept of mathematical transformations to simplify the solution to a problem is not new. For example, the mathematical operations of multiplication and division are transformed into the simpler operations of addition and subtraction by means of the logarithm transform. Once the addition/subtraction is performed, the solution to the problem is obtained by using the inverse transform, or antilog operation. The transformation is designed to create a new "domain" where mathematical manipulations are easier to carry out. Once the unknown is found in the new domain, it can be inverse-transformed back to the original domain.

In circuit analysis, the Laplace transform is used to transform the set of differential equations from the time domain (t) to a set of algebraic equations in the new domain called the complex frequency domain or, alternatively, the s-domain. The Laplace transform of a function is given by the expression

$$L\{f(t)\} = \int_0^\infty f(t)\, e^{-st}\, dt \qquad \text{(Eq 1)}$$

where the symbol $L\{f(t)\}$ is read as "the Laplace transform of $f(t)$." The Laplace transform is also denoted by the notation $F(s)$, that is,

$$F(s) = L\{f(t)\} \qquad \text{(Eq 2)}$$

This notation emphasizes that once the above integral has been evaluated, the resulting expression is a function of s. Since the exponent of the e in Eq 1 must be dimensionless, s must have the units of reciprocal time, hence the use of the alternate terms "frequency domain" and "s-domain" to describe the realm of the transformed function.

It can be shown that the "transformation" (or, more briefly, "transform") described by Eq 1 has special mathematical properties. Given an original expression involving both an unknown function (i.e., current, voltage, etc.), and operations on that function (i.e., derivatives, integrals, etc.), the s-domain expression that results when each term is transformed according to Eq 1 can be manipulated by ordinary algebraic procedures to yield a solution for the unknown function. The solution for the unknown function in the s-domain can then be transformed back to the time domain to produce the desired result. These mathematical methods can be used to greatly simplify the solution of complex differential equations.

The solution of a system problem involving a linear expression can then be determined in four simple steps, as follows:

(1) Formulate the differential time domain equations for the particular expression, which may contain terms like

$$Ri(\text{t}),\ C\ \frac{\mathrm{d}v}{\mathrm{d}t},\ \int_0^t i(\text{t})\ \mathrm{d}t,\ \text{etc.} \qquad\qquad (\text{Eq 3})$$

(2) Find the Laplace transform of the terms in the differential equation, including any initial conditions, according to the definition of the Laplace transform or using Laplace transform tables and equivalent circuit tables such as that shown in Table 1 and Table 2, respectively.

(3) Solve the transform for the unknown variable. The form of the s function should be manipulated into a form similar to those available in tables of Laplace transform pairs.

(4) From a table, find the inverse Laplace transform of the unknown.

An alternate approach is to interchange steps (1) and (2) above. In this case, the network is first transformed into the s-domain so that all the advantages of impedance and admittance operations and the use of network solution techniques such as mesh/nodal analysis and Thevenin/Norton equivalent circuits become available.

Applying this transformation to all elements in the system under study, the result will be a network consisting of s-domain equivalent elements. Then, usual circuit analysis will provide the quantities of interest in the s-domain. Finally, from a table of inverse Laplace transforms, the quantity of interest in the time domain can be obtained. In subsequent paragraphs, we will discuss the Laplace transform and its application to the solution of transient problems.

3.2.7.1 Transient Analysis by Laplace Transforms. The transient analysis of a circuit using Laplace transforms is a straightforward application of elementary algebra. Both the excitation sources and the impedance elements of the system under study must be replaced with their s-domain equivalent circuits. Table 1 contains Laplace transform pairs for commonly used excitation sources. The Laplace transform of the driving voltages and/or currents in the studied circuit must be determined for use in the s-domain equivalent circuit. Using Table 2, one of

45

Table 1
Laplace Transform Pairs

$f(t)$ $(t > 0^-)$	$F(s)$
$\delta(t)$	1
$u(t)$	$\dfrac{1}{s}$
t	$\dfrac{1}{s^2}$
e^{-at}	$\dfrac{1}{s+a}$
$\sin \omega t$	$\dfrac{\omega}{s^2 + \omega^2}$
$\cos \omega t$	$\dfrac{s}{s^2 + \omega^2}$
te^{-at}	$\dfrac{1}{(s+a)^2}$
$e^{-at} \sin \omega t$	$\dfrac{\omega}{(s+a)^2 + \omega^2}$
$e^{-at} \cos \omega t$	$\dfrac{s+a}{(s+a)^2 + \omega^2}$

two possible s-domain equivalent circuits must be chosen for each impedance element in the circuit. Once the s-domain network has been developed, the circuit equations can be solved algebraically.

The following examples will illustrate the procedure. They are simple and could very well be solved with less sophistication.

3.2.7.1.1 RL and RC Transients. The study of RL transients will begin with a series RL network with a constant (step function) voltage source $V(t)$ applied to the circuit through the closing of a switch. The network is illustrated in Fig 10.

According to Kirchoff's voltage law, at $t = 0^+$ the equation describing the circuit is

$$Vu(t) = Ri(t) + L\frac{di}{dt} \qquad \text{(Eq 4)}$$

Converting the terms of the above equation into the s-domain, according to either Table 1 or Table 2, yields

$$\frac{V}{s} = RI(s) + sLI(s) - LI(0^-) \qquad \text{(Eq 5)}$$

Since there could be no current flowing prior to the closing of the switch, the last term on the right can be ignored (initial conditions equal to zero). Solving for the current according to step (3), the result is

$$I(s) = \frac{V}{L}\left[\frac{1}{s(s + R/L)}\right] = \frac{V}{R}\left[\frac{1}{s} - \frac{1}{s + R/L}\right] \qquad \text{(Eq 6)}$$

Table 2
s-Domain Equivalent Circuits

Time Domain	Frequency Domain

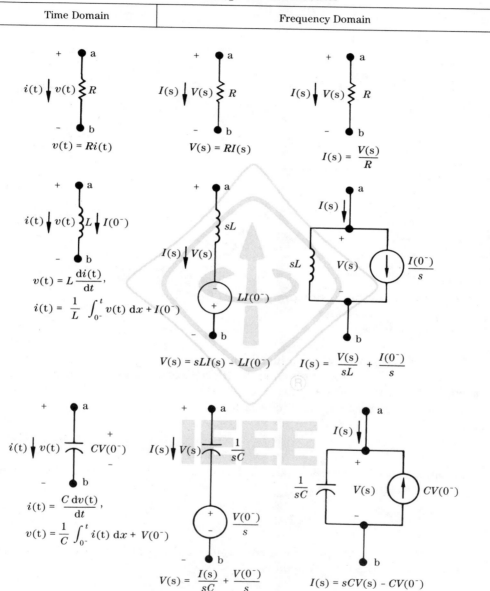

$$v(t) = Ri(t)$$

$$V(s) = RI(s)$$

$$I(s) = \frac{V(s)}{R}$$

$$v(t) = L\frac{di(t)}{dt},$$

$$i(t) = \frac{1}{L}\int_{0^-}^{t} v(t)\,dx + I(0^-)$$

$$V(s) = sLI(s) - LI(0^-)$$

$$I(s) = \frac{V(s)}{sL} + \frac{I(0^-)}{s}$$

$$i(t) = \frac{C\,dv(t)}{dt},$$

$$v(t) = \frac{1}{C}\int_{0^-}^{t} i(t)\,dx + V(0^-)$$

$$V(s) = \frac{I(s)}{sC} + \frac{V(0^-)}{s}$$

$$I(s) = sCV(s) - CV(0^-)$$

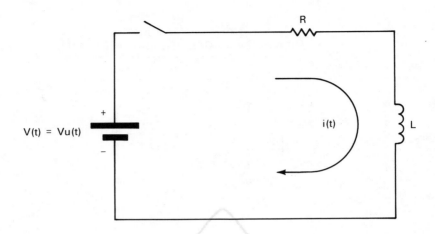

Fig 10
RL Network

The transient response is readily obtained from Table 1 as

$$i(\text{t}) = \frac{V}{R} \left[1 - e^{-tR/L}\right] \qquad \text{(Eq 7)}$$

The current $i(\text{t})$ is shown graphically in Fig 11.

The next circuit to be examined is the RC network depicted in Fig 12, already drawn in the symbols of the Laplace transform.

The equation describing the circuit is

$$\frac{V}{s} = RI(s) \times \frac{I(s)}{sC} - \frac{Vc(0^-)}{s} \qquad \text{(Eq 8)}$$

Assuming there is no initial charge in the capacitor ($Vc(0^-) = 0$) and solving for the current, yields

$$I(\text{s}) = \frac{V}{R} \left[\frac{1}{s + 1/RC}\right] \qquad \text{(Eq 9)}$$

Again, using the appropriate inverse Laplace transform from Table 1, the transient response in the time domain is

$$i(\text{t}) = \frac{V}{R} e^{-t/RC} \qquad \text{(Eq 10)}$$

The current response of the RC circuit is shown graphically in Fig 13 below.

If the quantity of interest was the voltage across the capacitor, then from Ohm's law we would have:

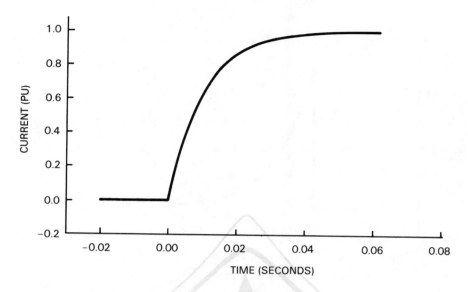

Fig 11
Current Response (Eq 7)

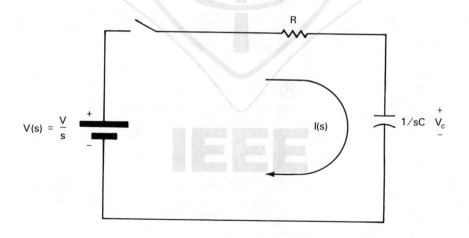

Fig 12
RC Network

$$Vc(s) = I(s)Z(s) \qquad \text{(Eq 11)}$$

Where $I(s)$ is defined by Eq 9 and $Z(s)$ is the capacitor impedance, namely

$$Z(s) = \frac{1}{sC} \qquad \text{(Eq 12)}$$

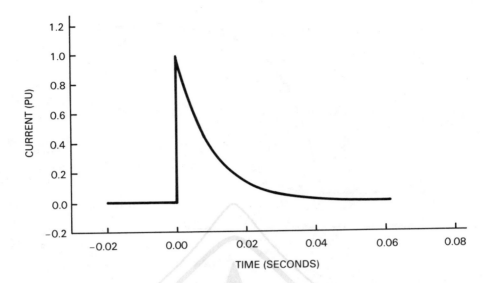

Fig 13
Current Response (Eq 10)

Since in a series circuit the current is common to all elements, we can substitute the above equation into Eq 11. After solving for $I(s)$, we substitute the result into Eq 9. Rearranging the terms, we obtain

$$Vc(s) = \frac{V}{RC}\left[\frac{1}{s(s+1/RC)}\right] = V\left[\frac{1}{s} - \frac{1}{s+1/RC}\right] \qquad \text{(Eq 13)}$$

From Table 1, the voltage across the capacitor in the time domain is

$$Vc(t) = V(1 - e^{-t/RC}) \qquad \text{(Eq 14)}$$

The capacitor voltage described by the above equation is shown graphically in Fig 14.

In cases involving more advanced circuits, the next step is the combination of all three components R, L, and C in a common network. This approach is considered in Chapter 11 of this book where additional circuits with higher degrees of difficulty are investigated using the same basics presented here.

3.2.8 The Single-Phase Equivalent Circuit. The single-phase equivalent circuit is a powerful tool for simplifying the analysis of balanced three-phase circuits, yet its restraints are probably most often disregarded. Its application is best understood by examining a three-phase diagram of a simple system and its single-phase equivalent, as shown in Fig 15. Also illustrated is the popular single-line diagram representation commonly used to describe the same three-phase system on engineering drawings.

If a three-phase system has a perfectly balanced symmetrical source excitation (voltage) and load, as well as equal series and shunt system and line impedances

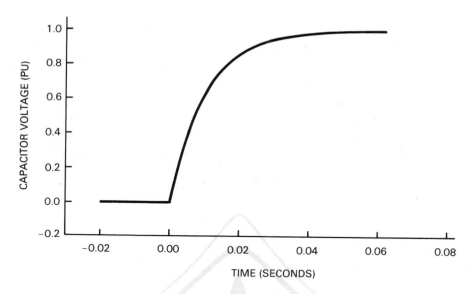

**Fig 14
Voltage Response**

connected to all three phases (see Fig 15(a)), imagine a conductor (shown dotted) carrying no current connected between the effective neutrals of the load and the source. Under these conditions, the system can be accurately described by either Fig 15(b) or Fig 15(c).

The single-phase equivalent circuit is particularly useful since the solution to the classical loop equations is much easier to obtain than for the more complicated three-phase network. To determine the complete solution, it is only necessary to realize that the other two phases will have responses that are shifted by 120° and 240° but are otherwise identical to the reference phase.

Anything that upsets the balance of the network renders the model invalid. A subtle way this might occur is illustrated in Fig 16. If the switching devices operate independently in each of the three poles, and for some reason the device in phase A becomes opened, the balance or symmetry of the circuit is destroyed. Neither the single-phase equivalent nor the single-line diagram representation is valid. Even though the single-phase and the single-line diagram representations would imply that the load has been disconnected, it continues to be energized by single-phase power. This can cause serious damage to motors and result in unacceptable operation of certain load apparatus.

More importantly, if only one switching device operates in response to a fault condition in the same phase, as depicted at location X, the system sources would continue to supply fault current from the other unopened phases through the impedance of the load. The throttling effect of the normally substantial load impedance, possibly in combination with additional arc impedance, can reduce the level of the current to a point where detection may not occur in phases (b) and (c). Needless to say, substantial damage can result before the fault finally burns enough

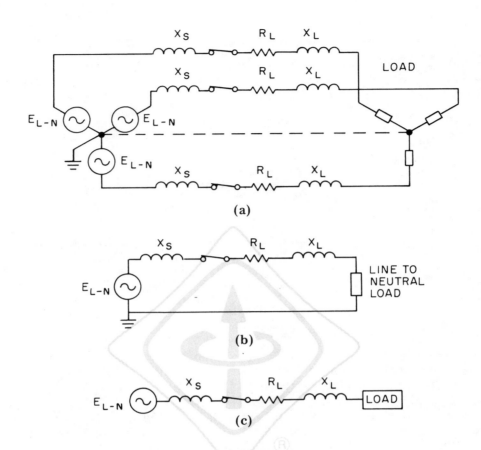

Fig 15
(a) Three-Phase Diagram, (b) Single-Phase
Equivalent, and (c) Single-Line Impedance Diagram

to involve the other phases directly and accomplish complete interruption. Meanwhile, both of the single-line representations fail to recognize the problem, and in fact, suggest that the condition has been safely disconnected. Therefore, the restraints of this calculating aid are:

(1) Symmetry of the electrical system, including all switching devices and applied load
(2) Any of the other previously described restraints which apply to the analytical technique being used in combination with the single-phase equivalent

3.2.9 The Symmetrical Component Analysis. This approach comes to the analyst's rescue when he is confronted with an unbalance, the most common circuit condition which invalidates the single-phase equivalent circuit solution method. The symmetrical component analysis allows the response to any unbalanced condition in a three-phase power system to be investigated and correctly synthesized by the sum of the responses to as many as three separate balanced system conditions.

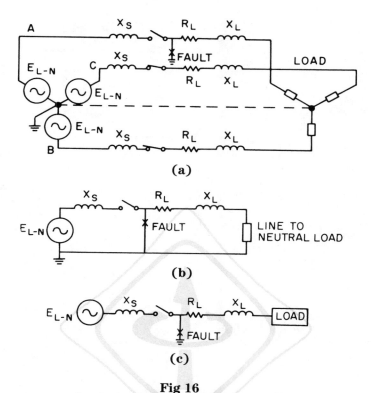

Fig 16
(a) Three-Phase Diagram, (b) Single-Phase
Equivalent, and (c) Single-Line Impedance Diagram

The application of an unbalanced set of voltage phasors, such as displayed in Fig 17, to a balanced downstream load is the sum of the responses of the balanced components which vectorially add to form the oiginal unbalanced set.

Similar conclusions apply when the voltages are balanced, but the connected phase impedances and/or loads, and the line currents are unbalanced. Here, the unbalanced current phasors are the sum of up to three balanced sets that flow through the balanced system impedances on either or both sides of the unbalance, producing voltage drops that satisfy the needs of the applied voltages and the boundary conditions at the point of unbalance.

The mathematical expression for three unbalanced phasors as a function of the balanced phasor components is as follows:

$$\overline{A} = \overline{A}_0 + \overline{A}_1 + \overline{A}_2$$

$$\overline{B} = \overline{B}_0 + \overline{B}_1 + \overline{B}_2$$

$$\overline{C} = \overline{C}_0 + \overline{C}_1 + \overline{C}_2$$

The positive (1), negative (2), and zero (0) sequence vector components of any phase always have the angular relationship with respect to one another as de-

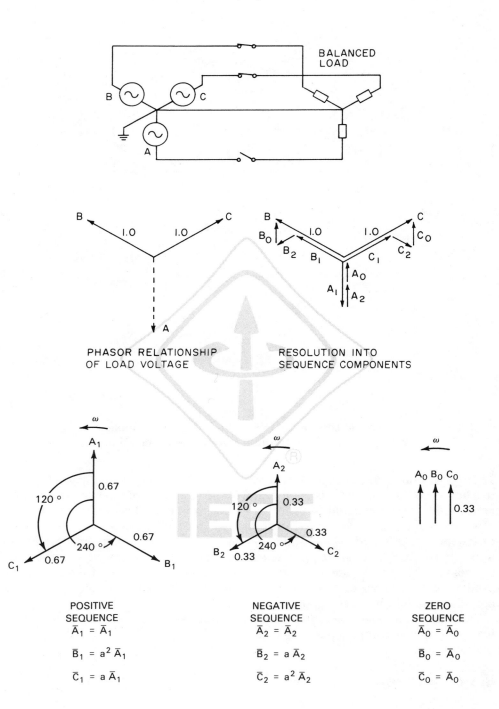

Fig 17
The Symmetrical Component Analysis

scribed by the vector diagram and defined by the identities of Fig 17. The operator (a) causes a counterclockwise rotation through an angle of 120° and is defined as

$$a = 1 \underline{/120°} = -0.5 + j0.866$$

These are assigned a counterclockwise direction of rotation in the *time domain* as illustrated. In the *space domain*, the negative sequence phasors will produce exactly the same results as a set of equal magnitude phasors that are displaced from one another by 120° and that rotate clockwise with time.

The proof that a set of N unbalanced vectors can be completely represented by N sets of balanced vectors is seldom presented in texts dealing with the subject of symmetrical components. The derivation of symmetrical component analysis is often generalized to three-phase systems, that is, systems where $N = 3$. First, it is postulated that it might be possible to describe three arbitrary but defined vectors, A, B, and C by summing their respective symmetrical components (as previously shown). Using the vector identities of Fig 17, the set of equations for the unbalanced vectors A, B, and C can be simplified into the linear set of three equations in three unknowns shown below:

$$\overline{A} = \overline{A}_0 + \overline{A}_1 + \overline{A}_2$$

$$\overline{B} = \overline{A}_0 + a^2\overline{A}_1 + a\overline{A}_2$$

$$\overline{C} = \overline{A}_0 + a\overline{A}_1 + a^2\overline{A}_2$$

Rearranging as shown below produces three independent equations for three unknowns (A_0, A_1, and A_2) which is everything required to uniquely and completely describe A_0, A_1, A_2 and, therefore, substantiate their existence. The corresponding B and C phase components are defined by the relationships shown in Fig 17.

$$\overline{A}_0 = \tfrac{1}{3}(\overline{A} + \overline{B} + \overline{C})$$

$$\overline{A}_1 = \tfrac{1}{3}(\overline{A} + a\overline{B} + a^2\overline{C})$$

$$\overline{A}_2 = \tfrac{1}{3}(\overline{A} + a^2\overline{B} + a\overline{C})$$

The merit of the symmetrical component analysis is that a relatively complicated, and often unwieldy problem, can be solved by simply vectorially summing the solution to no more than three balanced network problems. Several reference texts [1], [3], and [8] provide convenient tables showing the network interconnections that must be used to solve for the responses to many commonly encountered system unbalances, as well as certain balanced conditions. The three impedance networks (positive, negative, and zero sequence) are symbolically represented in shorthand fashion by an empty *block diagram* for the phase most definitive of the condition being studied up to the unbalance or other point of interest. Here, the final interconnections of the networks are shown which satisfy the necessary boundary conditions describing the system at the point of concern. The analyst can fill the *block diagrams* with the proper sources and impedances, including loads, in each sequence network and solve the single-phase loop equations. This produces the three sequence responses, that is, current or voltage, which add vectorially to produce the resultant phase responses. Similarly, the other phase responses can be obtained by adding vectorially the individual sequence solutions shifted by the appropriate multiple of 120°.

One curious and often confounding feature of this solution procedure is that the phase in the system which usually provides the best, and sometimes the only, approach to the solution for an unbalance is the one least actively involved in the event. The unbalance illustrated in Fig 17 is one such example where the solution is obtained through an analysis of the nonconducting phase A. A double line-to-ground fault is another, where examination of the open phase gives the most direct access to the network solution.

The symmetrical component analysis always involves the use of superposition as well as most of the other procedures previously discussed. The restraints which apply to these other procedures, therefore, must also govern the use of the symmetrical component analysis. In addition, due to mutual phase winding coupling and other effects, the impedance displayed by electrical machines will be different when excited by the different sequence sources. Hence, the *per phase* impedance of the negative and zero sequence networks will, in general, be different from the positive. Currents flowing in the zero sequence network are in phase and do not sum to zero as do the positive and negative sequence currents. Zero sequence currents must therefore flow through the ground circuit and are influenced by any impedance in this circuit path. When harmonic excitation sources are present (requiring the use of the Fourier representation), special care must be exercised in treating the sequence networks. Starting with the fundamental, the harmonic term progressively shifts from the positive to the negative to the zero sequence networks, and then the process repeats.

3.2.10 The Per Unit Method. This method of calculation and its close companion, the percentage method, are well documented in [1], [7], and [9] and are generally well known. As a result, they will only be mentioned in passing here.

Fundamentally, the per unit method and the percentage method amount to a shorthand calculating procedure where all equivalent system and circuit impedances are converted to a common kVA and kV *base*. This permits the ready combination of circuit elements in a network where different system voltages are present without the need to convert impedances each time responses are to be determined at a different voltage level.

Associated with each impedance element and its kVA base is a line-to-line kV base (usually the *nominal* line voltage at which the element is connected to the system), along with the resulting *base impedance* and *base current* related by the following expressions:

(1) *Three-Phase Network*

$$I_{base} = \frac{kVA_{base}}{\sqrt{3}\ kV_{base}}$$

$$Z_{base} = \frac{kV_{base}^2}{kVA_{base}} \cdot 1000$$

$$kVA_{base} = \sqrt{3}\, kV_{base}\, I_{base}$$

(2) *Single-Phase Network*

$$I_{base} = \frac{kVA_{base}}{kV_{base}}$$

$$Z_{base} = \frac{kV_{base}^2}{kVA_{base}} \cdot 1000$$

$$kVA_{base} = kV_{base}\, I_{base}$$

To illustrate the use of the per unit method, consider the example in Fig 18. The first step in a per unit calculation is the arbitrary selection of the system base kVA. Second, the choice of base kV must be made at one voltage level from which the base value at the other voltage levels is dictated by the turns ratios of the transformers in the network. For the circuit in Fig 18(a), the base values selected are 5000 kVA and 1.0 kV on the primary of the 10:1 transformer. The resultant base voltage on the secondary of the transformer is 1.0 kV/10 or 0.1 kV. The base impedance can next be calculated (both on the primary and secondary levels) and the per unit values of the primary and secondary impedances determined as shown.

Fig 18
(a) Classical Ohmic Representation

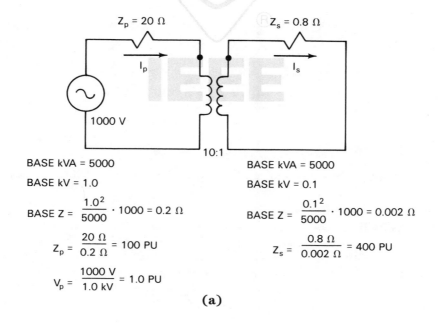

BASE kVA = 5000

BASE kV = 1.0

$$BASE\ Z = \frac{1.0^2}{5000} \cdot 1000 = 0.2\ \Omega$$

$$Z_p = \frac{20\ \Omega}{0.2\ \Omega} = 100\ PU$$

$$V_p = \frac{1000\ V}{1.0\ kV} = 1.0\ PU$$

BASE kVA = 5000

BASE kV = 0.1

$$BASE\ Z = \frac{0.1^2}{5000} \cdot 1000 = 0.002\ \Omega$$

$$Z_s = \frac{0.8\ \Omega}{0.002\ \Omega} = 400\ PU$$

(a)

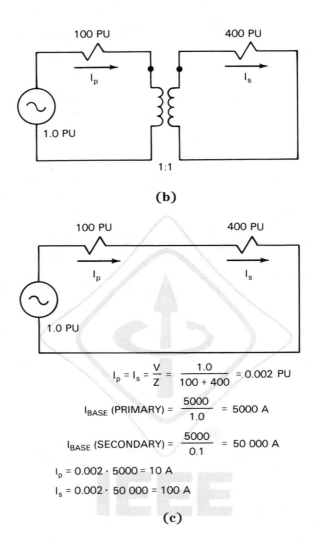

Fig 18 *(Continued)*
(b) Per Unit Representation, and (c) Simplified Per Unit Representation

Once the per unit impedance and excitation source values have been determined, the circuit can be simplified as shown in Fig 18(b). A key advantage of the per unit method, the transformer turns ratio becomes 1:1 thereby effectively removing it from the calculations as modeled in Fig 18(c). Working with the circuit in Fig 18(c), the primary current (I_p) and secondary current (I_s) are the same and can be easily calculated by simply applying Ohm's law. The last step in the procedure is to determine the actual current in amperes by multiplying each per unit value by the base current on the primary and secondary levels. Although the per unit values calculated for I_p and I_s are equal, the base currents are different, and therefore the solutions expressed in amperes are different. Since the per unit method of calculation is based on the existence of linearity and is always used in combination with one or more of the other principles, it is necessary to observe all of the associated restraints discussed earlier as they apply.

3.3 References and Bibliography

[1] BEEMAN, DONALD. *Industrial Power Systems Handbook*, New York: McGraw-Hill, 1955, chap. 2.

[2] CLOSE, C. M. *The Analysis of Linear Circuits*, New York: Harcourt, Brace and World, Inc., 1969.

[3] *Electrical Transmission and Distribution Reference Book*, East Pittsburgh, PA: Westinghouse Electric Corporation, 1964, chaps. 2 and 6.

[4] FITZGERALD, A. E. and KINGSLEY, Jr., CHARLES. *Electric Machinery*, New York: McGraw-Hill, 1961.

[5] HAYT, Jr., W. H., and KEMMERLY, J. E. *Engineering Circuit Analysis*, New York: McGraw-Hill, 1962.

[6] PUCHSTEIN, A. F. and LLOYD, T. C. *Alternating Current Machines*, New York: John Wiley and Sons, Inc., 1947.

[7] STEVENSON, Jr., W. D. *Elements of Power System Analysis*, New York: McGraw-Hill, 1962.

[8] WAGNER, C. F. and EVANS, R. D. *Symmetrical Components*, New York: McGraw-Hill, 1933.

[9] WEEDY, B. M. *Electric Power Systems*, New York: John Wiley and Sons, Inc., 1972, chap. 2.

A general bibliography that can be used to supplement the material is presented for reader consideration.

[B1] GOLDMAN, S. *Laplace Transform Theory and Electrical Transients*, New York: Dover Publications, 1966.

[B2] MILLER, R. *Algebraic Transient Analysis*, San Francisco: Reinhart Press, 1971.

[B3] NILSSON, J. W. *Electric Circuits*, Reading: Addison-Wesley Publishing Co., 1984.

Chapter 4
System Modeling

4.1 Introduction. This chapter addresses the following questions:
 (1) How can each component or each group of components of an industrial or commercial power system be represented so that an analysis of the system performance can be made?
 (2) Which of the several possible representations, or *models*, of the given components will best describe the system to meet the objectives of a given study?
 (3) What mathematical expressions will describe the characteristics of each system element so that it can be quantified and programmed for computer input?

There are an infinite number of possible power system configurations and a large variety of study types. Consequently, standards cannot be established to dictate specific models for all specific circumstances.[3] This text will therefore serve as a guide to help the reader make judicious trade-offs in selecting models for a study.

Derivations and proofs of mathematical expressions will not be given. References should be used for such purposes. However, several fundamental relationships of electrical and mechanical quantities will be mentioned in the text. This should save the reader the time needed to locate them in textbooks or handbooks. It should also help refresh the memories of those who have not been exposed to power studies or academic activities for a long time.

The material is intended to be as basic and simple as permitted by the subject. The emphasis is to correlate real-life systems with the abstraction of mathematics in order to communicate with the computer properly.

4.2 Modeling. Scale modeling of power systems as a means of analyzing their performance is impractical. However, scale models of certain mechanical components of power systems are used to evaluate their characteristics. This is often the case with hydraulic sections of hydroelectric plants, such as turbine runners, spiral cases, gates, draft tubes, etc. Nonetheless, much expertise is required to establish scaling and normalization factors, to construct the model, to gather meaningful data by measurement, and to interpret and extrapolate the results.

---

[3] An exception to this statement are the application standards for circuit breakers where models for sources of short-circuit current are specified. See 4.9.2.2 and 4.10.3.

Digital computers can be programmed to solve quickly and at relatively low cost a large number of simultaneous equations and can handle the algebra of large matrixes. This makes them particularly well suited for applications in power system analysis. An immense variety of programs have been written to study an ever-increasing number of problems in the electrical field. These programs are usually set up to receive the problem information in the form of numbers rather than analog settings. This then forces the power system analyst to model the system quantitatively. The programs, designed to maximize their usefulness, are written in general nonspecific terms. This exposes the analyst to a choice of program features and alternatives that require decisions to be made every step of the way. Finally, the programs are often structured to handle extensive power systems (3000 bus programs are not uncommon).

4.3 Review of Basics. Power network elements may be classified in two categories, passive elements and active elements.

4.3.1 Passive Elements. The passive elements comprise such components as transmission lines, transformers, reactors, and capacitors. They will, in general, be regarded as linear. They will be modeled by one or more of the following electrical quantities:

Name	Symbol	Unit
resistance	R	ohm
inductance	L	henry
capacitance	C	farad

The voltage across and the current through the element will be governed by these relationships:

$$v = Ri \qquad i = \frac{v}{R} \tag{Eq 15}$$

$$v = L\frac{di}{dt} \qquad i = \frac{1}{L}\int v\, dt \tag{Eq 16}$$

$$v = \frac{1}{C}\int i\, dt \qquad i = C\frac{dv}{dt} \tag{Eq 17}$$

where the lowercase letters represent the instantaneous values of voltage and current.

In dc circuits under steady-state conditions, these equations will reduce to:

$$V = RI \qquad I = \frac{V}{R}$$

$$V = 0 \qquad \left(\text{since } \frac{di}{dt} = 0\right)$$

$$I = 0 \qquad \left(\text{since } \frac{dv}{dt} = 0\right) \tag{Eq 18}$$

In ac circuits with sinusoidal waveshapes, the equations become:

$$V = RI \qquad\qquad I = \frac{V}{R}$$

$$V = jX_{L}I$$

where

$$X_{L} = 2\pi fL$$

$$= \text{inductive reactance} \qquad\qquad \text{(Eq 19)}$$

$$V = -jX_{C}I$$

where

$$X_{C} = \frac{1}{2\pi fC}$$

$$= \text{capacitive reactance} \qquad\qquad \text{(Eq 20)}$$

The capital letters for voltages and currents represent their rms values, f is the frequency in hertz, and j is the 90° operator (= $\sqrt{-1}$). Inverting and combining these elements in series or parallel will define the set of quantities of Table 3.

It should be noted here that it is customary in ac power circuits to use the R, X and Z quantities for the series (line) elements and the G, B and Y quantities for the shunt (line to neutral) elements.

Note also that Z and Y are complex quantities that can be expressed in the rectangular form above or the polar form $Z = |Z| \underline{/\theta}$ or $Y = |Y| \underline{/\theta}$. Most computer programs accept the Z and Y values in the rectangular form.

A final remark concerns the sign ahead of the reactances and susceptances. The four diagrams of Fig 19 are self-explanatory. The wise analyst will verify the program instructions to make sure that the computer will interpret the input data properly.

4.3.2 Active Elements. The active elements of a power system comprise such components as motors, generators, synchronous condensers, other loads like furnaces, adjustable speed drives, etc. The active elements will be regarded as nonlinear, although some of the components may behave linearly under certain circumstances.

Table 3
Equation References for Conductance, Susceptance,
Impedance and Admittance

Name	Symbol	Unit	Defining Expression
conductance	G	S (mho)	
inductive susceptance	B	S (mho)	$R + jX = \dfrac{1}{G + jB}$
capacitive susceptance	B	S (mho)	
impedance	Z	Ω (ohm)	$(R + jX)$
admittance	Y	S (mho)	$(G + jB)$

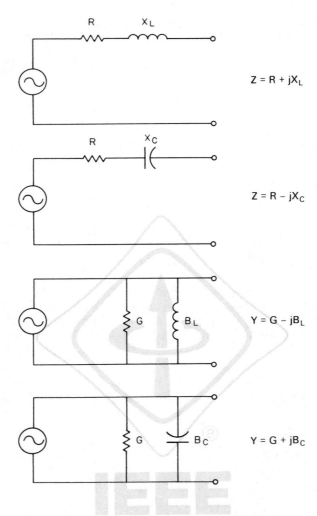

Fig 19
Equivalent Circuit Diagrams
Showing Sign Convention

One or more of the parameters of a model of an active element will vary as a function of time, phase angle, frequency, speed, etc.

The four expressions for power quantities given in Table 4 can be used to model nonlinear elements. Given any two of the four values, the remaining two can be defined. Power can also be expressed in polar form: $S = |S| \underline{/\theta}$ which yields these relationships: PF $= \cos\theta$, $P = S\cos\theta$ and $Q = S\sin\theta$.

Note that the signs of P or Q may be positive or negative. By convention, the positive sign of Q is used for inductive loads; that is, the current will lag the voltage applied to a load that *consumes* vars. In this sense, it is said that capacitors

Table 4
Four Defining Expressions for Power Quantities

Name	Symbol	Unit	Defining Expression
complex power	S	VA	$S = P + jQ$
active power	P	W	$P = \sqrt{S^2 - Q^2}$
reactive power	Q	var	$Q = \sqrt{S^2 - P^2}$
power factor	PF		$PF = \dfrac{P}{S}$

generate vars (current leads the voltage) and that squirrel cage induction motors *absorb* vars.

By convention also, the sign of P is positive for a load that consumes energy or a source that generates energy. Thus a load with a negative sign for P could be used to represent a generator, and vice versa.

Some of the power system components can best be expressed in terms of current or voltage. For instance, an infinite bus may be specified by a voltage source of constant magnitude and angle, and a particular load may be described as a constant current element. The current and voltage quantities may be complex numbers, in which case they have to be described in terms of a reference vector that may be a voltage or current quantity. This introduces the phase angle concept:

$$I = |I| \underline{/\theta} = I_x + jI_y = I(\cos\theta + j\sin\theta)$$

$$V = |V| \underline{/\theta} = V_x + jV_y = V(\cos\theta + j\sin\theta)$$

where the x axis of the coordinate system is taken as the reference shown in Fig 20.

So far, the review has included most of the quantities used in the type of studies that require network solutions at a single instant: a *snapshot* of the network in equilibrium. This instant shows the system in a steady state or during a transient, at an instant for which all network parameters are known, say ½ cycle after a short circuit has occurred.

Fig 20
Vector Diagram

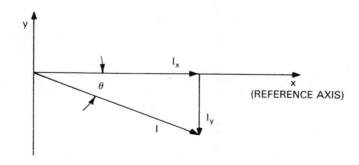

Some studies, of which motor starting and transient stability are examples, require that a complete period of time be covered to assess the effects of a disturbance on system performance. This requirement for a certain period of time to be considered introduces the need for mechanical quantities.

The fundamental quantities of mechanics are space, matter, and time. Length measures space, and mass measures matter. Other mechanical quantities are derived from these three. Some of the derived quantities officially recognized for electrical engineering work have been tabulated in Table 5, with the MKS system of units and the defining equations.

4.4 Power Network Solution. Before dealing with the detailed models of power system components, it is important to review what constitutes the solution of a network.

It can be said that a network is resolved if all the bus voltages and the relative phase angles between these voltages are known. This of course requires that the impedances between the buses be known. It also requires that a reference voltage and angle be specified for one bus.

Table 5
Fundamental Equations for Translation and Rotation

Name	Symbol	Unit	Defining Expression
Fundamental			
length	l	meter (m)	
mass	m	kilogram (kg)	
time	t	second (s)	
Translation			
velocity	v	m/s	$v = \dfrac{\mathrm{d}l}{\mathrm{d}t}$
acceleration	a	m/s^2	$a = \dfrac{\mathrm{d}^2 l}{\mathrm{d}t^2}$
force	F	newton (N)	$F = ma$
work	W	joule (J)	$W = \int F\,\mathrm{d}l$
power	P	watt (W)	$P = \dfrac{\mathrm{d}w}{\mathrm{d}t}$
momentum	M'	N/s	$M' = mv$
Rotation			
radius	r	m	
circular arc	s	m	
moment of inertia	I	kg-m^2	$I = \int r^2\,\mathrm{d}m$
angle	θ	radian (rad)	$\theta = \dfrac{s}{r}$
angular velocity	ω	rad/s	$\omega = \dfrac{\mathrm{d}\theta}{\mathrm{d}t}$
angular acceleration	α	rad/s^2	$\alpha = \dfrac{\mathrm{d}^2\theta}{\mathrm{d}t^2}$
torque	T	N·m	$T = rF$
work	W	J	$W = \int T\,\mathrm{d}\theta$
power	P	W	$P = T\omega$
angular momentum	M	Js/rad	$M = I\omega$

Consider for instance Fig 21, which shows a small simplified section of the typical plant from the single-line diagram in Chapter 1. Cable and transformer data has been shown to agree with the systems information in Fig 33. Assume that the voltages at buses 2, 4 and 24 (also called A, B, and C to simplify notations) and the

Fig 21
Section of Typical Single-Line Diagram (Simplified)

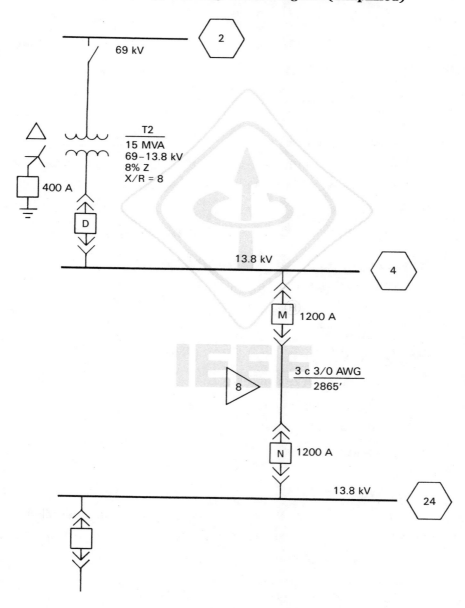

impedances of T2 and line 8 are known. They are summarized on Fig 22 and listed as follows:

$$V_2 = V_A = 69.00 \text{ kV} \underline{/0°}$$

$$V_4 = V_B = 13.60 \text{ kV} \underline{/-1.6°}$$

$$V_{24} = V_C = 13.57 \text{ kV} \underline{/-1.82°}$$

$$Z_{T2} = Z_{AB} = 1.0 + j8.0 \text{ (\%) on 15 MVA base}$$

$$= (0.01 + j0.08) \cdot \frac{69^2}{15}$$

$$= 3.174 + j25.39 \text{ } \Omega \text{ (at 69 kV)}$$

$$= (0.01 + j0.08) \cdot \frac{13.8^2}{15}$$

$$= 0.1270 + j1.016 \text{ } \Omega \text{ (at 13.8 kV)}$$

$$Z_{L8} = Z_{BC} = 0.19138 + j0.12119$$

The current from bus A to bus B can be found by:

$$I_{AB} = (V_B - V_A)/Z_{AB} \sqrt{3} \qquad\qquad\qquad \text{(Eq 21)}$$

$$= 10^3 \left(69.00 \underline{/0°} - 13.60 \cdot \frac{69}{13.8} \underline{/-1.6°}\right)/(3.174 + j25.39) \sqrt{3}$$

$$= 10^3 [69.00 + j0 - (67.97 - j1.90)]/(25.59 \underline{/82.87°}) \sqrt{3}$$

$$= 10^3 (1.03 + j1.90)/(25.59 \underline{/82.87°}) \sqrt{3}$$

$$= 10^3 (2.16 \underline{/61.54})/(25.59 \underline{/82.87°}) \sqrt{3}$$

$$= 10^3 \cdot 2.16/25.59 \sqrt{3} \underline{/61.54 - 82.87°}$$

$$= 48.73 \underline{/-21.33°} \text{ A (at 69 kV)}$$

The power flow from bus A to bus B is:

$$S_{AB} = \sqrt{3} V_A \cdot \hat{I}_{AB} \text{ (the caret on } \hat{I} \text{ means } \textit{conjugate}) \qquad \text{(Eq 22)}$$

$$= (\sqrt{3} \cdot 69 \underline{/0°}) (48.73 \underline{/+21.33°})$$

$$= 5823 \underline{/21.33°} \text{ kVA}$$

$$= P_{AB} + jQ_{AB}$$

$$= 5425 + j2118$$

The current I_{AB} on the 13.5 kV side is that at the 69 kV side multiplied by the transformation ratio:

$$I_{AB} = 48.73 \underline{/-21.33°} \cdot \frac{69}{13.8}$$

$$= 243.7 \underline{/-21.33°} \text{ (at 13.8 kV)}$$

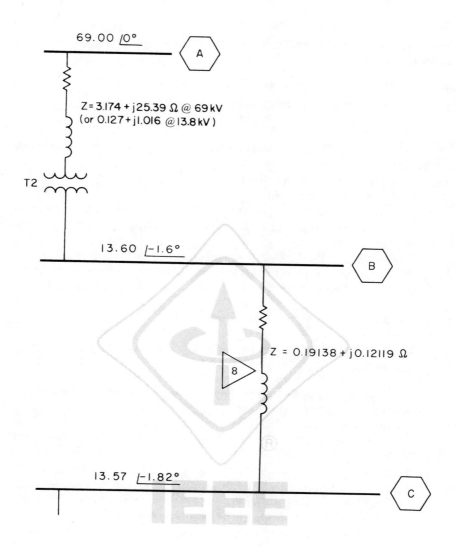

Fig 22
Impedance Diagram

Next find the power flow from bus B to bus A:

$$S_{BA} = -\sqrt{3}\, V_B \cdot \hat{I}_{AB} \text{ (at 13.8 kV)}$$

$$= -(\sqrt{3} \cdot 13.60 \,\underline{/-1.6°}) \cdot (243.7 \,\underline{/21.33°})$$

$$= -5741 \,\underline{/19.73°} \text{ kVA}$$

$$= P + jQ$$

$$= -5403 - j1938$$

The transformer losses can now be found by adding the power from A to B and that from B to A:

$$S_{\text{Losses}} = S_{\text{AB}} + S_{\text{BA}}$$

$$= (5425 + j2118) + (-5403 - j1938)$$

$$= 22 + j180$$

The losses are 22 kW and 180 kvar between buses A and B. Figure 23 summarizes these results.

This example illustrates that once the bus voltages are known, the remaining calculations are straightforward.

The real problem of analyzing even a modest size system is the determination of the bus voltages. The loads are known but mostly nonlinear and complicated further by the fact that the reference voltage is often many buses away from the loads. In order to find the bus voltages, one has to resort to a cut-and-try iterative method. The computer is an effective tool for this method since it can complete the set of calculations shown above much faster than if done by hand.

4.5 Impedance Diagram. Several things remain to be said about Figs 21, 22, and 23.

(1) All three diagrams show a single-phase equivalent of the three-phase system. The conditions for this equivalence to be true were covered in Chapter 3.

(2) Bus and line designators have been shown on all three diagrams. The impedance diagram of Fig 22 is the rigorously correct way to represent graphically resistances and reactances. However, it is felt that drawing the graphical symbols of resistances, inductances, and capacitances is superfluous since the expressions for impedances alongside a straight line sufficiently describe the line elements. With all this in mind, the graphical representation described by Fig 24 is suggested as one method for bridging the gap between the single- line diagram and the computer input document and illustrates use of the same bus and line designator conventions as shown in Figs 21, 22, and 23. Such a skeleton drawing of the power system showing buses, lines, generators, and loads, each with its assigned numbers, can be duplicated for multiple use as an impedance diagram and as a flow diagram. It should be noted that these diagrams are working tools and as such do not require standardization. However, the analyst should adopt a method suitable for keeping track of masses of data, for even small system studies require and generate a large amount of information.

(3) In power systems analysis, the term *bus* does not always have the meaning understood by a plant electrician, for instance. The analyst calls *bus* any point of the system where voltages are calculated. The term is interchangeable with *node*. Fictitious *buses* may be introduced on the network to obtain voltage solutions at certain points of interest. An example of this may be a 150 mile transmission line broken down in 5 sections of 30 miles (that is, a *bus* introduced every ⅕ of the length) in order to avoid the complicated but exact model of the long line.

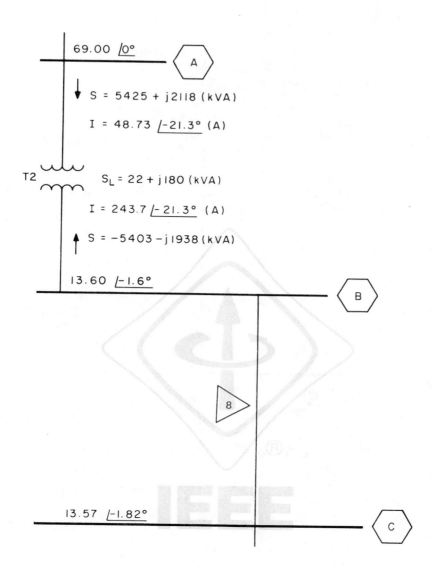

Fig 23
Flow Diagram

(4) In the same vein, the term *line* is often given the more general meaning of *branch*, that is, any element between two nodes. For example, the transformer data will be entered on a computer input document called *line data*.

4.6 Extent of the Model

4.6.1 General. No rigid rules can be established on how much of a power system should be modeled for a given study. System analysts have to exercise judgment and develop a feel for this as they gain experience.

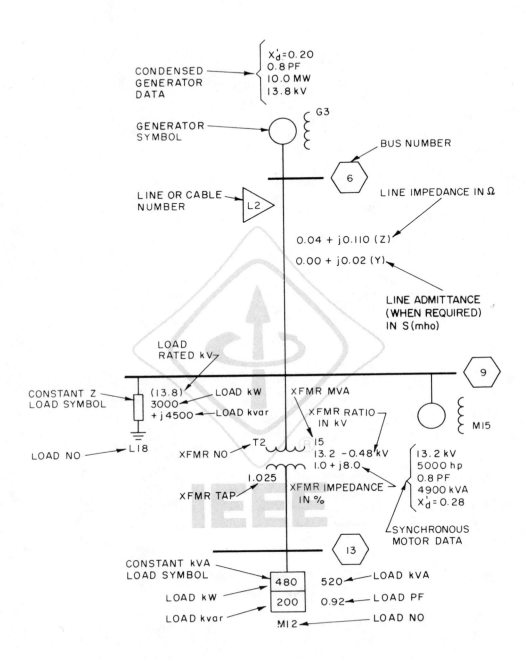

Fig 24
Data Presentation for Impedance and Other Diagrams

The objectives of the study should always be kept in focus. This will help in eliminating useless work.

4.6.2 Utility Supplied Systems. A large number of industrial and commercial establishments are supplied by stiff utility systems. Stiffness is a relative function of the size of the plant load and local generation. If the external power system or utility is large compared with that of the plant, disturbances within the plant do not affect the voltage at the point of connection. In such a case, the utility system is said to be an infinite system. The connection point will be an infinite bus.

This concept can be extended within the plant electrical distribution system when studies are concerned with small areas electrically remote from the utility supply. Conversely, sections of utility systems may require modeling in cases where this stiffness does not exist. It is, therefore, important that a sound knowledge of the utility supply systems be acquired before proceeding with the studies.

4.6.3 Isolated Systems. The question of whether an isolated system should be modeled in full or in part is easier to determine. These are usually relatively small and as such could be represented fully for most kinds of studies. The extra effort of gathering a set of data for the entire system, even though a smaller section would suffice, will not be lost since the additional data will be used in some future study. The nature of an isolated system is such that a modification or a disturbance is more apt to be felt throughout the system.

4.7 Models of Branch Elements

4.7.1 Lines. Four parameters affect the performance of a line connecting a source to a load: series resistance, series inductance, shunt capacitance, and shunt conductance. A short length of conductor can be modeled as in Fig 25. A line can be considered as many short lengths of conductors placed in series to yield the model of Fig 26(a). The individual lengths of conductor could be made shorter thus increasing the number of lengths for a given length of line. Continuing this process to the limit defines the model called *line with distributed constants*. This model has

**Fig 25
Equivalent Circuit of Short Conductor**

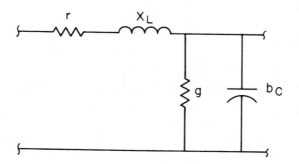

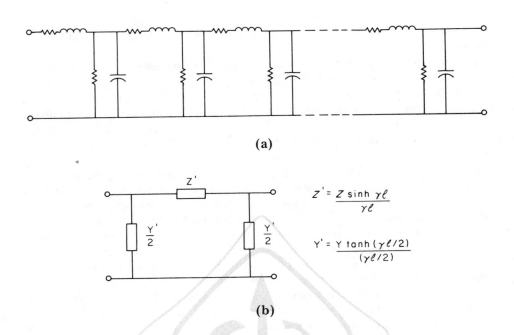

**Fig 26
(a) Line with Distributed Constants
(b) Long Line Equivalent Circuit**

been reduced to the equivalent circuit shown in Fig 26(b), where the series arm is defined by

$$Z' = Z \frac{\sinh \gamma\ell}{\gamma\ell} \tag{Eq 23}$$

and the shunt branches by:

$$\frac{Y'}{2} = \frac{Y}{2} \frac{\tanh(\gamma\ell/2)}{(\gamma\ell/2)}$$

$$= \frac{1}{Z_C} \frac{(\cosh \gamma\ell - 1)}{\sinh \gamma\ell} \tag{Eq 24}$$

Two figures of merit appear in these equations:

$Z_C = \sqrt{z/y}$, defined as the *characteristic or surge impedance* of the line (Eq 25)

$\gamma = \sqrt{y\,z}$, defined as the *propagation constant* (Eq 26)

Both Z_C and γ are complex numbers. The propagation constant γ can be expressed in the rectangular form:

$$\gamma = \alpha + j\beta \tag{Eq 27}$$

This defines:

α = attenuation constant
β = phase constant (in radians)

The other variables are:

l = total length of line
r = conductor effective resistance in ohms per unit of length
x = conductor series inductive reactance in ohms per unit of length
$x = 2\pi f L$
g = shunt conductance to neutral in siemens (mho) per unit of length
b = shunt capacitive susceptance in siemens (mho) per unit of length
$b = 2\pi f C$
L = conductor total inductance in henrys per unit of length
C = conductor shunt capacitance in farads per unit of length
z = r + jx = series impedance in ohms per unit of length
$y = g + jb$ = shunt admittance to neutral in siemens (mhos) per unit of length
$Z = zl$ = total series impedance of line in ohms
$Y = yl$ = total shunt admittance of line to neutral in siemens (mhos) per unit of length
sinh, cosh, tanh = hyperbolic functions

These functions of complex numbers can be evaluated by using the following relationships:

$$\sinh(\gamma\ell) = \sinh(\alpha\ell + j\beta\ell)$$
$$= \sinh(\alpha\ell)\cos(\beta\ell) + j\cosh(\alpha\ell)\sin(\beta\ell) \tag{Eq 28}$$

$$\cosh(\gamma\ell) = \cosh(\alpha\ell)\cos(\beta\ell) + \sinh(\alpha\ell)\sin(\beta\ell) \tag{Eq 29}$$

$$\tanh X = \frac{\sinh X}{\cosh X} \tag{Eq 30}$$

$$\sinh(\alpha\ell) = \frac{e^{\alpha\ell} - e^{-\alpha\ell}}{2} \tag{Eq 31}$$

$$\cosh(\alpha\ell) = \frac{e^{\alpha\ell} + e^{-\alpha\ell}}{2} \tag{Eq 32}$$

The surge impedance Z_C is approximately 400 Ω for single circuit overhead power lines, while cable circuits are about 30–40 Ω. The distributed constants model is valid for short or long lines at power or communication frequencies.

4.7.1.1 Long Lines. Power frequency overhead lines in excess of 150 miles should be represented by the distributed constants model reduced to an equivalent π as shown in Fig 26(b).

The shunt conductance g may be neglected since the dielectric is air (a good dielectric) and the conductor spacing is large. However, if the corona losses are important, they may be represented as a value for g. Computer programs will readily accept the data for Z' expressed in the rectangular form, that is, the equivalent series resistance R' and the equivalent series inductive reactance X'. It should be noted that even though g is neglected ($g = 0$), a nonzero value of G' will appear in the equivalent circuit of Fig 26(b). The values of $G'/2$ and $B'/2$ (in siemens) may have to be modified to MW or Mvar to suit the program input requirements. (The computer treats them as constant impedance loads.) If the computer requires a *line charging Mvar*, the value B' and not $B'/2$ must be used to calculate:

$$\text{Mvar} = 3 \, (\text{kV})^2 \cdot B' \tag{Eq 33}$$

The program will internally assign half of that value to the bus at each end of the line.

The value of $G'/2$ can be modified to:

$$\text{MW} = 3 \, (\text{kV})^2 \cdot \frac{G'}{2} \tag{Eq 34}$$

and input as a constant impedance bus load.

The voltage (kV) is the line-to-ground voltage corresponding to the base voltage used in the program input document.

4.7.1.2 Medium Lines. In the range of 50 to 150 miles approximately, little accuracy is lost by simplifying Eqs 23 and 24 to

$$Z' = Z$$

$$\frac{Y'}{2} = \frac{Y}{2}$$

Thus, neglecting the products $\dfrac{\sinh \gamma \ell}{\gamma \ell}$ and $\dfrac{\tanh (\gamma \ell / 2)}{(\gamma \ell / 2)}$ yields the model of Fig 27(a), called the nominal π circuit. In this model, the shunt branches are purely capacitive (no conductance).

The nominal π circuit can be thought of as being formed by the process described at the beginning of 4.7.1, with the exception that the unit length of conductor is made longer (instead of shorter), and the circuit made symmetrical (bilaterally). This results in the constants being lumped by an approximation process. The nominal T circuit is formed the same way, with the difference that all the shunt constants are lumped into one as compared with the π circuit where series constants are lumped into a single one.

Use of the nominal T model is not popular since it requires addition of a fictitious bus in the middle of the line. Entering data into the program, for the nominal π circuit follows the same requirements as for the long line model.

4.7.1.3 Short Lines. For overhead lines shorter than 50 miles, neglecting the shunt capacitance in the models presented earlier will not greatly affect the results of load flow, short-circuit, or stability calculations. This yields the model of Fig 28.

4.7.2 Cables. The overhead line models are equally applicable to cables. While the resistances are substantially the same, the relative values of reactances are vastly different. Table 6 compares two cases, one at 69 kV, the other at 13.8 kV. The cable inductive reactance is about ¼ that of the line; but the capacitive reactance is 30 to 40 times that of the line.

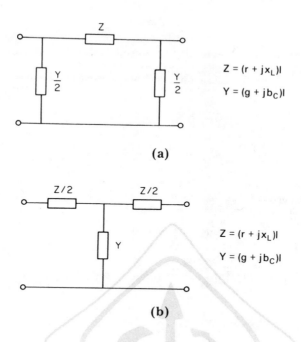

$$Z = (r + jx_L)l$$
$$Y = (g + jb_C)l$$

(a)

$$Z = (r + jx_L)l$$
$$Y = (g + jb_C)l$$

(b)

Fig 27
Medium Line Equivalent Circuits
(a) Nominal π (b) Nominal T

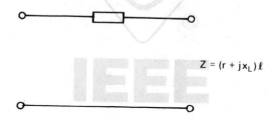

$$Z = (r + jx_L)\ell$$

Fig 28
Short Line Equivalent Circuit

This comparison suggests that the medium line model, the nominal π, should be used for cables in the order of one mile in length (approximately $\frac{1}{40}$ of 50 miles). The shorter the cable run, the better the accuracy when using this model.

It is doubtful that any medium voltage system will have feeder lengths requiring representation of the capacitive reactance.

4.7.3 Determination of Constants. Electrical conductor characteristics are available from numerous sources and need not be repeated here. A few general comments are in order.

Table 6
Comparison of Overhead Lines and Cable Constants

	Values in Ω/mile for 500 kcmil Cu Conductors			
	69 kV		13.8 kV	
	Overhead Line*	Cable**	Overhead Line†	Cable‡
Resistance	0.134	0.134	0.134	0.134
Inductive reactance	0.695	0.176	0.613	0.146
Capacitive reactance	0.162×10^6	0.005×10^6	0.142×10^6	0.003×10^6

*Open-wire equilateral conductor spacing of 8 ft
**Three-conductor oil-filled paper-insulated cable rated 69 kV
†Open-wire equilateral conductor spacing of 4 ft
‡Three-conductor oil-filled paper-insulated cable rated 15 kV

4.7.3.1 Resistance. The *effective resistance* of the conductors should be used. The effective resistance takes into account the conductor:

(1) Material
(2) Size
(3) Shape
(4) Temperature
(5) Frequency
(6) Environment

Copper and aluminum are the most used conductor materials for lines and cables. Soft annealed chemically pure copper has 100% conductivity (IACS standards). This is equivalent to 875.2 Ω for a one mile long round wire weighing one pound at 20 °C. All other materials can have their conductivities expressed as a percentage of the standard, a few of which are listed in Table 7.

The conductor resistance will vary with temperature according to the following formula:

$$R_2 = R_1 \left[1 + \alpha(t_2 - t_1) \right] \qquad \text{(Eq 35)}$$

where

R_2 = resistance at temperature t_2
R_1 = resistance at temperature t_1
α = temperature coefficient per degree at temperature t_1
At 20 °C, the coefficient α per degree Celsius is:
Copper: 0.00393
Aluminum: 0.00403
Galvanized steel:
SM: 0.0039
HS: 0.0035
EHS: 0.0032

It is not possible to predict the exact operating conductor temperature if the conductor current is not known. The analyst has the choice of either estimating the conductor temperature or assuming the worst case, which, in some studies, might

Table 7
Conductor Data

Material	% Conductivity	Application
Copper		
soft	100	cable construction
soft, tinned	93.15 to 97.3	cable construction
hard, shape	98.4	bus bar
hard, round	97.0	overhead line conductors
Aluminum	61.0	cables, bars, tubes
Aluminum alloys		
5005-H19	53.5	overhead line conductors
6201-T81	52.5	overhead line conductors
Galvanized steel		
Siemens-Martin	12.0	
high strength	10.5	
extra high strength	9.4	

be the maximum allowable temperature of the cable. Other studies might require that the minimum conductor temperature be used for the worst case.

The ac resistance of conductors is higher than the dc resistance due to skin effects. The effect is more pronounced as the conductor cross section or the operating frequency increases. Conductor data tables usually include ac resistances at power frequencies. The skin effect is a major factor in the design of high-current (several thousand amperes) ac bus systems, such as for electric furnaces.

The flux established by alternating current in a conductor may link other conductors or metallic masses in its proximity thus generating voltages in those parts. These voltages may cause currents to flow through closed circuits and thus cause I^2R losses other than those of the conductor itself. These losses can be represented as an additional component of resistance in series with the conductor resistance. The reader should consult References [13][4] and [25] for information on this subject.

4.7.3.2 Inductive Reactance. The inductive reactance of a circuit has two components: that due to its own circuit (self) and that due to other circuits in its vicinity (mutual). The inductance of a conductor also has two components: that caused by the current in itself and that caused by the currents in other conductors of the same circuit. Finally, the inductance of a conductor due to its own current is divided in two parts: the first part considers the flux internal to the conductor; the second part, the flux external to the conductor. This last division has been modified to simplify tables of conductor characteristics. There are two types of tables commonly used to determine the inductive reactance of a conductor. One table lists the conductor inductive reactance X_s at one foot spacing even if the actual spacing is larger or smaller than one foot.

---

[4]The numbers in brackets correspond to those in the References at the end of this chapter.

A second table, valid for any type or size of conductor, lists spacing factors X_d, which added to the one foot reactance will give the correct total reactance for the given circuit conductor spacing. The spacing factor table is calculated from the equation:

$$X_d = 4.657 \cdot 10^{-3} \cdot f \cdot \log \text{GMD}$$

$$= \Omega/(\text{conductor} \cdot \text{mile}) \qquad \text{(Eq 36)}$$

where

GMD = geometric mean distance of the conductors

For three conductors spaced d_1, d_2, d_3

$$\text{GMD} = \sqrt{d_1 \cdot d_2 \cdot d_3}$$

Note that in Eq 36 a GMD smaller than 1 yields a negative spacing factor.

Cables in steel conduit exhibit higher reactances than in free air. The calculations are too complex to develop by hand, hence, the tables in Chapter 1 of Reference [21] should be used for estimating purposes.

4.7.3.3 Shunt Capacitive Reactance. The determination of the capacitive reactance follows the same pattern as the inductive reactance. Conductor tables give the value of reactance X_s at one foot spacing. A spacing factor X_d is added to X_s to yield the total capacitive reactance of the conductor. Spacing factor tables are calculated from:

$$X_d' = \frac{4.099}{f} \cdot 10^{-6} \log \text{GMD}$$

$$= \Omega - \text{mile/conductor} \qquad \text{(Eq 37)}$$

The capacitive reactance of shielded cables is determined from:

$$X_c' = \frac{1.79 \, G \cdot 10^6}{f \cdot k}$$

$$= \Omega - \text{mile/conductor} \qquad \text{(Eq 38)}$$

where

G = geometric factor
k = dielectric constant of cable insulation
f = frequency

$$G = 2.303 \log \frac{2r}{d} \qquad \text{(Eq 39)}$$

where

r = inside diameter of shield
d = outside diameter of conductor

Typical values of k are 6.0 for rubber, 5.0 for varnished cambric, 2.6 for polyethylene, and 3.7 for paper.

4.7.4 Reactors. Reactors are used as branch elements in the following applications:

(1) To limit current during fault conditions
(2) To buffer cyclic voltage fluctuations caused by repetitive loads (in conjunction with condensers)
(3) To limit motor starting currents

They are modeled as impedances consisting of inductive reactance in series with a resistance expressed as $R + jX$. Manufacturers' design or test data should be obtained for existing applications.

4.7.5 Capacitors. Series capacitors are sometimes used on transmission and distribution lines to compensate for the inductive reactance drop or to improve the system stability by increasing the amount of power that can be transmitted on tie lines. They are represented by a negative reactance of the form $0 - jX$, in series with the line impedance.

For capacitors specified in microfarads per phase, the reactance may be expressed in the general form:

$$X = \frac{10^6}{2\pi fC} = \Omega/(\text{per phase}) \tag{Eq 40}$$

When specified in kilovars per phase (Q_C), the capacitor voltage rating (V_C) must also be known to calculate:

$$X = \frac{V_C^2}{Q_C} \cdot 10^3 = \Omega/(\text{per phase}) \tag{Eq 41}$$

It should be noted that the series capacitor voltage rating is a function of the amount of compensation of the design and will generally be a fraction of the system line-to-neutral voltage. The application of series capacitors should always be accompanied by thorough studies, since it is easy to create destructive overvoltage and ferroresonance conditions.

4.7.6 Transformers

4.7.6.1 Two-Winding Transformers. The equivalent circuit of a transformer is shown in Fig 29(a). The dashed rectangle represents an ideal voltage transformation ratio $n_s/n_p = N$, where n_s and n_p are the number of turns of the primary and secondary windings, respectively. R_p and R_s are the effective resistances of the windings; X_p and X_s are the leakage reactances. G_0, the shunt conductance, models the iron losses that remain constant when the transformer is energized at rated voltage and B_M, the shunt inductive susceptance, is equivalent to the quadrature magnetizing current at no load.

It can be demonstrated that Fig 29(a) is equivalent to Fig 29(b). In the latter, the secondary resistance and reactance have been *reflected* to the primary side of the ideal transformer by multiplication with the inverse of the square of the turns ratio N. A close approximation of this circuit is possible by moving the shunt branch and combining the primary and secondary impedances as shown on Fig 30(a). Another simplification consists in eliminating the shunt branch altogether to yield Fig 30(b). In many types of studies, the resistance R_T, being small with respect to X_T, is also

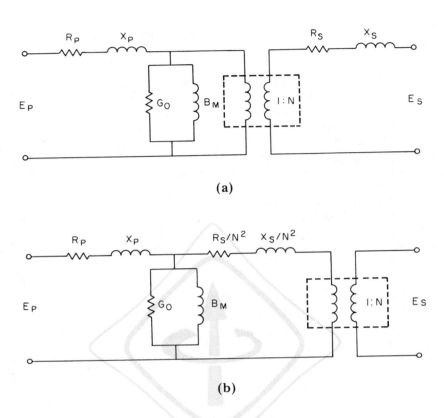

Fig 29
Two-Winding Transformer Equivalent Circuits

neglected, thus reducing the model of the transformer to a single series reactance.

Transformer nameplate specifies the impedance Z_T and the transformation ratio. An assumption may be made that $X_T \cong Z_T$ and that the single series reactance model was used.

Use of Fig 30(b) model requires that an estimate of R be made from typical data [1], and a value for X_T calculated from

$$X_T = \sqrt{Z^2 - R^2}$$

Transformer test data will usually be sufficient to calculate all the parameters for the circuits of Figs 29(a), 29(b) and 30(a). When maximum accuracy is needed, the effective resistance R_T should include the winding resistances corrected for the operating temperature and another series resistance to account for stray losses (see References [24] and [25]). The model of Fig 29 necessitates the creation of a fictitious bus for entry of the shunt admittance data in the program.

4.7.6.2 Transformer Taps. Thus far, only single ratio transformers have been dealt with. In real life, transformers have taps, normally on the high-voltage windings, to provide a voltage ratio best suited to the power system. The taps may

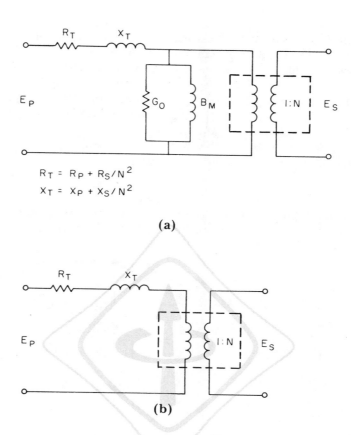

$$R_T = R_P + R_S/N^2$$
$$X_T = X_P + X_S/N^2$$

(a)

(b)

**Fig 30
Two-Winding Transformer
Approximate Equivalent Circuits**

be changeable automatically under load (LTC transformer) or fixed (manually changed in de-energized condition).

The resistance and leakage reactance of the tapped windings are slightly different at different taps. This may be ignored if the correct values are not known. On the other hand, transformer test data may specify values for the taps, in which case these values should be used. The main effect of changing taps is the change of voltage ratio and therefore the change of voltage base for which the impedance diagram should be prepared. This will be described in more detail in 4.8.

The analyst should pay particular attention to the specific requirements of programs for specifying taps. For instance, the tap value 1.05 per unit (105%), interpreted as an additional 5% to the voltage ratio, yields opposite results if applied to the opposite sides of the transformer.

Once the data is input for a given condition and a certain voltage on a specified bus is required, the computer program will, upon request, automatically adjust the taps and modify the system impedances as necessary for the new turns ratio.

4.7.6.3 Three-Winding Transformers. The circuits of Figs 29 and 30 apply to any two windings of the three-winding transformers.

The three possible combinations, put together using Fig 30 (b) as a basic model, form a delta as shown on Fig 31 (a), where the new subscript T denotes the tertiary winding. Elimination of the ideal transformer symbols is made possible by calling the voltages at nodes A, B and C, 1.0 per unit and remembering that those nodes have base voltage values E_p, E_s and E_t, respectively. Assuming that the three-phase system is balanced, the fictitious neutral will be common to all three voltage levels and can be eliminated to simplify the diagram.

Fig 31
Three-Winding Transformer Approximate Equivalent Circuits
(a) Simplified-Delta   (b) Simplified-Wye

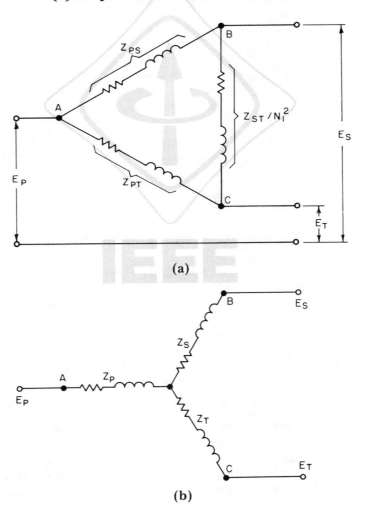

(a)

(b)

The loop formed by the delta (Δ) circuit can also be eliminated. This requires making a delta-wye transformation. The end result shown in Fig 31(b) is a format acceptable to computer programs. The new node D at the center of the Y creates a fictitious bus that cannot be identified physically but that is necessary to identify the impedance values to the computer.

The relationships for reduction from Δ to Y are:

$$Z'_\text{p} = \frac{1}{2}\left(Z_\text{ps} + Z_\text{pt} - \frac{1}{N_1^2}\,Z_\text{st}\right) \tag{Eq 42}$$

$$Z'_\text{s} = \frac{1}{2}\left(\frac{1}{N_1^2}\,Z_\text{st} + Z_\text{ps} - Z_\text{pt}\right) \tag{Eq 43}$$

$$Z'_\text{t} = \frac{1}{2}\left(Z_\text{pt} + \frac{1}{N_1^2}\,Z_\text{st} - Z_\text{ps}\right) \tag{Eq 44}$$

The relationship as a function of the Y values is:

$$Z_\text{ps} = Z'_\text{p} + Z'_\text{s} \tag{Eq 45}$$

$$Z_\text{pt} = Z'_\text{p} + Z'_\text{t} \tag{Eq 46}$$

$$Z_\text{st} = N_1^2\,(Z'_\text{s} + Z'_\text{t}) \tag{Eq 47}$$

4.7.6.4 Phase-Shifting Transformers. Consider again the example in 4.4. Assume this time that the voltage at bus B has the same magnitude but that the angle is shifted back another 2° by the winding arrangement within the transformer, and that the rest of the information on Fig 23 is the same as before. The new impedance diagram will appear as shown in Fig 32(a).

Now resolve as before with $V_\text{B} = 13.60\ \underline{/-1.6° - 2.0°}$. The results are shown in Fig 32(b). Note that the phase shift of a mere 2° has caused the real power from A to B to jump from 5425 to 11 803 kW; but the reactive power decreased from 2118 to 1605 kvar.

Fig 32
Impedance and Flow Diagrams
(a) Impedance Diagram (b) Flow Diagram

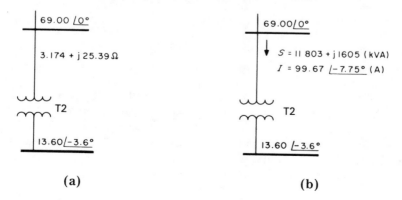

(a)

(b)

This example illustrates the main purpose of phase-shifting transformers to control the flow of real power between two buses.

These transformers usually have load tap changing mechanisms that will vary the phase angle between primary and secondary automatically or manually. Thus, computer programs will be set up to vary the amount of angle shift (plus or minus), within limits specified in the input, to achieve the desired amount and the direction of real power flow.

The data required by computer programs will generally include the following:
(1) Center (0° shift) position impedance
(2) Positive limit position impedance
(3) Negative limit position impedance
(4) Angle shift interval between taps
(5) Number of taps

Program subroutines allow the computer to automatically estimate the value of the impedances at the intermediate taps.

4.7.6.5 Other Transformer Models. The foregoing discussion has been intended to give the reader an introduction to the subject of transformer modeling. It is far from being complete. Reference [13] contains a considerable number of models that can be useful to the reader.

4.8 Power System Data Development

4.8.1 Per Unit Representations. Network calculations can be made with impedances, voltages and currents expressed either in actual ohms, volts or amperes or using the dimensionless per unit method. While both representations will yield identical results, the per unit method is generally preferred as it will do the job much more conveniently, especially if the system being studied has several different voltage levels. Also, the impedances of electric apparatus are usually given in per unit or percent by manufacturers.

The per unit value of any quantity is its ratio to the chosen base quantity of the same dimensions, expressed as a dimensionless number. For example, if base voltage is taken as 4.16 kV, voltages of 3740 V, 4160 V, and 4330 V will be 0.9, 1.00, and 1.04, respectively, when expressed in per unit on the given base voltage. The chosen base voltage, 4.16 kV, is referred to as base voltage, 100% voltage or unit voltage.

There are four base quantities in the per unit system: base kVA, base volts, base ohms, and base amperes. They are so related that the selection of any two of them determines the base values of the remaining two. It is a common practice to assign study base values to kVA and voltage. Base amperes and base ohms are then derived for each of the voltage levels in the system.

In power system studies, the base voltage is usually selected as the nominal system voltage at one point of the system, such as the voltage rating of a generator or the nominal voltage at the delivery point of the utility supply. The base kVA can be similarly taken as the kVA rating of one of the predominant pieces of system equipment, such as a generator or a transformer; but usual practice is to choose a convenient round number such as 10 000 for the base kVA. The latter selection has some advantage of commonality when several studies are made, while the former choice means that at least one significant component will not have to be converted to a new base.

The basic relationships for the electrical per unit quantities are as follows:

$$\text{Per unit volts} = \frac{\text{actual volts}}{\text{base volts}} \qquad \text{(Eq 48)}$$

$$\text{Per unit amperes} = \frac{\text{actual amperes}}{\text{base amperes}} \qquad \text{(Eq 49)}$$

$$\text{Per unit ohms} = \frac{\text{actual ohms}}{\text{base ohms}} \qquad \text{(Eq 50)}$$

For three-phase systems, the nominal line-to-line system voltages are normally used as the base voltages. The base kVA is assigned the three-phase kVA value. The derived values of the remaining two quantities are:

$$\text{Base amperes} = \frac{\text{base kVA}}{\sqrt{3}\ \text{base kV}} \qquad \text{(Eq 51)}$$

$$\text{Base ohms} = \frac{(\text{base kV})^2}{\text{base MVA}} \qquad \text{(Eq 52)}$$

It is convenient in practice to convert directly from ohms to per unit ohms without first determining base ohms according to the following expression:

$$\text{Per unit ohms} = \frac{\text{ohms} \cdot \text{base MVA}}{(\text{base kV})^2} \qquad \text{(Eq 53)}$$

For a three-phase system, the impedance is in ohms per phase and the base kVA is the three-phase value.

Where two or more systems with different voltage levels are interconnected through transformers, the kVA base is common for all systems; but the base voltage of each system is determined by the turns ratio of the transformer connecting the systems, starting from the one point for which the base voltage has been declared. Base ohms and base amperes will thus be correspondingly different for systems of different voltage levels.

Once the system quantities are expressed as per unit values, the various systems with different voltage levels can be treated as a single system and the necessary variables can be solved. Only when reconverting the per unit values to actual voltage and current values is it necessary to recall that different base voltages exist throughout the system.

When impedance values of devices are expressed in terms of their own kVA and voltage ratings which differ from the base values of a circuit, it is necessary to refer these values to the system base values. This may also happen when machines rated at one voltage may actually be used in a circuit at a different voltage. In such cases, the per unit impedance of the device must be changed to either a new base kVA or new base voltage, or both, by the equation:

$$\text{Per unit } Z_2 = \text{per unit } Z_1 \cdot \frac{(\text{base kV}_1)^2}{(\text{base kV}_2)^2}\ \frac{\text{base kVA}_2}{\text{base kVA}_1} \qquad \text{(Eq 54)}$$

where subscripts 1 and 2 refer to the old and new base conditions, respectively.

4.8.2 Applications Example. A section of the power system described in 4.5 has been repeated in Fig 33 as an illustration of the per unit system. The transformer ratios were changed slightly to improve this example. The steps in reducing the data to per unit are as follows:

Fig 33
Impedance Diagram Raw Data

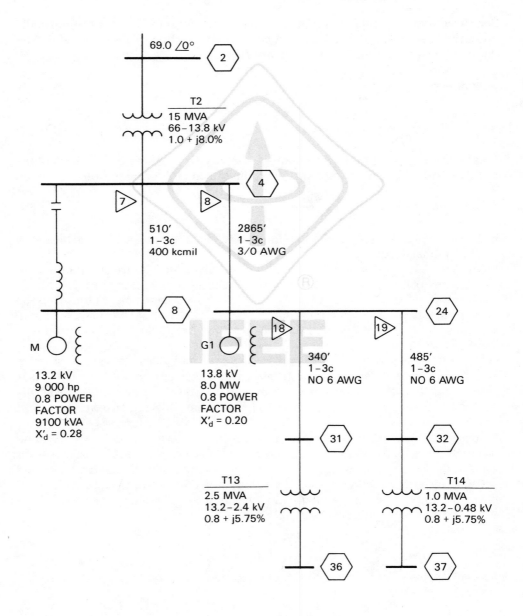

(1) Select base power: S = 10 000 kVA
(2) Determine base voltages
   (a) Select bus 2 nominal voltage of 69 kV as base

NOTE: This base voltage is selected here and is shown in Fig 33 to illustrate the full range the following calculations might take. The calculations are simplified considerably when base voltages and nominal transformer ratings correspond to each other, i.e., 66 kV at the bus 2 incoming service or 69 kV as the transformer primary rating. In order to fully explore this fact, the reader is encouraged to rework the calculations using 66 kV as the starting point (base voltage) at bus 2. In Figs 21, 22, and 23 of this chapter and in other figures in this book, 69 kV is used both as the incoming service bus voltage and as the incoming service transformer primary winding rating.

   (b) Calculate base voltages at other system levels

$$\text{Bus 4:} \quad kV = 69.0 \cdot \frac{13.8}{66}$$

$$= 14.427 \text{ kV}$$

$$\text{Bus 36:} \quad kV = 14.427 \cdot \frac{2.4}{13.2}$$

$$= 2.623 \text{ kV}$$

$$\text{Bus 37:} \quad kV = 14.427 \cdot \frac{0.48}{13.2}$$

$$= 0.525 \text{ kV}$$

(3) Calculate base impedances using Eq 52
   (a) 69 kV system:

$$Z = \frac{69^2 \cdot 10^3}{10\,000}$$

$$= 476.1 \ \Omega$$

   (b) 13.8 kV system:

$$Z = \frac{14.427^2 \cdot 10^3}{10\,000}$$

$$= 20.82 \ \Omega$$

   (c) 2.4 kV system:

$$Z = \frac{2.623^2 \cdot 10^3}{10\,000}$$

$$= 0.688 \ \Omega$$

   (d) 0.48 kV system:

$$Z = \frac{0.525^2 \cdot 10^3}{10\,000}$$

$$= 0.02756 \ \Omega$$

(4) Calculate base currents using Eq 51
(a) 69 kV system:

$$I = \frac{10\,000}{\sqrt{3 \cdot 69.0}}$$

$$= 83.67 \text{ A}$$

(b) 13.8 kV system:

$$I = \frac{10\,000}{\sqrt{3 \cdot 14.427}}$$

$$= 400.2 \text{ A}$$

(c) 2.4 kV system:

$$I = \frac{10\,000}{\sqrt{3 \cdot 2.623}}$$

$$= 2201 \text{ A}$$

(d) 0.48 kV system:

$$I = \frac{10\,000}{\sqrt{3 \cdot 0.525}}$$

$$= 11\,000 \text{ A}$$

(5) Summarize the base data in Table 8
(6) Convert transformer impedances to per unit using Eq 54
(a) T2:

$$Z = \frac{1.0 + j8.0}{100} \cdot \frac{66^2}{69^2} \cdot \frac{10}{15}$$

$$= 0.006\,10 + j0.048\,80$$

Table 8
System Base Values
(Base Power 10 000 kVA)

Bus	Base kV	Base Z	Base I
2	69.00	476.1	83.67
4	14.427	20.82	400.2
8	14.427	20.82	400.2
24	14.427	20.82	400.2
31	14.427	20.82	400.2
32	14.427	20.82	400.2
36	2.623	0.688	2201.0
37	0.525	0.027 56	11 000.0

or

$$Z = \frac{1.0 + j8.0}{100} \cdot \frac{13.8^2}{14.427^2} \cdot \frac{10}{15}$$

$$= 0.006\,10 + j0.048\,80$$

(b) T13:

$$Z = \frac{0.8 + j5.75}{100} \cdot \frac{13.2^2}{14.427^2} \cdot \frac{10}{2.5}$$

$$= 0.026\,79 + j0.192\,54$$

or

$$Z = \frac{0.8 + j5.75}{100} \cdot \frac{2.4^2}{2.623^2} \cdot \frac{10}{2.5}$$

$$= 0.026\,79 + j0.192\,55$$

(c) T14:

$$Z = \frac{0.8 + j5.75}{100} \cdot \frac{13.2^2}{14.427^2} \cdot \frac{10}{1}$$

$$= 0.066\,97 + j0.481\,35$$

or

$$Z = \frac{0.8 + j5.75}{100} \cdot \frac{0.48^2}{0.525^2} \cdot \frac{10}{1}$$

$$= 0.066\,87 + j0.480\,65$$

(7) Calculate line impedance in ohms
   (a) Lines 7, 8, 18 and 19 are 3/C, copper cables, paper insulated, shielded
       conductors; dielectric constant: 3.7. See Table 9.

Table 9
Cable Data

Line	Length (ft)	Conductor Size	$r(\Omega/1000')$ (ft)	$x(\Omega/1000')$ (ft)	Conductor Outside Diameter (inches)	Insulation Thickness (inches)
7	510	400 kcmil	0.0297	0.0370	0.728	0.175
8	2865	3/0	0.0668	0.0423	0.470	0.185
18	340	6	0.4960	0.0610	0.184	0.220
19	485	6	0.4960	0.0610	0.184	0.220

91

(b) Line 7:

R = 0.0297 · 510/1000

   = 0.015 15 Ω

X_L = 0.0370 · 510/1000

   = 0.018 87 Ω

X_C = (neglect due to short
   length)

(c) Line 8:

R = 0.0668 · 2865/1000

   = 0.191 38 Ω

X_L = 0.0423 · 2865/1000

   = 0.121 19 Ω

From Eqs 38 and 39

$$X_C' = \frac{1.79 \cdot 10^6}{60 \cdot 3.7} \cdot 2.303 \log \cdot \frac{(0.470 + 2 \cdot 0.185)}{0.470}$$

$$= 4683 \ \Omega\text{-mi}$$

X_C = 4683 · 5280/2865

   = 8630 Ω

(d) Line 18:

R = 0.4960 · 340/1000

   = 0.168 64 Ω

X_L = 0.0610 · 340/1000

   = 0.020 74 Ω

X_C = (neglect due to short
   length)

(e) Line 19:

R = 0.4960 · 485/1000

   = 0.240 56 Ω

X_L = 0.0610 · 485/1000

   = 0.029 59 Ω

X_C = (neglect due to short
   length)

(8) Calculate line impedances in per unit with Eq 50

(a) Line 7:

$$Z = \frac{0.015 \ 15 + j0.018 \ 87}{20.82}$$

$$= 0.000 \ 728 + j0.000 \ 906 \text{ per unit}$$

(b) Line 8:

$$Z = \frac{0.191 \ 38 + j0.121 \ 19}{20.82}$$

$$= 0.009 \ 192 + j0.005 \ 821 \text{ per unit}$$

$$Y = \frac{Z_C}{-jX_C} = \frac{20.82}{-j8630}$$

$$= 0 + j0.002\,413 \text{ per unit}$$

(c) Line 18:

$$Z = \frac{0.168\,64 + j0.020\,74}{20.82}$$

$$= 0.00810 + j0.00100 \text{ per unit}$$

(d) Line 19:

$$Z = \frac{0.240\,56 + j0.029\,59}{20.82}$$

$$= 0.011\,55 + j0.00142 \text{ per unit}$$

(9) Calculate X'_d of two synchronous machines in per unit with Eq 54

    (a) Synchronous motor on bus 8

$$X'_d = j0.28 \cdot \frac{13.2^2}{14.427^2} \cdot \frac{10\,000}{9800}$$

$$= j0.2392$$

    (b) Generator $G1$

$$X'_d = j0.20 \cdot \frac{13.8^2}{14.427^2} \cdot \frac{10\,000}{8500/0.8}$$

$$= j0.1722$$

The per unit data and the base voltages have been transferred to the impedance diagram of Fig 34, in readiness for the preparation of computer input document and as a record of the basic information for the study.

The above calculation method illustrates the use of the per unit system by selection of base voltages on the different parts of the system using transformer turns ratios. While this method is straightforward for hand calculations, it can lead to errors in interpretation of results when used in computer studies. In computer studies, it is usually preferred to select the base voltages for each part of the system to be equal to the nominal voltage. In the above example, the base voltages would be 13.8 kV for buses 4, 8, 24, 31 and 32, 2.4 kV for bus 36 and 0.48 kV for bus 37. The difference between the transformer turns ratio and the ratio of the bases selected for the buses on each side of the transformer is accounted for by using an off-nominal tap in the computer program model of the transformer. Modeling the tap on the high side of the transformer for the case being considered, the values of the tap would be

$$t = \frac{66}{69} = 0.956 \text{ per unit}$$

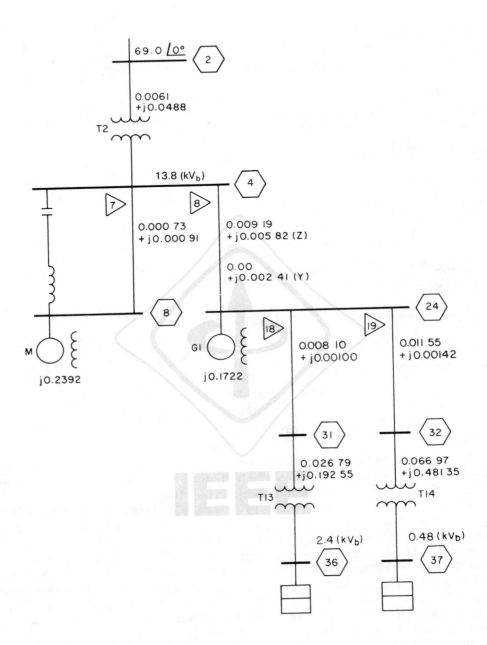

Fig 34
Impedance Diagram Per Unit Data
(Base MVA = 10)

Thus the off-nominal turns ratio is modeled in a manner similar to the modeling of transformers with movable taps as explained in 4.7.6.2.

The use of nominal voltages as base voltages has the following advantages:

- The chance of errors in interpretation of results is reduced: it is clear to all what 1.0 per unit means.
- It is much easier to spot abnormally high or low voltages. One only has to scan for voltages outside a certain range, say 0.95 to 1.05 per unit.
- Changes in network configuration will not require changes in network impedances. For example, if two parts of the 13.8 kV system are modeled using base voltages of 14.427 and 14.0 kV due to different transformer turns ratios, the study of the outage of one transformer and the connection of the two systems via a normally open breaker could not be performed until the impedances of one of the systems was converted to the base of the other.

The user's manual of most computer programs will contain information on the modeling of transformer taps and off-nominal ratios.

4.9 Models of Bus Elements

4.9.1 Loads in General. Power system loads may be classified by one or a combination of the following types, to account for their voltage dependence:

Constant power S
Constant impedance Z
Constant current I

Figure 35 shows their respective power/voltage relationships. A single expression:

$$\left(\frac{S}{S_i}\right) = \left(\frac{V}{V_i}\right)^k \qquad \text{(Eq 55)}$$

can represent the three load types by making $k = 0$ for constant power, $k = 1$ for constant current, and $k = 2$ for constant impedance loads. S_i is the initial power at V_i, the initial voltage, S the power at voltage V.

A more general expression can be formulated by expanding Eq 55 for the real and reactive power:

$$P + jQ = P_i \left(\frac{V}{V_i}\right)^{k_1} + jQ_i \left(\frac{V}{V_i}\right)^{k_2} \qquad \text{(Eq 56)}$$

The subscript i has the same meaning as above. The exponents k_1 and k_2 could be different and nonintegers.

A load or group of loads could also be expressed in a more restrictive way by:

$$P + jQ = \left[A + B \frac{V}{V_i} + C \left(\frac{V}{V_i}\right)^2 \right] P_i + j \cdot \left[D + E \frac{V}{V_i} + F \left(\frac{V}{V_i}\right)^2 \right] Q_i \qquad \text{(Eq 57)}$$

again to reflect voltage dependency. In this case A, B, C and D, E, F represent fractions of P and Q, respectively. The sums $A + B + C$ and $D + E + F$ must equal 1.

In stability studies, frequency, like voltage, may become an important factor in the modeling of loads. Linear frequency dependence would take the form

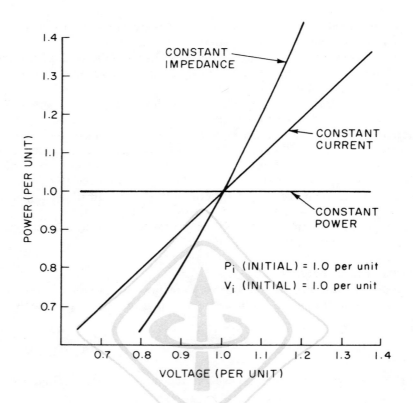

Fig 35
Effect of Voltage Variations for Three Types of Loads

$$P + jQ = (1 + G\Delta f)P_i + j(1 + H\Delta f)Q_i \qquad \text{(Eq 58)}$$

where G and H are the fractions of P_i and Q_i, respectively, being affected by the frequency deviation f from the steady-state frequency.

The problem of assigning correct values to the constants $(k_1, k_2, A$ through $H)$ is very difficult when studying utility type systems because the nature of the load is not known accurately. Additionally, the tasks of simulating the load in all its details would require a computer program of such size and cost that the effort might be prohibitive (see References [10], [19], [20], and [22]).

Industrial and commercial power systems are relatively modest in size. Moreover bus loads are often arranged in such a way that grouping by type is easy to do, thus facilitating the preparation of computer input data and offering the possibility of combining large sections of the system to reduce the overall size of the study sections.

4.9.2 Induction Motors. The induction motor single phase equivalent circuit is shown in Fig 36 (see References [14], [24], and [25]). The solution of this circuit for values of slip s at a given input voltage and frequency will yield, amongst other

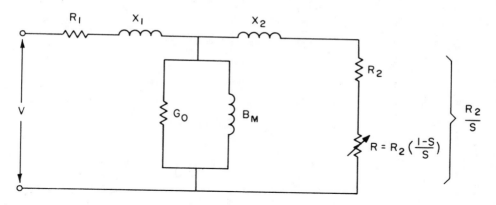

Fig 36
Induction Motor Equivalent Circuit

performance characteristics, three sets of curves: torque, current and power factor, typified in Figs 37, 38, and 39.

If the three curves are known, the shaft output power can be calculated for any speed using the slip and torque with the equation:

$$\tau = 7.05 \cdot \frac{P_o}{(1 - S)N_s} \qquad \text{(Eq 59)}$$

where

τ = torque in lb-ft
P_o = output power in watts
N_s = synchronous speed in rev/min

Input power can also be calculated using the current and power factor curves for the desired voltage.

$$S = 3VI\,(\cos\theta + j\sin\theta) \qquad \text{(Eq 60)}$$

$$\theta = \arccos PF \qquad \text{(Eq 61)}$$

where V is the line-to-neutral voltage.

In some cases, the equivalent circuit constants will be known along with the input or output power. The analytical solution consists of finding the correct value of slip that will match the specified power. A cut-and-try method may be used here; but this will prove to be a tedious process if resolved by hand because of the repeated reduction of the complex network for each trial of slip value. Computers excel at iterative methods, and for this reason programs will generally be written to accept this kind of input. The *equivalent circuit* model finds its major applications in motor starting and stability studies.

The curves of Figs 37, 38, and 39 are also used in motor starting studies. The computer input data will consist of the motor equivalent circuit and/or sets of values of torque, current and power factor taken at different intervals of slip along

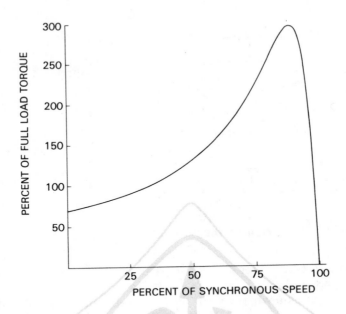

Fig 37
Induction Motor Torque versus Speed

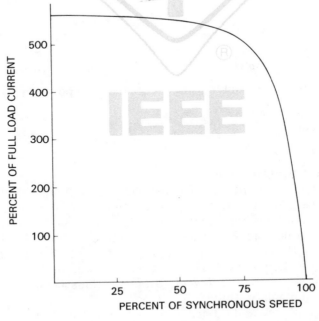

Fig 38
Induction Motor Current versus Speed

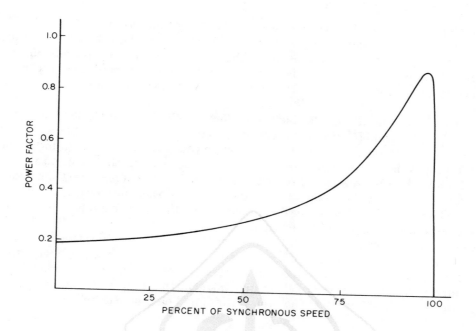

Fig 39
Induction Motor Power Factor versus Speed

these curves. The mechanical load characteristics in the form of a torque-speed curve (in the same format as for the motor) will normally be required as well by the program. The motor and load moments of inertia will complement the description of the mechanical system.

NOTE: All inertias must be calculated on the same base. In addition, inertias of couplings, pulleys, and flywheels should also be considered.

The equivalent circuit constants can be calculated from motor test data or obtained from the manufacturer (calculated or test values). References [6], [24] and [25] describe the methods used for the calculations from test data.

4.9.2.1 Constant kVA Model. It may be appropriate in some studies to represent a motor as a constant kVA load:

$$S = P + jQ = \text{constant}$$

This requires that the shaft horsepower (BHP), nameplate voltage, motor efficiency η, and power factor be known or estimated and substituted in the following equations:

$$P \text{ (kW)} = 0.746 \cdot \frac{\text{BHP}}{\eta}$$

(Eq 62)

$$\theta = \arccos \text{PF}$$

(Eq 63)

$$Q \text{ (kvar)} = P \tan \theta$$

(Eq 64)

Both the efficiency and power factor are functions of motor voltage and percent load. An example is shown in Fig 40. Consequently, if the objectives of the study warrant, it may be necessary to estimate the voltage in order to determine appropriate values for η and PF. Manufacturers publish motor characteristics (current, efficiency, power factor) at various loadings for a wide spectrum of motor sizes and types.

This model is generally applicable to load flow studies or other studies where the effect of model simplification is of secondary importance.

It should be emphasized that the reactive power Q is positive for an induction motor, even if the motor is operating in the generating mode, that is, negative slip.

4.9.2.2 Models for Short-Circuit Studies. In this area of power system analysis, standards specify the model to be used for motors to account for their contribution to short-circuit currents (see References [6] and [15]). The model is invariably a voltage source in series with an impedance as in Fig 41.

The voltage source is justified by the fact that at the very instant of a short circuit, the flux that exists at the air gap cannot change instantly. The energy that this flux represents has to dissipate itself in the form of current through resistance. This current flows in the electrical circuits linked by that flux, and is limited by the inductance of these circuits. The stator winding, being linked by that flux and

**Fig 40
Effect of Voltage Variations on
Typical Induction Motor Characteristics**

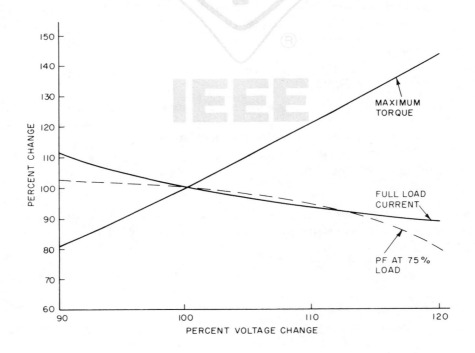

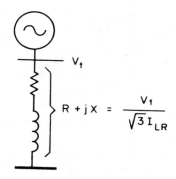

**Fig 41
Model of Induction Motor for Short-Circuit Study**

having inductance, will therefore act as a limited source of current to the network connected to it. The flux will decay rapidly, there being no source of current to maintain it. Thus, the time constant will be in the order of a few cycles.

The value of impedance in series with the voltage source is nearly the same value as the impedance which limits the locked rotor current when voltage is applied to the motor at rest. Accordingly, the impedance in Fig 41 can be calculated from:

$$R + jX = \frac{V}{\sqrt{3}\,I_{LR}}$$

(Eq 65)

where V is the motor nameplate line-to-line voltage and I_{LR} is the motor locked rotor current.

The resistance R is usually small relative to the reactance X. Moreover, short-circuit currents have a low power factor. These two factors justify neglecting R in short-circuit studies, where only the current magnitude is required.

When the short-circuit study objective is to check circuit breaker capabilities, it is necessary to consider different conditions such as first cycle duty, momentary duty, interrupting duty, and breaker speed, in order to determine an appropriate value of motor series impedance that will account for the decay in current it contributes to the fault. This is spelled out in Reference [1] and takes the form of impedance multiplying factors that vary as a function of time from the moment the fault occurred. The rationale for neglecting small motor contributions is based on their remoteness, electrically, from the power circuit breakers and their very short time constants.

4.9.2.3 Constant Impedance Model. It is sometimes desirable to determine the system voltages at the instant a motor at rest is energized. The appropriate model is the equivalent circuit given earlier with the slip s equal to 1. This circuit is a constant impedance. It is generally simplified to a single impedance to ground and calculated from Eq 65.

4.9.3 Synchronous Machines. Synchronous machine models vary tremendously whether a study considers steady-state or transient conditions. In this section, the text will include both the simpler model for steady-state conditions, and the more complicated models for transient conditions.

4.9.3.1 Steady-State Models

4.9.3.1.1 Generators. Once a power system has settled down after a change of any kind, generator output voltages will have been automatically brought back to the desired output values by the voltage regulator. The prime mover governor will also have taken a steady position to maintain the generator speed, consistent with the rest of the system, and to supply the amount of power programmed by its setting.

The generator output voltage will be a function of the field current. Its output power will be a function of the mechanical power applied to its shaft which will also determine the rotor angle that the field poles maintain relative to the revolving magnetic field in the stator.

Thus, in a load flow study for steady-state conditions, the generator can be modeled as a constant voltage source and a scheduled amount of kW. The system operating conditions may demand that a generator voltage output be adjusted automatically so that it supplies an amount of reactive power such that a bus, somewhere on the system, maintains a specified voltage. In this situation, the analyst specifies a range of reactive power (kvar) within which the machine will operate. The generator voltage will be adjusted by the program.

Other possibilities are:

(1) Specify a voltage and let the kW and kvar fall where they may.
(2) Specify fixed kW and kvar and let the voltage fall where it may.

The last alternative suggests that a generator may be considered analytically as a negative constant power load. Many computer programs will accept negative power signs. Thus negative kW input to the bus load data would model a generator.

4.9.3.1.2 Synchronous Condenser. One difference between this machine and a generator is that the condenser consumes a small amount of real power corresponding to its losses. It will generally be equipped with a voltage regulator similar to a generator and have its reactive power output, specified within certain limits, adjusted by the computer program to maintain a specified voltage at its own terminals or elsewhere in the system.

4.9.3.1.3 Synchronous Motors. Synchronous motors may or may not be equipped with regulators to control their excitation. Those equipped with regulators may control voltage, power factor, reactive power or even current at their terminals or elsewhere. The real power in kW will be a function of the load driven by the motor, and will not be adjustable for the given set of conditions under study.

The analyst may, therefore, resort to modeling the motors as negative generators with reactive power limits, if a voltage or current control device is supplied. In the case of power factor or reactive power regulators, the specified kW will define a value of kvar using Eqs 66 through 68.

For motors with fixed excitation and a given fixed load, vee-curves must be used to calculate the equivalent reactive power. Figure 42 shows the vee-curve for a particular motor. The reactive power is calculated as follows:

(1) Draw a vertical line for the fixed excitation current

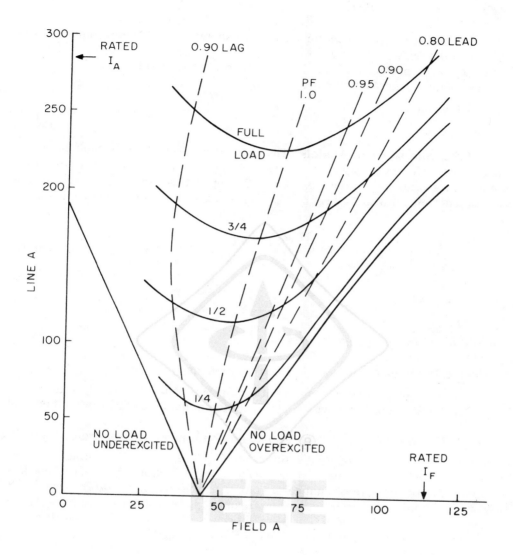

Fig 42
Vee-Curves: Synchronous Motor, 2000 hp,
4000 V, 180 rev/min, 0.8 Lead Power Factor

(2) Read the armature current I, at the intersection of the load line representing the motor running load, and the I line

(3) Estimate the motor terminal voltage (line to line)

(4) Calculate the reactive power

$$kvar = \sqrt{3(VI)^2 - kW^2}$$

(Eq 66)

Alternately, if the power factor curves are shown on the graph, replace above steps 3 and 4 with:

(3) Read the power factor at the intersection of the load line and I line

(4) Calculate the reactive power

$$\text{kvar} = \text{kW} \tan \theta \qquad \qquad \text{(Eq 67)}$$

$$\theta = \arccos \text{PF} \qquad \qquad \text{(Eq 68)}$$

4.9.3.2 Short-Circuit Models. The current contributed to a fault by a synchronous machine varies with time, from a high initial value to a moderate final steady-state value. Equations 69, 70, and 71 depict this variation of current as a function of time for a short circuit at the terminals of a machine operating initially at no load.

$$I_{ac} = E \left[\frac{1}{X_d} + \left(\frac{1}{X_d'} - \frac{1}{X_d} \right) e^{-t/T_d'} + \left(\frac{1}{X_d''} - \frac{1}{X_d'} \right) e^{-t/T_d''} \right] \qquad \text{(Eq 69)}$$

$$I_{dc} = \frac{\sqrt{2} \, E \cos \alpha}{X_d''} \, e^{-t/T_d} \qquad \qquad \text{(Eq 70)}$$

$$I_T = \sqrt{I_{ac}^2 + I_{dc}^2} \qquad \qquad \text{(Eq 71)}$$

The total current I is made up of two components:

(1) A power frequency ac component I_{ac}, the rms value of which decreases with time t, in accordance with Eq 69. This component is called the symmetrical current.

(2) A dc component I_{dc}, which decreases with time in accordance with Eq 70.

The initial magnitude of I_{dc} is a function of the angle α of the voltage wave at which the short circuit occurred. It is the amount of offset of the current wave. It is proportional to the rate of change of voltage at the instant of short circuit. The offset is maximum at $\alpha = 0$ because the rate of change of voltage is maximum when the voltage is 0. The offset is 0 at $\alpha = 90°$ because the rate of change is 0 at the positive or negative peak voltage values.

In the above equations:

E = the open circuit voltage

X_d, X_d', X_d'' = the direct-axis synchronous, transient, and subtransient reactances, respectively

T_a, T_d', T_d'' = the armature and the direct-axis transient and subtransient short-circuit time constants, respectively

For typical values of reactances and time constants, and with the maximum offset condition ($\cos \alpha = 1$), the equations will reduce to:

At time $t = 0$ (under subtransient conditions)

$$I_{ac}'' = \frac{E}{X_d''} \qquad \qquad \text{(Eq 72)}$$

$$I''_{\text{dc}} = \frac{\sqrt{2}E}{X''_{\text{d}}}$$

(Eq 73)

$$I''_{\text{T}} = \frac{E}{X''_{\text{d}}} \sqrt{1+2} = \frac{E}{X''_{\text{d}}} \sqrt{3}$$

(Eq 74)

At time $t = \infty$

$$I_{\text{ac}} = \frac{E}{X_{\text{d}}}$$

$$I_{\text{dc}} = 0$$

$$I_{\text{T}} = I_{\text{ac}} = \frac{E}{X_{\text{d}}}$$

(Eq 75)

Since T'_{d} is much larger than T''_{d}, there is a time t' smaller than T'_{d} and larger than T''_{d}, that will make $e^{-t'/T'_{\text{d}}}$ approximately equal to 1 and $e^{-t'/T''_{\text{d}}}$ approximately equal to 0. In this case, the original equations reduce to:

$$I'_{\text{ac}} = \frac{E}{X'_{\text{d}}}$$

(Eq 76)

$$I'_{\text{dc}} = \frac{\sqrt{2}\,E\,e^{-t'/T_{\text{a}}}}{X''_{\text{d}}}$$

(Eq 77)

Should there be an impedance $Z_{\text{L}} = R_{\text{L}} + jX$ between the machine terminal (still at no load) and the point of fault, Eqs 72, 75 and 76 will become:

$$I''_{\text{ac}} = \frac{E}{R_{\text{L}} + j(X_{\text{L}} + X''_{\text{d}})}$$

(Eq 78)

$$I'_{\text{ac}} = \frac{E}{R_{\text{L}} + j(X_{\text{L}} + X'_{\text{d}})}$$

(Eq 79)

$$I_{\text{ac}} = \frac{E}{R_{\text{L}} + j(X_{\text{L}} + X_{\text{d}})}$$

(Eq 80)

The armature time constant T_{a} will be shortened appreciably by addition of resistance R_{L} in the external circuit. So Eq 77 will become:

$$I'_{\text{dc}} = \frac{\sqrt{2}\,E\,e^{-t'/T'_{\text{a}}}}{X''_{(T'_{\text{a}} < T_{\text{a}})}}$$

(Eq 81)

The offset current will decay more rapidly the farther away (electrically) the machine is from the point of fault.

The voltage E, in all the above equations, is equal to the terminal voltage V_{t}, since it was assumed that the machine was carrying no load before the short circuit. If the machine was carrying a current I_{L} before the short circuit, the voltage E will be different in each equation, to satisfy prefault conditions. For the case of a generator, the voltage in Eqs 72, 75 and 76 will be:

$$E'' = V_t + I_L X_d'' \tag{Eq 82}$$

$$E' = V_t + I_L X_d' \tag{Eq 83}$$

$$E = V_t + I_L X_d \tag{Eq 84}$$

respectively. These voltages have been called voltage behind subtransient reactance (E''), voltage behind transient reactance (E'), and voltage behind synchronous reactance (E). It is not practical, in short-circuit studies, to calculate the system currents for the entire period — from the time of fault to the time that the current reaches a steady-state value. The normal procedure is to resolve the network at times $t = 0$ or $t = t'$, or both, using the models of Fig 43. Network solutions at $t = \alpha$ are meaningless since the machine field excitation has likely been changed at that time.

Depending on the study objectives, the effect of the offset current I_{dc} may or may not be important. In power circuit breaker applications, however, it is a very important consideration. To obviate the difficulties in resolving Eq 77, the breaker standards specify multipliers for the X_d'' and X_d' current components. These are a function of machine type and of the time from the inception of the short circuit. Reference [9] lists those multipliers, gives examples of their use and expands on this important aspect of short-circuit studies.

The models of Figs 43 and 44 are also applicable to synchronous motors and synchronous condensers, the difference being that the E'', E' and E voltages are calculated with:

$$E'' = V_t - I_L X_d'' \tag{Eq 84}$$

$$E' = V_t - I_L X_d' \tag{Eq 85}$$

$$E = V_t - I_L X_d \tag{Eq 86}$$

4.9.3.3 Stability Models. In stability studies, it is necessary to resolve the electrical networks at a series of intervals starting from the time of inception of a disturbance and for as long as necessary to establish a trend in the system perform-

Fig 43
Models of Synchronous Machines for Short-Circuit Studies

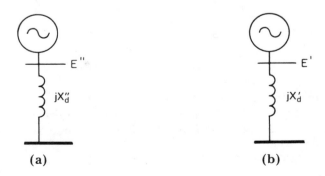

(a)   (b)

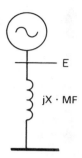

MF = Multiplier specified
in IEEE C37.010-1979 (ANSI)

Fig 44
General Model for AC Machines in Short-Circuit Studies

ance. This may cover a period shorter than 1 second for transient stability or several seconds for dynamic stability studies.

4.9.3.3.1 Classical Model. The classical model for a transient stability study consists of a simple constant voltage source behind a constant transient reactance as in Fig 43 (b). This model neglects the following factors by assuming that:

(1) The shaft mechanical power remains constant
(2) Field flux linkages remain constant
(3) Damping is nonexistent
(4) The constant voltage and reactance are not affected by speed variations
(5) The rotor mechanical angle coincides with the phase angle of the internal voltage

Ignoring dynamic stability effects sometimes associated with high-speed exciter systems, a system found stable under these assumptions will likely be stable if any or all of the above factors are taken into consideration. However, due to exciter effects and in cases where the above analysis suggests instability, there is a need to account for the neglected factors along with regulator action in certain circumstances. The various models, in increasing degree of sophistication, will be presented with simple explanations, if possible, or without explanations in which case references will be given for the reader to consult.

4.9.3.3.2 The H Constant. Stability studies are concerned with relative speed variations of rotating masses. The kinetic energy of a rotating mass, using units of Table 5, is:

$$KE = \tfrac{1}{2} I\omega^2 = \tfrac{1}{2} M\omega \text{ J} \qquad \text{(Eq 87)}$$

The rotating mass associated with a generator includes the rotor, shaft, coupling, turbine and exciter, if the rotating type is used. Since this mass is designed to rotate at a fixed (synchronous) speed, the stored energy, at synchronous speed, of a given machine is used as a constant. This constant is usually normalized by defining a quantity H that expresses the stored energy per unit of machine rated power. If kinetic energy is in megajoules and the rated power S in megavolt-amperes the H constant will be:

$$H = \frac{KE}{S} = \frac{J}{VA}$$

Since 1 joule is equal to 1 watt-second, the H constant is sometimes stated in equivalent kilowatt-second per kVA units.

To calculate the H constant use:

$$H = 0.231 \frac{(WR^2)\,(\text{rev/min})^2 \cdot 10^{-3}}{\text{MVA}} \qquad \text{(Eq 88)}$$

where WR^2 is the moment of inertia in pounds-feet squared and rev/min is the speed in revolutions per minute. The H constant values will fall within the narrow range of approximately 1 to 15, irrespective of the size of the machines.

4.9.3.3.3 Stability Model Variations. In the discussion of 4.9.3.2, only the direct-axis parameters were considered on the basis that short circuits produce currents of low power factor (quadrature currents predominate). This assumption may not be acceptable for disturbances considered in stability analysis. Therefore, additional synchronous machine parameters are required to more accurately model the behavior and account for the differences in the magnetic construction types, such as salient poles, smooth rotor, laminated rotor, solid iron rotor with or without dampers.

Quadrature-axis reactances and open-circuit time constants are defined for that purpose. Chapter 1 of Reference [23] is especially recommended as a clear and basic text on the subject.

The classical model may be improved one step by taking account of the variation of machine impedance with time from its initial value, X_d', to a steady-state value of X_d. The variation will be an exponential described by a time constant T_{do}' (transient, open-circuit time constant). The three parameters X_d, X_d' and T_{do}' will ignore the major effect of dampers.

Another improvement will involve adding the effect of dampers which predominate during fast changing conditions, that is, the subtransient state. The additional parameters X_d'' and T_{do}'' will take care of this effect.

The saliency of the rotor will be represented by the quadrature-axis parameters, X_q, X_q' and X_q'' and the associated time constants T_{qo}' and T_{qo}''.

Adding the parameters X_q' and T_{qo}' to those of the first improved model will increase the accuracy in the case of a generator with a solid iron rotor. But, as before, the damping effect will have been mostly neglected. The solid iron rotor will be fully represented by all direct- and quadrature-axis, synchronous, transient and subtransient reactances and associated time constants.

The transient quadrature-axis reactance of salient pole machines has the same value as the equivalent synchronous reactance. Thus salient pole machines can be fully modeled as the solid iron rotor machine by omission of the X_q' and T_{qo}' parameters.

4.9.3.4 Exciter Models

(1) *Saturation.* The field poles saturate as the excitation current exceeds a certain level. (See Fig 45.) Computer programs will usually account for the related nonlinearity of air-gap voltage and field current from input data representing two points on the saturation curve. Refer to program instructions for which two points to use.

(2) *Standard Models.* An IEEE committee has developed a number of models to represent excitation systems and the dynamic characteristics of synchronous machines for stability studies (see Reference [16]). A tutorial paper [18] supplements Reference [16] by discussing the transfer function blocks and their practical meanings as well as other topics related to excitation system response. Only type (1) is repeated here (Fig 46) to illustrate the following points.

Consider the simple circuit of Fig 47. If the input voltage V is a step function (voltage changes suddenly at $t = 0$ from 0 to 1.0 pu V), the output voltage V will be an exponential function of time:

$$V_\text{o} = V_\text{i}(1 - e^{-t/RC}) \qquad \text{(Eq 89)}$$

that may be rearranged

$$\frac{V_\text{o}}{V_\text{i}} = 1 - e^{-t/RC} \qquad \text{(Eq 90)}$$

Fig 45
Saturation Curves

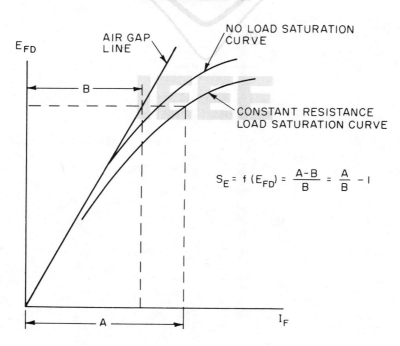

$$S_E = f(E_{FD}) = \frac{A-B}{B} = \frac{A}{B} - 1$$

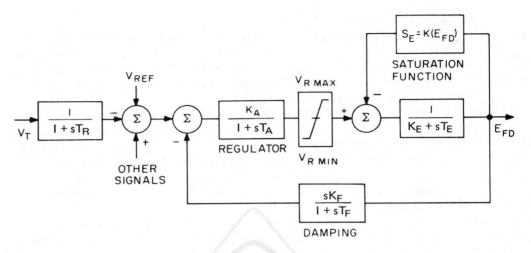

Fig 46
IEEE Type 1 Excitation System

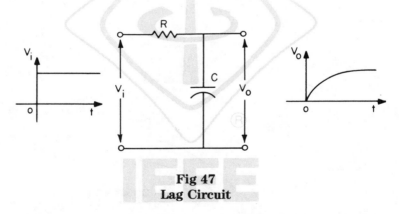

Fig 47
Lag Circuit

If the input is sinusoidal voltage, the output will also be sinusoidal but its amplitude will be reduced by the factor

$$\frac{1}{1 + j\omega CR}$$

and phase shifted by a lagging angle

$$\phi = \tan^{-1} \frac{R}{\omega C}$$

since:

$$V_o = V_i \frac{1}{1 + j\omega CR} \tag{Eq 91}$$

where

$\omega = 2\pi f$ and f is frequency.

Dividing both sides of Eq 91 by V_i yields:

$$\frac{V_o}{V_i} = \frac{1}{1 + j\omega CR}$$

(Eq 92)

This equation has the same form as the equation $1/(1 + sT_R)$ (transfer function) in the first block of the type 1 excitation system ($s = j\omega$ and $T_R = RC$). Analysis of the response of this simple RC *lag* circuit can therefore be made in the time domain (Eq 89) or the frequency domain (Eq 91). It is common practice to use the frequency domain representation for both the excitation (as in Fig 46) and the prime mover systems. It *pictorializes* the system.

The lag transfer function just introduced has many applications, a few of which are representations of:

(1) The delay for the field current to change to a steady-state value following a change of input field voltage
(2) The delay of water flow rate changes in penstocks
(3) The delays inherent in components of control mechanisms

A similar analysis with the circuit of Fig 48 would yield the *lead* transfer function which has the form $s/(1 + sT)$. Its response time can be compared with the previous one. In the type 1 excitation system, this function serves the purpose of damping the effect of the amplification K_A of the regulator that forces the field for faster response. K_F is the fraction of the output field voltage that is fed back to accomplish this damping.

4.9.3.5 Prime Movers and Governor Models. Basic models for speed-governing systems and turbines in power system stability studies have been presented in an IEEE Committee Report [17]. As mentioned earlier, the models are in the form of block diagrams with transfer functions describing the system components' performance. Two more papers [12] and [26] cover some of the basics and

**Fig 48
Lead Circuit**

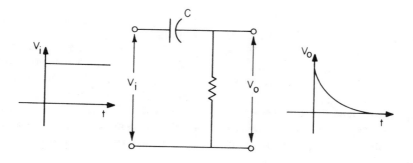

will help the novice to understand the relationships between the physical elements and the transfer functions.

Typical parameter values are also available in these references. Of course, analysts will be well advised to seek from the manufacturer the data applicable to their equipment before compromising with typical data.

4.10 Miscellaneous Bus Element Models

4.10.1 Lighting and Electric Heating. Lighting and electric heating often constitute a large section of a plant load, particularly in commercial buildings. This type of load can be modeled as constant admittance as suggested by Figs 4.10, 4.11, and 4.12 of Reference [21]. The constant admittance model would seem appropriate for fluorescent and mercury vapor lighting as well as for incandescent lighting. In utility type networks where substations are equipped with voltage regulators, lighting and heating can be represented as constant $P + jQ$.

To calculate the admittances, determine the watts P and vars Q at rated voltage V and resolve

$$Y \text{ (siemens)} = \frac{P + jQ}{E^2}$$

The fluorescent and mercury vapor lighting power factors will be determined from manufacturers' data in order to calculate the vars Q. Incandescent lights and electric heating will have unity power factors.

4.10.2 Electric Furnaces. In load flow studies, this load will usually be represented by constant power which will reflect a desired controlled operating condition to be analyzed. In the case of short-circuit and stability studies, the electric furnaces may behave like a constant impedance load. It is unlikely that automatic load tap changers and electrode position controls will have had time to change from the prefault condition to the end of the period covered in transient stability studies. This may very well be the case also in dynamic stability studies extending to several seconds. Since furnace loads can be a large section of a plant load, it is important that the constant impedance model be specified in stability problems because of its damping effect on the power system.

4.10.3 Shunt Capacitors. Banks of capacitors are used extensively to correct power factors and as a result, improve voltage regulation at the point of connection. They are modeled simply by a constant shunt capacitive susceptance.

$$jB = j2\pi fC$$

where C is in farads, f in hertz and B in siemens.

Often the capacitors will be specified in kvars. The rated voltage of the bus where the capacitors are connected must be known whether the susceptance or the equivalent kvars at the base voltage are used in the study. Remember that kvars vary proportionally as the square of the voltage for a constant susceptance.

4.10.4 Shunt Reactors. These are seldom used in industrial power systems. A constant admittance would be the correct model for reactors that do not saturate. The same voltage considerations as in the last paragraph would apply.

4.11 References

[1] IEEE C37.010-1979, Application Guide for AC High-Voltage Circuit Breakers Rated on a Symmetrical Current Basis (ANSI).[5]

[2] IEEE C37.04-1979 (R 1988), American National Standard Rating Structure for AC High-Voltage Circuit Breakers Rated on a Symmetrical Current Basis (ANSI).

[3] IEEE C57.12.80-1978 (R 1986), Terminology for Power and Distribution Transformers (ANSI).

[4] IEEE Std 58-1978 (R 1983), Induction Motor Letter Symbols (ANSI).

[5] IEEE Std 86-1987, Recommended Practice: Definitions of Basic Per-Unit Quantities for AC Rotating Machines (ANSI).

[6] IEEE Std 112-1984, Standard Test Procedure for Polyphase Induction Motors and Generators (ANSI).

[7] IEEE Std 115-1983, Test Procedures for Synchronous Machines (ANSI).

[8] IEEE Std 115A-1987, Standard Procedures for Obtaining Synchronous Machine Parameters by Standstill Frequency Response Testing (ANSI).

[9] IEEE Std 122-1985, Recommended Practice for Functional Performance Characteristics of Control Systems for Steam Turbine Generator Units (ANSI).

[10] ADLER, R. B. and MOSHER, C. C. Steady State Power Characteristics for Power System Loads, *IEEE Paper 70CP 706-PWR*, 1970.

[11] *Aluminum Electrical Conductor Handbook*, The Aluminum Association, 750 Third Ave., New York, 1971.

[12] EGGENBERGER, M. A. A Simplified Analysis of the No-Load Stability of Mechanical–Hydraulic Speed Control Systems for Steam Turbines, *ASME Paper 60-WA-34.*

[13] *Electrical Transmission and Distribution Reference Book*, East Pittsburgh, PA, Westinghouse Electric Corporation, 1964.

[14] FITZGERALD, A. E. and KINGSLEY, Jr., C. *Electric Machinery*, New York, McGraw-Hill, 1961.

[15] HUENING, Jr., W. C. Interpretation of New American National Standards for Power Circuit Breaker Applications, *IEEE Transactions on Industry and General Applications*, Sept./Oct. 1969.

[16] IEEE COMMITTEE REPORT. Computer Representation of Excitation Systems, *IEEE Transactions on Power Apparatus and Systems*, June 1968.

---

[5] IEEE publications are available from the Institute of Electrical and Electronics Engineers, IEEE Service Center, 445 Hoes Lane, P.O. Box 1331, Piscataway, NJ 08855-1331.

[17] IEEE COMMITTEE REPORT. Dynamic Models for Steam and Hydro Turbines in Power System Studies, *IEEE Transactions on Power Apparatus and Systems*, Nov./Dec. 1973.

[18] IEEE COMMITTEE REPORT. Excitation System Dynamic Characteristics, *IEEE Transactions on Power Apparatus and Systems*, Jan./Feb. 1973.

[19] IEEE COMMITTEE REPORT. System Load Dynamics — Simulation Effects and Determination of Load Constants, *IEEE Transactions on Power Apparatus and Systems*, Mar./Apr. 1973.

[20] ILICITO, F., CEYHAN, A. and RUCKSTUHL, G. Behavior of Loads During Voltage Dips Encountered in Stability Studies — Field and Laboratory Tests, *IEEE Transactions on Power Apparatus and Systems*, Nov./Dec. 1972.

[21] *Industrial Power Systems Handbook*, BEEMAN, D., editor, New York, McGraw-Hill, 1955.

[22] KENT, M. H., SCHMUS, W. R., McCRACKIN, F. A., and WHEELER, L. M. Dynamic Modeling of Loads in the Stability Studies, *IEEE Transactions on Power Apparatus and Systems*, May 1969.

[23] KIMBARK, E. W. Power System Stability: Synchronous Machines, vol. 3, New York, Dover Publications, Inc., 1968.

[24] McFARLAND, T. C., and VAN NOSTRAND, D. *Alternating-Current Machines*, New York, John Wiley & Sons, 1948.

[25] PUCHSTEIN, A. F., and LLOYD, T. C. *Alternating-Current Machines*, New York, John Wiley & Sons, 1974.

[26] RAMEY, D. G. and SKOOGLUND, J. W. Detailed Hydrogovernor Representation for System Stability Studies, *IEEE Transactions on Power Apparatus and Systems*, Jan. 1970.

[27] *Standard Handbook for Electrical Engineers*, FINK and BEATTY, editors, New York, McGraw-Hill, 12th edition.

[28] STEVENSON, Jr., W. D. *Elements of Power System Analysis*, New York, McGraw-Hill, 1975.

Chapter 5
Computer Solutions and Systems

5.1 Introduction. Serious power system analysis work requires the use of computers and specialized programs. Hand calculations are suitable for estimating the operating characteristics of a few individual circuits; but accurate calculation of voltages, power flows, or short-circuit currents throughout an industrial or commercial power system would be impractical without the use of computer programs. The various types of power system studies, as described in the following chapters, generally involve developing mathematical models in terms of algebraic or differential equations and then solving the equations for numerous design conditions. Computer-aided analysis permits the engineer to focus on power system operation rather than on numerical manipulations.

Success in selecting and applying computer techniques requires the engineer to be familiar with the power system problem as well as many computer hardware and software considerations. The purpose of this chapter is to acquaint prospective users of computer methods in power system analysis with the available types of hardware and software resources. In particular, commonly used numerical methods are introduced to provide insight into how computer programs manipulate large quantities of power system data and determine accurate solutions for various kinds of equations. This chapter is not intended to explain in detail how to write computer programs for power system analysis. For those so inclined, numerous books are devoted to power system analysis programming [B2],[6] [B14], [B15], [B27], [B30].

Digital computers were first used for small power system problems in the late 1940's; but it was not until the availability of large-scale computers in the mid-1950's that power system analysis programs became really useful. Before this time, system studies were performed using special-purpose analog computers, called network analyzers or network calculators, which were introduced in 1929 [B8]. Power system studies were limited to hand calculations, assisted by slide rule, before 1929. The development of digital computer technology has been remarkable. Currently, it is common to find computers many times more powerful than the room-filling machines of the 1950's on the desks of individual engineers. Steady advancements have also been made in numerical methods and computer programming.

---

[6]The bibliographic references (capital B plus number in brackets) are listed at the end of this chapter.

The advancements in computer technology, particularly as reflected in shrinking computing costs, have brought many engineers into direct contact with computers for the first time. Most find the experience rewarding, if approached with patience. Computers and power system analysis programs have become more readily available and easier to use. Unfortunately, it is still easy to get erroneous results from computer programs. Technology has done little to invalidate the maxim, "garbage in, garbage out." Data used in a study must be carefully assembled and checked for input errors. The impact of errors in any assumptions made by the user or by the program (e.g., default values) should be considered. Modeling and solution techniques should be understood. In general, it is important to exercise engineering judgment when reviewing computer results and avoid the tendency to accept numbers on a printout blindly.

5.2 Numerical Solution Techniques. Computer programs for power system analysis use efficient numerical methods that permit a standardized step-by-step approach to setting up and solving equations. Methodical procedures are required in general-purpose computer programs designed for solving complex engineering problems. Such programs adapt, within limits, to the size of the problem at hand. For example, a load flow analysis program might be designed to handle systems with only two buses or as many as 1000. Matrix methods are essential for achieving this flexibility and orderliness.

5.2.1 Matrix Algebra Fundamentals. A *matrix* is a rectangular array of numbers arranged in rows and columns enclosed by square brackets or large parentheses. The numbers in a matrix, called *elements*, may be real or complex. The *dimension* of a matrix is expressed as m × n (read "m by n") if the matrix has m rows and n columns. If a matrix has only one row, it is called a *row matrix* or *row vector*. Similarly, a matrix with a single column is known as a *column matrix* or *column vector*. A matrix is often denoted by a capital letter such as **A**, and an element in the matrix is represented as a_{ij}, where i is the row number and j is the column number. For example:

$$\mathbf{A} = \begin{bmatrix} a_{11} & a_{12} & \cdots & a_{1n} \\ a_{21} & a_{22} & \cdots & a_{2n} \\ \cdot & \cdot & & \cdot \\ \cdot & \cdot & & \cdot \\ \cdot & \cdot & & \cdot \\ a_{m1} & a_{m2} & \cdots & a_{mn} \end{bmatrix}$$

If m = n, the matrix is called a *square matrix* of *order* n. The elements of a square matrix for which i = j are called *diagonal elements*. The other elements in a square matrix are referred to as *off-diagonal elements*. If corresponding off-diagonal elements are equal (i.e., $a_{ij} = a_{ji}$). the matrix is a *symmetric matrix*. For example,

$$\begin{bmatrix} 3 & 6 & 2 & 7 \\ 6 & 5 & 1 & 5 \\ 2 & 1 & 4 & 8 \\ 7 & 5 & 8 & 9 \end{bmatrix}$$

is a symmetric matrix of order 4 (dimension 4 × 4).

Two matrixes are *equal* if they have the same dimension and the corresponding elements in each matrix are equal. If the rows and columns of a matrix **A** are interchanged, the resulting matrix is called the *transpose* of **A** and is denoted by $\mathbf{A}^T$. If **A** is a symmetric matrix, then $\mathbf{A} = \mathbf{A}^T$.

Addition and *subtraction* of matrixes are valid only for matrixes of the same dimension. The *sum*, $\mathbf{A} + \mathbf{B} = \mathbf{C}$, is carried out for each element in **C** as $c_{ij} = a_{ij} + b_{ij}$. The order of addition does not matter (i.e., $\mathbf{A} + \mathbf{B} = \mathbf{B} + \mathbf{A}$). Similarly, the *difference*, $\mathbf{A} - \mathbf{B} = \mathbf{C}$, is determined by calculating each element in **C** as $c_{ij} = a_{ij} - b_{ij}$.

Multiplication of a matrix by a number (or *scalar*) is valid for any matrix. The *product*, $k\mathbf{A} = \mathbf{B}$, is found by multiplying each element in **A** by the number k (i.e., $b_{ij} = ka_{ij}$). Multiplication of two matrixes, $\mathbf{AB} = \mathbf{C}$, is valid only if the number of columns in **A** equals the number of rows in **B**. If **A** is a matrix of dimension m $\times$ n and **B** is a matrix of dimension n $\times$ p, then the product **C** is a matrix of dimension m $\times$ p. Each element of **C** is calculated as follows:

$$c_{ij} = a_{i1}b_{1j} + a_{i2}b_{2j} + \ldots + a_{in}b_{nj}$$

For example, if

$$\mathbf{A} = \begin{bmatrix} 1 & 2 & 3 \\ 4 & 5 & 6 \\ 7 & 8 & 9 \end{bmatrix} \text{ and } \mathbf{B} = \begin{bmatrix} u & x \\ v & y \\ w & z \end{bmatrix} \text{ then }$$

$$\mathbf{AB} = \begin{bmatrix} u+2v+3w & x+2y+3z \\ 4u+5v+6w & 4x+5y+6z \\ 7u+8v+9w & 7x+8y+9z \end{bmatrix} = \mathbf{C}$$

If the product **AB** is valid, then **BA** is valid only if **A** and **B** are square matrixes of the same order. In general, the products **AB** and **BA** are not equal. An exception is when **B** is equal to the *unit matrix* **U**. The unit matrix is a square matrix of the same order as **A** in which all diagonal elements equal 1 and all off-diagonal elements are 0. In this case, $\mathbf{AU} = \mathbf{UA} = \mathbf{A}$. For example:

$$\begin{bmatrix} 8 & 3 \\ 5 & 2 \end{bmatrix}\begin{bmatrix} 1 & 0 \\ 0 & 1 \end{bmatrix} = \begin{bmatrix} 1 & 0 \\ 0 & 1 \end{bmatrix}\begin{bmatrix} 8 & 3 \\ 5 & 2 \end{bmatrix} = \begin{bmatrix} 8 & 3 \\ 5 & 2 \end{bmatrix}$$

The only form of division defined in matrix algebra is division of a matrix by a scalar k. This is equivalent to multiplication of a matrix by the reciprocal, or *inverse*, of the number k^{-1}. Given the matrix expression $\mathbf{AX} = \mathbf{B}$ in which the elements of **X** are unknown, a solution $\mathbf{X} = \mathbf{A}^{-1}\mathbf{B}$ may exist where $\mathbf{A}^{-1}$ is the inverse of the square matrix **A**. If $\mathbf{A}^{-1}$ exists, it is a square matrix of the same order as **A**, and it satisfies the condition $\mathbf{A}^{-1}\mathbf{A} = \mathbf{AA}^{-1} = \mathbf{U}$, where **U** is the unit matrix. This definition implies a method for determining the inverse of a matrix. For example:

If $\mathbf{A} = \begin{bmatrix} 8 & 3 \\ 5 & 2 \end{bmatrix}$, let $\mathbf{A}^{-1} = \begin{bmatrix} w & y \\ x & z \end{bmatrix}$

Then $\mathbf{AA}^{-1} = \begin{bmatrix} 8w+3x & 8y+3z \\ 5w+2x & 5y+2z \end{bmatrix} = \begin{bmatrix} 1 & 0 \\ 0 & 1 \end{bmatrix}$

The resulting equations $\quad 8w+3x=1 \quad\quad 8y+3z=0$
$\quad\quad\quad\quad\quad\quad\quad\quad\quad\quad 5w+2x=0 \quad\quad 5y+2z=1$
have the solution $\quad w = 2,\ x = -5,\ y = -3,\ z = 8.$

Therefore $\mathbf{A}^{-1} = \begin{bmatrix} 2 & -3 \\ -5 & 8 \end{bmatrix}$

Matrix inversion can be used to solve simultaneous algebraic equations. This will be described later along with a more practical method for finding the inverse of a matrix.

An operation that is useful for simplifying computations with a large matrix or for showing the specific structure of a matrix is called *partitioning*. A partitioned matrix is divided by horizontal and vertical lines into *submatrixes* that are treated as single elements in addition, subtraction, and multiplication. For example, the 3×3 matrix $\mathbf{A}$ shown below can be partitioned into four submatrixes $\mathbf{A}_1, \mathbf{A}_2, \mathbf{A}_3$, and $\mathbf{A}_4$.

$$\mathbf{A} = \begin{bmatrix} 1 & 2 & \vdots & 3 \\ 4 & 5 & \vdots & 6 \\ \cdots & \cdots & + & \cdots \\ 7 & 8 & \vdots & 9 \end{bmatrix} = \begin{bmatrix} \mathbf{A}_1 & \mathbf{A}_2 \\ \mathbf{A}_3 & \mathbf{A}_4 \end{bmatrix}$$

A column vector $\mathbf{B}$ can be partitioned to facilitate multiplication by $\mathbf{A}$.

$$\mathbf{B} = \begin{bmatrix} 3 \\ 2 \\ \hline 1 \end{bmatrix} = \begin{bmatrix} \mathbf{B}_1 \\ \mathbf{B}_2 \end{bmatrix}$$

$$\mathbf{AB} = \begin{bmatrix} \mathbf{A}_1 & \mathbf{A}_2 \\ \mathbf{A}_3 & \mathbf{A}_4 \end{bmatrix} \begin{bmatrix} \mathbf{B}_1 \\ \mathbf{B}_2 \end{bmatrix} = \mathbf{C} = \begin{bmatrix} \mathbf{C}_1 \\ \mathbf{C}_2 \end{bmatrix}$$

The submatrix $\mathbf{C}_2$, which consists of a single element for this partitioning, can be determined as follows:

$$\mathbf{C}_2 = \mathbf{A}_3\mathbf{B}_1 + \mathbf{A}_4\mathbf{B}_2 = \begin{bmatrix} 7 & 8 \end{bmatrix} \begin{bmatrix} 3 \\ 2 \end{bmatrix} + \begin{bmatrix} 9 \end{bmatrix} \begin{bmatrix} 1 \end{bmatrix}$$

$$= 21 + 16 + 9 = 46$$

The original matrixes and corresponding submatrixes must be compatible for multiplication, addition, or subtraction if one of these operations is to be performed on partitioned matrixes. Special techniques for finding the transpose and the inverse of a matrix by operating on its submatrixes are also available [B27].

Only the fundamental definitions in matrix algebra have been presented here. Comprehensive information is available in numerous books on mathematics and electrical circuits [B5], [B11], [B26], [B27], [B29].

5.2.2 Power System Network Matrixes. The matrixes used in computer programs for several types of power system analysis are based on the mesh-current and node-voltage analysis methods described in introductory electrical circuit theory texts [B5], [B13], [B24]. In power systems work, the terms *loop* and *bus* are frequently used in place of mesh and node, respectively. The simple, three-bus power system shown in the single-line diagram of Fig 49 will be used to explain these methods. Resistance is neglected in the discussion in order to keep the arithmetic easier to follow, although computer programs are normally written to use both resistance and reactance in calculations.

A single-phase equivalent impedance diagram is the basis for the loop current analysis method. Generators and motors are represented as series reactances and emf's connected to the neutral bus. The values of the series reactances will vary depending on whether a transient or steady-state analysis is required. A per unit impedance diagram for the system of Fig 49 is shown in Fig 50. Three loop currents, I_1, I_2, and I_3, are shown circulating clockwise within each loop. Kirchoff's voltage law, which states that the sum of the potential differences around a closed circuit loop is 0, is applied to the loop indicated by I_1. This can be repeated for the other two loops; but the equations can be expressed directly in standard matrix form. The voltage matrix $\mathbf{V}_{\text{LOOP}}$ includes the total emf rise around each loop in the direction assumed positive for the loop current. The impedance matrix $\mathbf{Z}_{\text{LOOP}}$ consists of *self-impedances* and *mutual impedances*. The impedances Z_{11}, Z_{22}, and Z_{33} are the self-impedances that equal the sum of the impedances in loops 1, 2, and 3, respectively. The other impedances are the mutual impedances that equal the

**Fig 49
Single-Line Diagram**

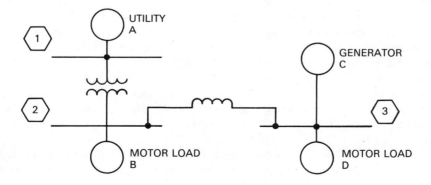

119

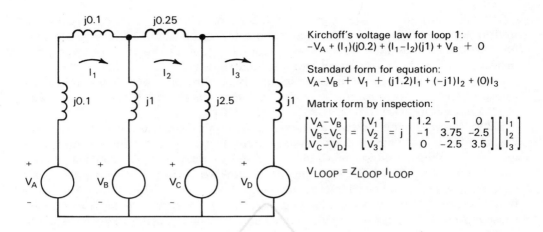

Fig 50
Impedance Diagram and Mesh (Loop) Current Analysis

negative of the impedance common to two loops. For example, $Z_{23} = Z_{32} = -j2.5$, the negative of the impedance common to loops 2 and 3. If the direction assumed positive for *one* of the currents had been counterclockwise, then the loop currents would flow in the same direction through the common impedance, and the mutual impedance would be positive. It is important to realize that the voltage, current, and impedance subscripts refer to loops, not buses.

An analysis based on the node-voltage method directly references bus quantities. Most power system analysis programs work with bus, rather than loop, quantities. Since the node-voltage method uses admittances instead of impedances, the impedance diagram in Fig 50 has been converted to the admittance diagram in Fig 51. Voltage sources with series impedances are converted to equivalent current sources with shunt impedances using well-known methods. The shunt impedances equal the series impedances, and currents are calculated as the voltages divided by the series impedances (i.e., $I_1 = V_A/j0.1$, $I_2 = V_B/j1$, $I_3 = V_C/j2.5 + V_D/j1$). Individual admittances are calculated as reciprocals of the corresponding impedances. Kirchoff's current law, which states that the sum of the currents entering a node equals the sum of the currents leaving the node, can now be applied to bus 1. This is shown in Fig 51. Bus voltages, V_1, V_2, and V_3, are with respect to the neutral reference bus. Currents leaving a node are expressed as the potential difference across a branch multiplied by the admittance of the branch. The resulting current equations can be written in a standard form suitable for matrix methods. The *bus admittance matrix* $\mathbf{Y}_{BUS}$ consists of *self-admittances* and *mutual admittances*. The self-admittances, Y_{11}, Y_{22}, and Y_{33}, are each equal to the sum of the admittances connected to the node identified by the double subscripts. The mutual admittances are each equal to the negative of the admittance between the two nodes indicated by the subscripts (e.g., $Y_{12} = Y_{21} = j10$).

The admittances in the bus admittance matrix are not reciprocals of the impedances in the loop impedance matrix, nor are the two matrixes inverses of each

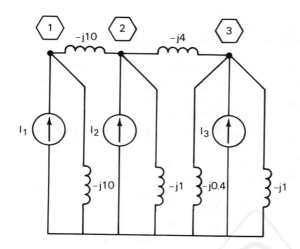

Kirchoff's current law at bus 1:
$I_1 + (V_1)(-j10) + (V_1-V_2)(-j10)$

Standard form for equation:
$I_1 + (-j20)V_1 + (j10)V_2 + (0)\ V_3$

Matrix form by inspection:

$$\begin{bmatrix} I_1 \\ I_2 \\ I_3 \end{bmatrix} = j \begin{bmatrix} -20 & 10 & 0 \\ 10 & -15 & 4 \\ 0 & 4 & -5.4 \end{bmatrix} \begin{bmatrix} V_1 \\ V_2 \\ V_3 \end{bmatrix}$$

$I_{BUS} = Y_{BUS}\ V_{BUS}$

**Fig 51
Admittance Diagram and Node (Bus) Voltage Analysis**

other. However, the *bus impedance matrix* $\mathbf{Z}_{BUS}$ is defined to be the inverse of the bus admittance matrix $\mathbf{Y}_{BUS}$. The diagonal elements of $\mathbf{Z}_{BUS}$ are called *driving-point impedances*, and the off-diagonal elements are known as *transfer impedances*. Although the bus impedance matrix cannot be determined from a simple impedance diagram, efficient algorithms for developing $\mathbf{Z}_{BUS}$ directly from a list of impedance elements are available ([B3], [B27]). The bus impedance matrix is widely used in short-circuit calculation programs. The bus admittance matrix is often used in programs for load flow analysis.

Certain properties of power system network matrixes are exploited to improve the efficiency of computer programs [B14]. The fact that impedance and admittance matrixes are symmetric allows a reduction in computer memory for this data of nearly 50%. Only the diagonal elements and the elements above the diagonal need be stored since $Y_{ji}=Y_{ij}$ and $Z_{ji}=Z_{ij}$. Another matrix property that can be used to reduce storage requirements is known as *sparsity*. A matrix is sparse if most of its elements are 0. The bus admittance matrix is usually very sparse since each bus is connected through a branch (nonzero mutual admittance) to only a few other buses. A row or column of $\mathbf{Y}_{BUS}$ contains 3 to 4 nonzero entries, on average. Consequently, at least 96% of the elements in the bus admittance matrix for a typical 100 bus system are 0. The bus impedance matrix, in contrast, frequently contains no zero elements. Programs which take advantage of sparsity may store only the nonzero elements along with the row-column locations of those entries. Also, numerical techniques have been developed that are especially effective in performing operations on sparse matrixes.

Matrix partitioning is also useful for reducing computer storage and processing requirements. One particularly powerful application of partitioning is *network reduction*. Buses that are not connected to current sources can be grouped together in submatrixes within the expression $\mathbf{I}_{BUS} = \mathbf{Y}_{BUS}\,\mathbf{V}_{BUS}$. These buses can be eliminated from the expression through a straightforward series of matrix operations [B29].

There is often a trade-off of increased processing time associated with techniques to reduce computer memory requirements. Extra steps may be required to locate an element to be used in a calculation if it is not stored in standard matrix form. However, some memory reduction schemes are commonly used in power system analysis programs in order to maximize the number of buses that can be handled on a given size computer.

5.2.3 Solution of Simultaneous Algebraic Equations.

Modeling of many problems in science, engineering, or business results in systems of simultaneous algebraic equations that must be solved for several unknowns. Simultaneous equations typically exist in cases of interdependent quantities. For example, the current that one conductor in an underground duct bank can carry is dependent on the heat produced by the current flowing in all of the conductors in the duct bank. Similarly, all the voltages and currents of a power system are mutually dependent. The development of the equations and the associated matrix representation (i.e., $\mathbf{I}_{BUS} = \mathbf{Y}_{BUS}\,\mathbf{V}_{BUS}$) for modeling this relationship has been described. The solution of these equations is the basis for load flow analysis as described in Chapter 6.

Simultaneous algebraic equations may be classified as *linear* or *nonlinear*. In linear equations, the sum of one or more variables of degree 1, each multiplied by a constant coefficient, is equal to a constant. Nonlinear equations may contain variables raised to powers other than 1, division of an expression by a variable, multiplication of two or more variables, or expressions with nonalgebraic functions of variables (e.g., trigonometric). In this context, a linear equation is not just the equation of a line in the xy plane. Systems of linear equations are common in many kinds of analysis even though, by definition, they represent a small subset of all types of equations. Sometimes a linear system of equations becomes nonlinear due to constraints placed on the solution. For example, in the load flow problem, real and reactive power for each bus, rather than bus current, are typically known. When the required substitution $(\mathbf{I} = (\mathbf{P} - j\mathbf{Q})/\mathbf{V}^{*})$ is made, the equations become nonlinear.

Techniques for solving simultaneous algebraic equations may be described as either *direct* or *iterative* [B27]. Direct methods are applicable to linear systems of equations. Iterative methods may be employed to solve either linear or nonlinear systems of equations. Sometimes a combination of direct and iterative methods is used. Regardless of method, some type of computer-determined solution becomes almost essential for systems of more than three or four equations.

5.2.3.1 Direct Methods.

A direct method determines a solution in a predictable number of arithmetic operations. Such methods are also called "exact" since all steps are carried out according to strict mathematical definitions without assuming values for any unknowns. Unfortunately, these methods can be far from exact unless measures are taken to minimize the cumulative effect of round-off errors.

Several direct methods, useful for solving systems of linear equations, involve the use of *determinants*. A determinant is similar to a matrix in that it consists of entries arranged in rows and columns. A determinant is denoted by vertical bars instead of the brackets or parentheses used to indicate a matrix. The main difference between a determinant and a matrix is that a single value can be calculated for a determinant, while the number of entries in a matrix cannot be changed. A determinant is defined based on a system of two linear equations as follows [B26]:

$$a_1 x + b_1 y = c_1$$

$$a_2 x + b_2 y = c_2$$

This system of equations, representing two lines in the xy plane, can be solved simultaneously by various algebraic manipulations to find the unknowns x and y, which are the coordinates where the two lines intersect.

$$x = \frac{c_1 b_2 - b_1 c_2}{a_1 b_2 - b_1 a_2} \qquad y = \frac{a_1 c_2 - c_1 a_2}{a_1 b_2 - b_1 a_2}$$

Given the definition of a determinant of order 2,

$$\begin{vmatrix} a & b \\ c & d \end{vmatrix} = ad - bc$$

we can write the solutions for x and y in determinant form.

$$x = \frac{\begin{vmatrix} c_1 & b_1 \\ c_2 & b_2 \end{vmatrix}}{\begin{vmatrix} a_1 & b_1 \\ a_2 & b_2 \end{vmatrix}} \qquad y = \frac{\begin{vmatrix} a_1 & c_1 \\ a_2 & c_2 \end{vmatrix}}{\begin{vmatrix} a_1 & b_1 \\ a_2 & b_2 \end{vmatrix}}$$

Examination of these expressions shows that the denominator for both x and y is a determinant containing the coefficients of x and y in the original system of equations. The numerator for x is in the same form, except the coefficients of x are replaced by the constants c_1 and c_2. Similarly, to write the numerator of y, the coefficients of y are replaced by the constants. This procedure, known as *Cramer's rule*, is applicable to systems of any number of linear equations.

To evaluate a determinant of order greater than 2, some additional definitions are required. Given a square matrix **A** of order n, there is an associated determinant, det(**A**). Taking any element a_{ij} in **A**, a new determinant of order $(n-1)$ can be obtained by removing row i and column j. This new determinant is called the *minor* of a_{ij}. If the minor of a_{ij} is multiplied by $(-1)^{i+j}$, the result is called the *cofactor* of the element a_{ij} and is denoted A_{ij}. The value of a determinant can be calculated by adding the products $a_{ij}A_{ij}$ for all the elements in any row or column. This rule, called the *Laplace expansion*, is illustrated below to find the determinant associated with the bus admittance matrix developed in Fig 51.

$$\det(\mathbf{Y}) = j \begin{vmatrix} -20 & 10 & 0 \\ 10 & -15 & 4 \\ 0 & 4 & -5.4 \end{vmatrix}$$

If this is expanded in terms of cofactors for the first column, the result is as follows:

$$\det(\mathbf{Y}) = -j20(-1)^{1+1} \begin{vmatrix} -15 & 4 \\ 4 & -5.4 \end{vmatrix} + j10(-1)^{2+1} \begin{vmatrix} 10 & 0 \\ 4 & -5.4 \end{vmatrix} + 0$$

$$= -j20(81-16) - j10(-54-0) = -j760$$

Note that the 0 in the first column that was picked for expansion eliminated the need to evaluate the third determinant. A technique for systematically introducing zeros by algebraic manipulations is known as the *Gauss elimination method* [B27]. This technique can also be applied directly to a system of linear equations. The result is a coefficient matrix with all diagonal elements equal to 1 and all elements below the diagonal equal to 0. The solution can then be determined by a simple process of *back substitution*. While the Gauss elimination method is too involved to present here, it is actually much more efficient computationally than Cramer's rule.

Matrix inversion was previously described as a useful technique for solving systems of linear equations. However, the method shown for inverting a matrix only generates more equations that must be solved simultaneously using another method. Now that determinants and cofactors have been introduced, an effective matrix inversion procedure can be presented. If $\mathbf{A}$ is a square matrix and $\det(\mathbf{A})$ is not zero (i.e., $\mathbf{A}$ is *nonsingular*), then a unique inverse $\mathbf{A}^{-1}$ exists that can be expressed as follows:

$$\mathbf{A}^{-1} = \frac{[A_{ij}]^T}{\det(\mathbf{A})}$$

where $[A_{ij}]$ is the matrix of the cofactors of the corresponding elements a_{ij} in $\mathbf{A}$. $[A_{ij}]^T$ indicates its transpose (i.e., its rows and columns are interchanged).

To illustrate this method, a system of equations will be solved by matrix inversion. The basis of the method will be clarified by solving the same equations by Cramer's rule. The system of equations

$$8x_1 + 3x_2 = 14$$
$$5x_1 + 2x_2 = 9$$

can be expressed in matrix form $\mathbf{AX} = \mathbf{B}$ as:

$$\begin{bmatrix} 8 & 3 \\ 5 & 2 \end{bmatrix} \begin{bmatrix} x_1 \\ x_2 \end{bmatrix} = \begin{bmatrix} 14 \\ 9 \end{bmatrix}$$

The solution matrix $\mathbf{X}$ can be determined as $\mathbf{X} = \mathbf{A}^{-1}\mathbf{B}$:

$$[A_{ij}]^T = \begin{bmatrix} (-1)^{1+1}(2) & (-1)^{1+2}(5) \\ (-1)^{2+1}(3) & (-1)^{2+2}(8) \end{bmatrix}^T = \begin{bmatrix} 2 & -3 \\ -5 & 8 \end{bmatrix}$$

$$\det(\mathbf{A}) = \begin{vmatrix} 8 & 3 \\ 5 & 2 \end{vmatrix} = (8)(2) - (3)(5) = 1$$

$$\mathbf{A}^{-1} = \frac{[A_{ij}]^T}{\det(\mathbf{A})} = \begin{bmatrix} 2 & -3 \\ -5 & 8 \end{bmatrix}$$

Notice that $\mathbf{A}^{-1}$ matches the determinant determined previously using the fundamental definition of an inverse matrix. Now the solution of the equations can be determined by matrix multiplication.

$$\begin{bmatrix} x_1 \\ x_2 \end{bmatrix} = \begin{bmatrix} 2 & -3 \\ -5 & 8 \end{bmatrix} \begin{bmatrix} 14 \\ 9 \end{bmatrix} = \begin{bmatrix} (2)(14)+(-3)(9) \\ (-5)(14)+(8)(9) \end{bmatrix} = \begin{bmatrix} 1 \\ 2 \end{bmatrix}$$

Using Cramer's rule:

$$x_1 = \frac{\begin{vmatrix} 14 & 3 \\ 9 & 2 \end{vmatrix}}{\begin{vmatrix} 8 & 3 \\ 5 & 2 \end{vmatrix}} = \frac{(14)(2) - (3)(9)}{(8)(2) - (3)(5)} = \frac{1}{1} = 1$$

$$x_2 = \frac{\begin{vmatrix} 8 & 14 \\ 5 & 9 \end{vmatrix}}{1} = \frac{(8)(9) - (14)(5)}{1} = \frac{2}{1} = 2$$

The similarity in the calculations for the two methods shows the common basis for the techniques. Additional direct methods for solving systems of linear equations are described in bibliographic references [B14], [B17], [B21], [B27].

5.2.3.2 Iterative Methods. Iterative methods for solving linear and nonlinear simultaneous algebraic equations are sometimes called "approximate." Generally, however, a solution determined by an iterative method can be as accurate as required. Also, round-off errors tend to be corrected from one iteration to the next. Useful iterative methods should have the following features [B10]:

(1) A means for making a satisfactory first guess
(2) A means for systematically improving on previous approximations
(3) A criterion for stopping the iterations when sufficient accuracy has been achieved

The number of successive approximations, or iterations, needed to arrive at a solution of the desired accuracy cannot be predicted in advance. Depending on the nature of the equations, the choice of initial approximations, and the specific solution technique, the successive approximations may converge to an accurate solution quickly or slowly. Sometimes each iteration will cause the approximations to oscillate or become less accurate (i.e., diverge). For some types of problems, one solution technique may fail consistently, while another may provide reliable results. Two commonly used iterative methods, *Gauss-Seidel* and *Newton-Raphson*, are described here in general terms. These methods are widely used in programs for load flow analysis. Specific application considerations and comparison information for these methods, as applied to load flow analysis, are included in Chapter 6.

The *Gauss-Seidel iterative method* is a simple substitution technique in which a variable calculated using one equation is substituted in the following equations to calculate other variables. The calculations are repeated until the results match the previous iteration within a specified tolerance. To implement the Gauss-Seidel method, each equation is rewritten to isolate a different unknown variable. This is shown below for a system of three linear equations.

$$a_{11}x_1 + a_{12}x_2 + a_{13}x_3 = b_1$$

$$a_{21}x_1 + a_{22}x_2 + a_{23}x_3 = b_2$$

$$a_{31}x_1 + a_{32}x_2 + a_{33}x_3 = b_3$$

The variable x_1 can be isolated in the first equation and x_2 and x_3 in the second and third equations, respectively. Provided that the equations are processed sequentially in the order listed, we can also assign a superscript of k or $k+1$ to indicate the iteration number of the variable being calculated or the variable used in a calculation.

$$x_1^{k+1} = 1/a_{11}(b_1 - a_{12}x_2^k - a_{13}x_3^k)$$

$$x_2^{k+1} = 1/a_{22}(b_2 - a_{21}x_1^{k+1} - a_{23}x_3^k)$$

$$x_3^{k+1} = 1/a_{33}(b_3 - a_{31}x_1^{k+1} - a_{32}x_2^{k+1})$$

The iteration process begins with calculations for iteration number 1 (i.e., $k = 0$). An initial guess for a variable is used if its iteration number is 0. In calculating x_2 for the first iteration, the value just calculated for x_1 is used; but a calculation for x_3 has not yet been made so its initial value is used. Unless the nature of the problem provides a means of guessing initial values, the following is generally used:

$$x_i^0 = b_i/a_{ii} \quad \text{for } i = 1, 2, \dots, n$$

When x_3 is calculated, an iteration is finished. If the changes in the variables from the previous iteration are within the specified tolerance, the procedure is stopped. Otherwise, k is incremented by 1, and calculations for another iteration are performed.

Although this procedure may seem involved, it is quite easy to implement in a computer program. A simple BASIC language program for solving the same system of equations previously solved using direct methods is shown in Fig 52. Comments are included in the program listing to clarify the main functions. The program prints iteration numbers x_1 and x_2 after each iteration. The beginning and ending sections of the printout are also shown in Fig 52. Changes in x_1 and x_2 of less than 0.00001 between iterations indicate that the program has converged to a solution

```
100 ' --------------------------------------------------------------
110 ' |      This BASIC program uses the Gauss-Seidel iterative       |
120 ' |      method to solve the following system of equations:       |
130 ' |                                                               |
140 ' |                  8(X1) + 3(X2) = 14                            |
150 ' |                  5(X1) + 2(X2) = 9                             |
160 ' --------------------------------------------------------------
170 DEFINT I                    ' Define iteration no. to be an integer
180 TOLER = .00001              ' Set tolerance for convergence check
190 X1 = 14/8                   ' Set initial guesses
200 X2 = 9/2                    '
210 FOR I = 1 TO 1000           ' Top of Loop  (allow 1000 iterations)
220 X1OLD = X1                  ' Save results of previous iteration
230 X2OLD = X2                  '   to check for convergence
240 X1 = (1/8) * (14-3*X2)      ' Calculate new X1
250 X2 = (1/2) * ( 9-5*X1)      ' Calculate new X2
260 LPRINT I,X1,X2              ' Print iteration no. X1 and X2
270                             ' Check for convergence
280 IF ABS(X1-X1OLD) < TOLER THEN GOTO 290 ELSE GOTO 300
290 IF ABS(X2-X2OLD) < TOLER THEN END
300 NEXT I                      ' Bottom of Loop
```

1	.0625	4.34375
2	.1210938	4.197266
3	.1760254	4.059937
4	.2275238	3.931191
5	.2758036	3.810491
6	.3210659	3.697335
7	.3634993	3.591252
8	.4032805	3.491799
9	.4405754	3.398562
10	.4755395	3.311152
.	.	.
.	.	.
.	.	.
140	.9998808	2.000298
141	.9998882	2.000279
142	.9998952	2.000262
143	.9999018	2.000246
144	.9999079	2.00023
145	.9999136	2.000216
146	.999919	2.000202
147	.9999241	2.00019
148	.9999288	2.000178
149	.9999333	2.000167
150	.9999376	2.000156
151	.9999414	2.000146

Fig 52
Computer Program Using Gauss-Seidel Method

of sufficient accuracy. Notice that 151 iterations were required before the solution converged. *Acceleration factors* are often used to speed up convergence with the Gauss-Seidel method. This is done by replacing each new value of x_i with the following accelerated value:

accelerated $x_i^{k+1} = x_i^k + a(x_i^{k+1} - x_i^k)$

where

Acceleration factor $a = 1.0$ to 1.8
$i = 1, 2, \dots, n$

When this was added to the program of Fig 52, the optimum acceleration factor was found to be 1.61, which reduced the required number of iterations to 27. Factors greater than 1.8 caused the solution to diverge. The optimum acceleration factor will vary depending on the nature of the equations. The value 1.6 is commonly used in Gauss-Seidel load flow solutions.

The characteristics of the *Newton-Raphson iterative method* are quite different from those of the Gauss-Seidel method. While the Gauss-Seidel method tends to require a large number of simple iterations, the Newton-Raphson method needs relatively few iterations; but the calculations needed to complete an iteration are extensive. The Newton-Raphson method is only useful for systems of nonlinear equations since solution of a linear system of equations by another method is part of each Newton-Raphson iteration.

Since the Newton-Raphson method is also applicable to one equation in one variable, a graphical development of the procedure based on Fig 53 is helpful [B10]. Given an equation in the form $f(x) = 0$, we can graph the function as $y = f(x)$. The solution of the equation is the x-coordinate where the curve crosses the x-axis (i.e., $y = 0$). If an initial guess at the solution of x_0 is made, the corresponding y-coordinate can be calculated as $y_0 = f(x_0)$. A line tangent to the curve can be drawn at (x_0, y_0). The x-coordinate (x_1) at which this line crosses the x-axis will be a closer approximation to the actual solution. Analytically, we can take the equation of the tangent line and solve it to find the point $(x_1, 0)$:

$$y - y_0 = \left. \frac{df}{dx} \right|_0 (x - x_0)$$

$$0 - f(x_0) = \left. \frac{df}{dx} \right|_0 (x_1 - x_0)$$

where $\left. \dfrac{df}{dx} \right|_0 = $ the derivative (slope) of the function $f(x)$ with respect to x evaluated at the point (x_0, y_0)

To show the similarity of this equation to that presented later for the case of two equations in two unknowns, it can be rewritten as:

$$f(x_0) + \left. \frac{df}{dx} \right|_0 (x_1 - x_0) = 0$$

128

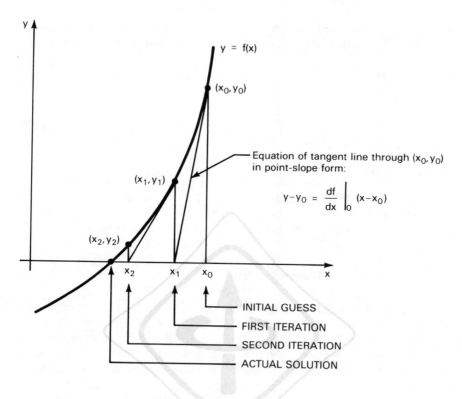

Fig 53
Newton-Raphson Method to Solve an Equation of the Form f(x) = 0

The function and its derivative can now be evaluated for $x = x_0$ to calculate an improved approximation of the solution:

$$x_1 = x_0 - \frac{f(x_0)}{\frac{df}{dx}\Big|_0}$$

This process can be repeated to find x_2, the next approximation.

This method has been developed based on a graphical approach; but the same result can be obtained by expanding $f(x)$ as a Taylor series about x_0 and truncating the series after the first derivative term. This approach is required to rigorously develop the Newton-Raphson method for multiple equations in multiple unknowns. Here, we will take the case of two equations in two unknowns and simply write the Newton-Raphson equations, using the similarity to the case of one equation in one unknown as justification. Given a system of two equations in the form

$$f(x,y) = 0$$
$$g(x,y) = 0$$

and an initial guess at the solution (x_0, y_0), the following linear system of equations can be solved simultaneously to find an improved approximation (x_1, y_1):

$$f(x_0, y_0) + \left.\frac{\partial f}{\partial x}\right|_0 (x_1 - x_0) + \left.\frac{\partial f}{\partial y}\right|_0 (y_1 - y_0) = 0$$

$$g(x_0, y_0) + \left.\frac{\partial g}{\partial x}\right|_0 (x_1 - x_0) + \left.\frac{\partial g}{\partial y}\right|_0 (y_1 - y_0) = 0$$

These equations contain additional terms because of the second variable; but the form of the equations matches that for the case of one equation in one unknown. Partial derivatives are required since we are now dealing with functions of more than one variable. Although the equations look formidable, everything reduces to simple numbers except the variables x_1 and y_1.

Iterative methods provide the only means for solving many types of simultaneous algebraic equations. Additional information on this subject may be found in numerous texts about numerical methods [B1], [B10], [B17], [B21], [B27].

5.2.4 Solution of Differential Equations. The dynamic performance of a physical system is described by one or more differential equations. The physical system may be mechanical, electrical, fluidal, thermal, or a combination of these [B4]. For example, in a complex power system transient stability study, differential equations describing synchronous machine rotor positions and speeds, internal voltages, exciter control systems, and prime mover speed governors are solved simultaneously [B27].

The Laplace transform technique is useful for evaluating the dynamics of a system since easily manipulated algebraic equations are substituted for the more difficult differential equations [B4], [B5], [B13], [B23], [B24], [B26]. The technique permits a complex system to be divided into smaller elements, each described by a *transfer function*. The transfer function of a system, or part of a system, is defined as the Laplace transform of its output variable divided by the Laplace transform of its input variable. The transfer function for each part of a system may be represented within a block of a diagram like that shown for an exciter control system in Fig 54. Numerous block diagram transformations are available to reduce such a diagram to a single transfer function [B4]. For a computer program solution, however, the individual transfer functions are usually replaced by differential equations that are solved simultaneously. For example, the differential equation for the first block in Fig 54 can be determined as follows:

$$\frac{E_2}{E_1} = \frac{1}{1 + sT_R}$$

Cross-multiplying yields the frequency-domain equation:

$$sE_2 = \frac{1}{T_R} (E_1 - E_2)$$

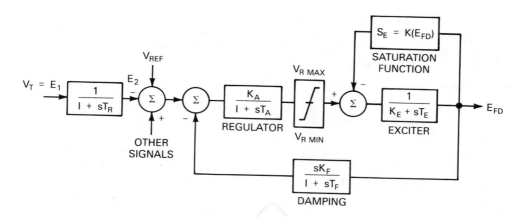

Fig 54
Block Diagram of an Exciter Control System

Since the Laplace operator s can be treated as the differential operator d/dt, we can write the time-domain differential equation:

$$\frac{dE_2}{dt} = \frac{1}{T_R}(E_1 - E_2)$$

First-order differential equations of this type can be solved at desired time increments using a mathematical approach similar to that used in the graphical development of the Newton-Raphson equation. A differential equation of order n can be represented as n first-order equations so that the methods described below can be used. For each time increment, the value of the dependent variable is determined. The calculated values approximate the smooth curve, which is the actual solution of the differential equation. For example, if we know E_1 and E_2 at time zero, we can calculate the slope (dE_2/dt) of a line tangent to the solution curve at time zero. If the time increment Δt is sufficiently small, we can assume that a point on the tangent line also falls on the solution curve. Therefore, a new value of E_2 after the first time increment can be calculated as:

$$E_2^1 = E_2^0 + \left.\frac{dE_2}{dt}\right|_0 \Delta t$$

This value for E_2 may be used in a similar manner to calculate a value for E_2 after the second time interval. This method for solving a differential equation is known as *Euler's method*.

Euler's method is not recommended for serious use because more accurate methods are available. The *modified Euler method* uses an iteration scheme to find the slope of the tangent line at the end of each time interval. The average slope over the time interval is used, instead of the beginning slope, to calculate each new solution. One of the most commonly used methods for the numerical solution of differential equations is the *fourth-order Runge-Kutta method*. In this method, the slope is evaluated four times during each interval: once at the initial point, twice at trial midpoints, and once at a trial end point. These four slope values are used in a special formula to calculate the average slope over the time interval without using iteration. More information on these and other methods for solving differential equations is available in several texts [B21], [B23], [B27].

5.3 Computer Systems. A few of the numerical techniques used in power system analysis programs have been described in order to eliminate some of the mystery in how computers generate answers to problems too big to tackle manually. Many other aspects of computer programs, or software, should be understood before selecting or using computer techniques. The computer itself and peripheral equipment (i.e., hardware) must be considered in conjunction with software. A program is designed to be used on a particular type of computer, although alternate versions for different computers are sometimes available. Factors important in choosing the most appropriate hardware and software for power system analysis are described in this section. The improvements in computer technology continue to be rapid and dramatic. For this reason, the discussion will be kept somewhat general. Still, some of the information will probably seem antiquated in a few years.

5.3.1 Computer Terminology. The following is a brief, informal glossary of terms frequently used in discussing hardware and software [B16], [B25], [B28]:

ASCII — A 7 bit digital code used by many computers, including most personal computers, to represent 128 alphanumeric and device control characters. ASCII stands for American Standard Code for Information Interchange. Many computers use an eighth bit to provide 128 more nonstandard special characters (e.g., graphics).

baud — A unit of data transmission speed that corresponds to "bits per second."

binary — A number system using only the digits 0 and 1 (also called base 2). Binary is employed in digital computers since the use of off and on to represent 0 and 1 is easily implemented in electronic circuits.

bit — An acronym for binary digit (i.e., a single 0 or 1). Usually, bit is abbreviated as lowercase b.

bug — An error in a program or a hardware flaw.

bus — In a computer, a group of conductors carrying related signals between devices. Bus width, expressed in bits (typically 8, 16, 32, or 64), affects the processing speed of a computer (wider = faster).

byte — A string of 8 bits treated as a unit. One byte is required to store one character. Usually, byte is abbreviated as uppercase B.

Central Processing Unit (CPU) — The part of a computer that fetches and executes instructions. It consists of an Arithmetic and Logic Unit (ALU), a control unit, and a number of registers for storing data currently being processed.

copy protection — A method for preventing a program from being used on more computers than it was purchased for.

cursor — A pointer that appears on a monitor, usually to show where characters will appear as they are typed on the keyboard.

default — A value or option assumed by a program when none is specified by the user.

disk (or disc) — A revolving, flat, circular plate coated with magnetic material upon which data is recorded in concentric tracks. A "hard" or "fixed" disk holds large quantities of data and is permanently installed. A "floppy" disk or diskette is removable, and it stores a much smaller volume of data on a thin, flexible magnetic disk inside a flexible or rigid cover.

K — In reference to data storage, used as a prefix meaning 1024 (2^{10}). For example, 2 KB (kilobytes) is 2048 bytes. If K is used alone in this context, bytes are implied.

M — In reference to data storage, used as a prefix meaning 1 048 576 (2^{20}). For example, 10 MB (megabytes) is 10 485 760 bytes. If M is used alone in this context, bytes are implied.

microprocessor — A single integrated circuit, or "chip", which performs the functions of the Central Processing Unit (CPU). A microprocessor is used as the main "engine" in personal computers and many other small computers. Microprocessor type is a key factor determining the processing power of such computers. A related factor is the microprocessor clock speed (in MHz) that sets the rate at which internal functions are performed.

modem — Acronym for modulator/demodulator. Converts digital signals to analog and vice versa, so that two computers can communicate over a telephone line.

monitor — A video display containing a Cathode Ray Tube (CRT), which is used to display program output and data entered by the user on the keyboard. A wide range of types, from monochrome to high-resolution color, are available.

operating system — Software that controls the execution of other programs and the interaction with disk drives, printers, and other devices. Operating systems may be described as single user, multitasking, or multiuser.

pixel (or pel) — A picture element or dot used to make up the display on a monitor. The more pixels that a monitor displays, the higher its resolution will be.

RAM — Random Access Memory that can be read from and written into by a user program. The amount of memory required by a program is an important factor for matching hardware to software.

ROM — Read-Only Memory that can only be read from. Usually, ROM contains programs that control certain basic computer functions. A program stored in ROM is called firmware (software directly implemented in hardware).

terminal — A keyboard and monitor for accessing a remote computer. Also called VDT, Video Display Terminal, or simply CRT, Cathode Ray Tube. A teletypewriter can also be considered a terminal.

word—The natural unit of memory in a computer consisting of a series of bits treated as a unit and stored at one memory address. The number of bits in a word generally matches the internal bus width.

5.3.2 Computer Hardware. There is probably no area of technology that changes as quickly or as dramatically as that related to computer hardware. In particular, the steadily increasing power of desktop computers has changed many basic ideas about data processing. Even the categories of computer types have blurred as computers of one type begin to match, or even exceed, the capabilities associated with a more powerful category of computers [B7], [B9], [B18], [B20]. Because of these advancements, even the smallest engineering organizations can now afford some kind of genuinely useful computer equipment. In fact, the need to remain competitive pushes most companies toward expanded computer usage in all areas. Current computer periodicals should be reviewed for help in selecting hardware.

There may still be valid reasons, however, for not using an in-house computer for all power system studies. For example, the purchase cost for some complicated analysis programs (e.g., transient stability) may make use of a program from a computer time-sharing service more economical for an occasional study. Even in this case, a personal computer with a 1200 baud (or faster) modem would typically be used to access the program running on the time-sharing computer. Computer time-sharing was a big business in the early 1980's; but the availability of software and inexpensive hardware has forced many such companies to disband. Several companies, however, still offer time-sharing programs for power system analysis. Also, comprehensive studies can be performed by engineering consultants, either as part of an expansion project design contract or as a stand-alone service. Since certain studies are already required for significant expansion projects, it is a good time to request a thorough review of the plant distribution system. A stand-alone study by a consultant is justified when qualified plant engineering personnel are unavailable. Such studies are more economical if the client assists in developing the input data.

There are many available options to consider if in-house hardware is to be used for power system studies. If the prospective user is already doing other work on a computer, then obtaining suitable analysis software for the existing equipment would be a natural first approach. It's best not to jump to this conclusion, however, since analysis programs may be much more effective with a faster processor, higher resolution color monitor for graphics, and better printing/plotting equipment. Generally, hardware can be obtained either for lease or for purchase, depending on company financial considerations. Although distinctions are sometimes vague, four types of computer systems suitable for engineering applications are described below. In order of increasing cost (although there is some overlap), the categories are: personal computer, engineering workstation, minicomputer, and mainframe. A fifth category, supercomputer, is not described here since these processors, the fastest and most expensive number crunchers available, are not necessary for analyzing even the largest industrial or commercial power systems. Supercomputers can be effectively applied, however, for modeling interconnected utility systems with thousands of buses.

(1) *Personal computers*, or PC's, have been available since the late 1970's. Personal computers are described as desktop computers, although some PC system units are placed on the floor next to a desk supporting a monitor and keyboard. Until 1984, these computers were used mainly for simple, isolated tasks like word processing and spreadsheet calculations. Since then, more powerful personal computers have been introduced that can be used effectively for power system analysis. More advanced microprocessors, faster clock speeds, and math coprocessor chips (which work with the microprocessor to speed up arithmetic) have been responsible for this improvement in processing power. As performance improves, personal computers are also being connected through Local Area Networks (LAN's) to PC file servers and larger computers to share expensive peripherals and to access common databases. Advanced graphics capabilities suitable for demanding Computer-Aided Design and Drafting (CADD) have become available for use with PC's. New operating systems, which allow a PC to perform several tasks at the same time and use large amounts of memory, are also being implemented. These improvements are bringing top-of-the-line personal computers into the same class as engineering workstations.

(2) An *engineering workstation* can be described as a high-powered, multitasking, networked, desktop computer with large memory, designed for scientific, engineering, and CADD applications. This description is essentially the same as for high-end PC's. Also, the cost of some engineering workstations is dropping into the personal computer range. Terms like "personal workstation" are being used to describe the apparent merging of the two categories. The term *microcomputer* is sometimes used for personal computers and engineering workstations because both use microprocessors for CPU's. This terminology is misleading, however, since numerous computers with microprocessor CPU's are generally considered to be minicomputers.

(3) The term *minicomputer* was first used in the 1960's to describe a new class of computer that was small enough to fit into an office, yet capable of serving about five people working at terminals. The capabilities of these computers have increased so that they are now able to support a few dozen users. These computers generally work with 16 bit words. When a more advanced computer using 32 bit words was introduced in the late 1970's, the term *superminicomputer* was used to distinguish it from 16 bit processors.

(4) A *mainframe* is a large, general-purpose computer supporting perhaps hundreds of users, sometimes in worldwide locations. Other computers also called mainframes are no more powerful than a superminicomputer. Again, there is a blurring of the distinction in the categories. Mainframes, superminicomputers, and minicomputers are all supported by a data processing staff. Usually, there is some type of intracompany accounting and billing for the use of time on these types of computers.

5.3.3 Power System Analysis Software. Programs for power system analysis are available for all of the computer types described in the previous section. Usually, a program for a larger computer will cost more than an equivalent program for a smaller computer. This is not unreasonable because development costs are usually more and because more people can use the one copy of the program installed on a central mainframe or minicomputer. Alternatively, a company can develop its own analysis software. In-house development was more common before a wide range of PC-based software became available. Development usually cannot be justified when a usable program is available for purchase, since, even with qualified engineering and programming personnel, development costs will probably be at least ten times the purchase cost. Available programs for power system analysis are frequently discussed and advertised in periodicals [B6], [B9], [B22].

A careful evaluation is always required before purchasing any kind of engineering software. A trial use period for the complete program is preferred over use of a limited demonstration disk. The following points should be considered [B19], [B22]:

(1) Type of computer required — Manufacturer, model, amount of RAM needed, graphics board and monitor, math coprocessor, and operating system should be checked.

(2) Program accuracy — A description of benchmark tests to verify program results should be available. The user should also run the program for a case previously solved by another program or method.

(3) Vendor — The vendor should be investigated to determine length of time in business, ability to support the program, and policy on program upgrades.

(4) Documentation — The user's manual should be clear, concise, and thorough.

(5) Limitations — The program should model all items required for a particular power system including a sufficient number of buses, local generation, three-winding transformers, load tap changing transformers, industry standard adjustment factors for short-circuit currents, etc.

(6) Ease of use — "User friendly" programs can be operated without constant reference to the user's manual. Programs with on-screen help provide input assistance when requested. Menu-driven programs allow selection of the desired program option from a displayed list.

(7) Input — Full screen input on labeled screens as shown in Fig 55 is preferred over the 80 column input file approach of Fig 56. Programs that use input screens, menus, and user prompts are called *interactive* programs. *Batch* programs work using an input data file without interacting with the user during execution. Both Fig 55 and Fig 56 are for exciter control system data like indicated in Fig 54. Ideally, input data is available in a common database for use by a family of analysis programs (e.g., load flow, short circuit). The input data should be easy to edit after each run in order to make changes for different design cases. Data should be validated, to the extent possible, when entered. The program should allow input in the form normally available (e.g., transformer impedance in percent on kVA base).

(8) Output — The output report should list all input data in a readable manner so that the user can verify the basis for the run at a later date. The pages in each section of a solution report should be in a logical, visually pleasing format with appropriate titles. Some programs have the ability to plot results on the system single-line diagram.

The chapters that follow discuss many aspects of software that are specific to particular studies.

```
┌─────────────────────────────────────────────────────────────────────────┐
│ File: EXAMPLE        │    MACHINE DATA (STABILITY)     │ Study #: MS-100-A │
├─────────────────────────────────────────────────────────────────────────┤
│ Machine#   Name    Rated MVA  Rated kV   X/R      % Xd'       % Xd         │
│   101     Gen. #1    35.300    13.800    89.50     18.60      180.00       │
│   Tdo'      H       D       % Xq     % Xl     S100   S120   %BUS P  %BUS Q  │
│   3.50    1.200   0.030   160.00    20.0     1.07   1.18   100.0   100.0   │
│              Synchronous Generator Parameters                             │
├─────────────────────────────────────────────────────────────────────────┤
│ Type      KA        KF        KE      VRmax    VRmin    SE max   SE .75    │
│   1      250.0     0.060     1.00     17.5     -15.5     1.65     1.13     │
│          E fd       TR        TA       TE       TF                         │
│          6.60      0.005     0.030    1.250    1.000                       │
│  Exciter Parameters; IEEE Type 1 (Cont. Acting Regulator, Rotating Exc.)  │
├─────────────────────────────────────────────────────────────────────────┤
│ Type       % Droop    P max     P min     Tc        Tch                    │
│   ST         5.0      35.30      0.00     0.10       0.15                   │
│             Mode                 Tsr       Trh       Fhp                    │
│             DROOP               0.10      5.00       0.70                   │
│  Governor - Turbine Parameters; Steam - Turbine (with Reheater)           │
├─────────────────────────────────────────────────────────────────────────┤
│  Enter an existing machine number.  This must be a machine bus.           │
└─────────────────────────────────────────────────────────────────────────┘
```

F3:Calculator F5:Add F6:Const.kVA F7:Typ Data F8:Save F9:Return F10:Browse

**Fig 55
Full-Screen Data Input**

5.4 Bibliography. This chapter has presented some of the most important concepts of matrix methods, numerical solution techniques, computer hardware, and software as used for power system analysis. More detailed information is available in the following:

[B1] ARBENZ, KURT and WOHLHAUSER, ALFRED. *Advanced Mathematics for Practicing Engineers*, Norwood, MA, Artech House, 1986.

[B2] ARRILLAGA, J., ARNOLD, C. P., and HARKER, B. J. *Computer Modeling of Electrical Power Systems*, New York, John Wiley & Sons, 1983.

[B3] BROWN, H. E., PERSON, C. E., KIRCHMAYER, L. K., and STAGG, G. W. "Digital Calculation of Three-phase Short Circuits by Matrix Method," *Trans. AIEE*, vol. 79, pt. III, pp. 1277–1281, 1960.

[B4] DORF, RICHARD C. *Modern Control Systems*, Reading, MA, Addison-Wesley, 1974.

[B5] EDMINISTER, JOSEPH A. *Theory and Problems of Electric Circuits*, New York, Schaum's Outline Series, McGraw-Hill, 1965.

[B6] *Electrical Construction & Maintenance*, New York, McGraw-Hill (monthly).

[B7] GLEASON, BILL. "A New Culture—32-bit Technology on Your Desk," *Computer*, IEEE Computer Society, July 1988, pp. 74–76.

[B8] GONEN, TURAN. *Modern Power System Analysis*, New York, John Wiley & Sons, 1988.

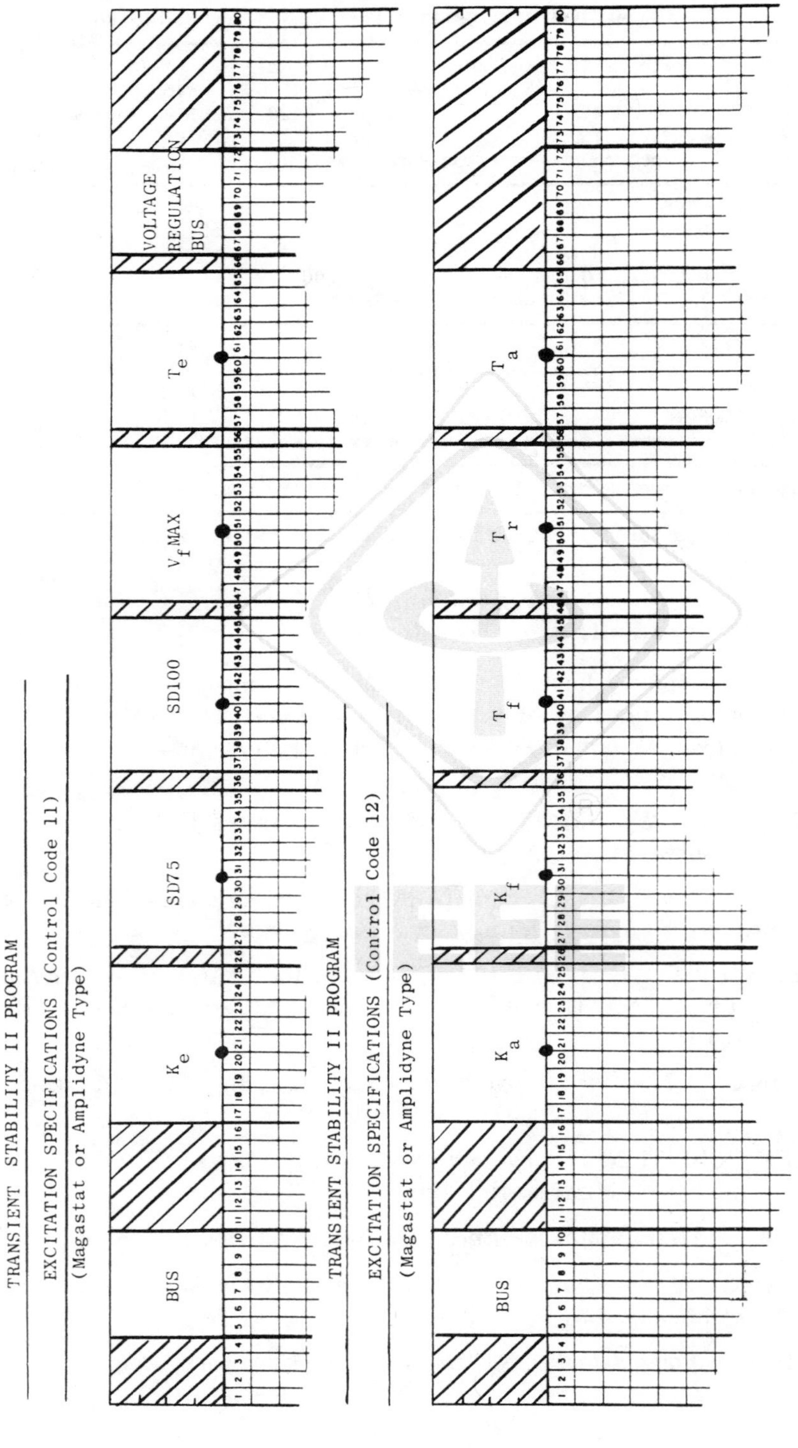

Fig 56
Forms for 80-Column File Input

[B9] GRANT, IAN S., LASKOWSKI, T. F., and WEEKLEY, A. R. "Small Computer Capabilities vs. Large Power-Planning Program Requirements," *IEEE Computer Applications in Power*, IEEE Power Engineering Society, Jan. 1988, pp. 34–36.

[B10] GROVE, WENDELL E. *Brief Numerical Methods*, Englewood Cliffs, NJ, 1966.

[B11] GROZA, V. S. and SHELLEY, S. *Precalculus Mathematics*, New York, Holt, Rinehart and Winston, 1972.

[B12] HASHEMI, N., LOVE, D. J., and TAJADDODI, F. Y. *A Relational Database Approach to Design of Power Plant and Large Industrial Electrical Facilities*, 1987 I&CPS Conference Paper ICPSD-87-01, Presented at Nashville, TN, May 5, 1987.

[B13] HAYT, Jr., WILLIAM H. and KEMMERLY, JACK E. *Engineering Circuit Analysis*, New York, McGraw-Hill, 1978.

[B14] HEYDT, G. T. *Computer Analysis Methods for Power Systems*, New York, Macmillan, 1986.

[B15] KUSIC, GEORGE L. *Computer Aided Power System Analysis*, Englewood Cliffs, NJ, Prentice-Hall, 1986.

[B16] MARKOWSKY, GEORGE. *A Comprehensive Guide to the IBM Personal Computer*, Englewood Cliffs, NJ, Prentice-Hall, 1984.

[B17] MARON, W. J. *Numerical Analysis*, New York, Macmillan, 1987.

[B18] MURPHY, ERIN E. "Minisuper, Supermini: What's in an Adjective?", *IEEE Spectrum*, May 1989, p. 22.

[B19] OLIVER, W. K. "Evaluating CAE Software," *Hydrocarbon Processing*, Feb. 1985.

[B20] POLLACK, ANDREW. "Sun, Apple on Collision Course," *Houston Chronicle*, Mar. 12, 1989.

[B21] PRESS, W. H., FLANNERY, B. P., TEUKOLSKY, S. A., and VETTERLING, W. T. *Numerical Recipes in C*, Cambridge, England, Cambridge University Press, 1988.

[B22] "Problem Solving Software," *Plant Engineering*, Cahners, June 9, 1988, pp. 62–102.

[B23] RAINVILLE, EARL D. and BEDIENT, PHILLIP E. *Elementary Differential Equations*, New York, Macmillan, 1974.

[B24] SCOTT, RONALD E. *Elements of Linear Circuits*, Reading, MA, Addison-Wesley, 1965.

[B25] SOMERSON, PAUL. *PC Magazine DOS Power Tools*, New York, Bantam Books, 1988.

[B26] SPIEGEL, MURRAY R. *Theory and Problems of Advanced Mathematics for Engineers and Scientists*, New York, Schaum's Outline Series, McGraw-Hill, 1971.

[B27] STAGG, GLENN W. and EL-ABIAD, AHMED H. *Computer Methods in Power System Analysis*, New York, McGraw-Hill, 1968.

[B28] STALLINGS, WILLIAM. *Computer Organization and Architecture*, New York, Macmillan, 1987.

[B29] STEVENSON, Jr., WILLIAM D. *Elements of Power System Analysis*, New York, McGraw-Hill, 1982.

[B30] WALLACH, Y. *Calculations and Programs for Power System Networks*, Englewood Cliffs, NJ, Prentice-Hall, 1986.

Chapter 6
Load Flow Studies

6.1 Introduction. One of the most common computational procedures used in power system analysis is the load flow calculation. The planning, design, and operation of power systems require such calculations to analyze the steady-state (quiescent) performance of the power system under various operating conditions and to study the effects of changes in equipment configuration. These load flow solutions are performed using computer programs designed specifically for this purpose. The basic load flow question is: Given the load power consumption at all buses of a known electric power system configuration and the power production at each generator, find the power flow in each line and transformer of the interconnecting network and the voltage magnitude and phase angle at each bus.

Analyzing the solution of this problem for numerous conditions helps ensure that the power system is designed to satisfy its performance criteria while incurring the most favorable investment and operation costs.

Some examples of the uses of load flow studies are to determine:
- Component or circuit loadings
- Steady-state bus voltages
- Reactive power flows
- Transformer tap settings
- System losses
- Generator exciter/regulator voltage set points
- Performance under emergency conditions

Modern systems are complex and have many paths or branches over which power can flow. Such systems form networks of series and parallel paths. Electric power flow in these networks divides among the branches until a balance is reached in accordance with Kirchoff's laws.

Computer programs to solve load flows are divided into two types — static (offline) and dynamic (real time). Most load flow studies for system analysis are based on static network models. Real time load flows (online) that incorporate data input from the actual networks are typically used by utilities in automatic Supervisory Control And Data Acquisition (SCADA) systems. Such systems are used primarily as operating tools for optimization of generation, var control, dispatch, losses, and tie line control. This discussion is concerned with only static network models and their analysis.

Because the load flow problem pertains to balanced, steady-state operation of power systems, a single-phase, positive sequence model of the power system is used. Three-phase load flow analysis software is available; but it is not normally needed for routine studies.

A load flow calculation determines the state of the power system for a given load and generation distribution. It represents a steady-state condition as if that condition had been held fixed for some time. In actuality, line flows and bus voltages fluctuate constantly by small amounts because loads change constantly as lights, motors, and other loads are turned on and off. However, these small fluctuations can be ignored in calculating the steady-state effects on system equipment.

As the load distribution, and possibly the network, will vary considerably during different time periods, it may be necessary to obtain load flow solutions representing different system conditions such as peak load, average load, or light load. These solutions will be used to determine either optimum operating modes for normal conditions, such as the proper setting of voltage control devices, or how the system will respond to abnormal conditions, such as outages of lines or transformers. Load flows form the basis for determining both when new equipment additions are needed and the effectiveness of new alternatives to solve present deficiencies and meet future system requirements.

The load flow model is also the basis for several other types of studies such as short-circuit, stability, motor starting, and harmonic studies. The load flow model supplies the network data and an initial steady-state condition for these studies.

6.2 System Representation. Utility and industrial plant electrical systems can be extensive. A simplified visual means of representing the complete system is essential to understanding the operation of the system under its various possible operating modes. The system single-line diagram serves this purpose. The single-line diagram consists of a drawing identifying buses and interconnecting lines. Loads, generators, transformers, reactors, capacitors, etc., are all shown in their respective places in the system. It is necessary to show equipment parameters as well as their relationship to each other. Figure 57 is a single-line diagram of the sample industrial plant system that will be used later to illustrate some aspects of load flow studies. It shows the operating condition to be studied in terms of which breakers are open or closed.

Buses are usually named as well as numbered. As suggested in Chapter 4, interconnecting lines are shown with their impedance values entered or cross referenced with tables of values. Equipment associated with a bus is shown connected to that bus. For instance, generators are shown connected to their bus with their equipment parameters specified as illustrated in Fig 58. Similarly a load is shown connected to bus 19 in Fig 59. Each line originates on a bus and terminates on a different bus also as depicted in Fig 59. For example, a line runs from bus 6 (FDR 34) to bus 14 (FDR 61) and is shown to be 2193 feet long, 3-conductor no. 2 cable, with $R = 0.3596 \ \Omega$ and $X = 0.1096 \ \Omega$.

Transformers, like lines, are shown between two buses with the primary connected to one bus and the secondary to the other. Information to convey an off-nominal turns ratio should be given when applicable.

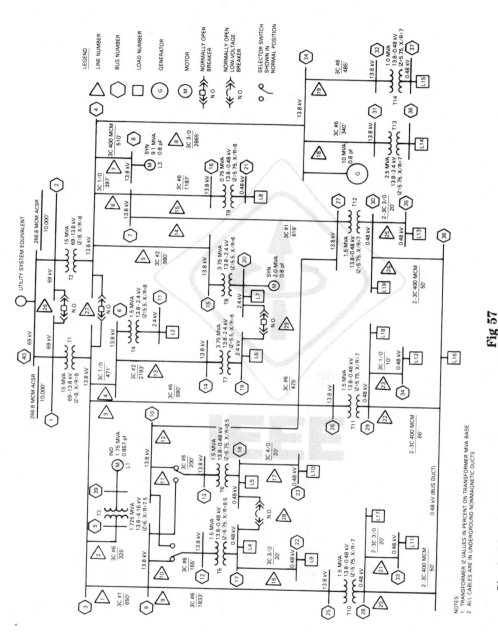

Fig 57

Single-Line Diagram of Typical Industrial Power System for Load Flow Study Example

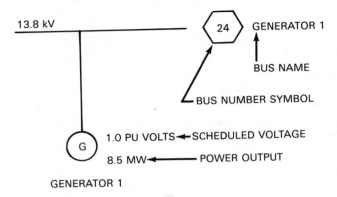

Fig 58
Bus and Generator Representation

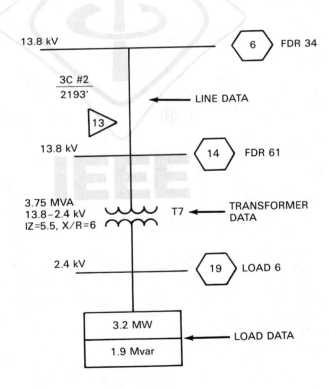

Fig 59
Representation of Loads, Lines, and Transformers

Note that these single-line diagrams show network data. The transfer of this data to the load flow program for analysis is discussed in the next section.

6.3 Input Data. The system information, shown on the single-line diagram, defines the system configuration and the location and size of loads, generation, and equipment. It is organized into a list of data that defines the mathematical model for each power system component and how the components are connected together. The preparation of this data file is the foundation of all load flow analysis, as well as other analysis requiring the network model, such as short-circuit and stability analysis. It is therefore essential that the data preparation be performed in a consistent, thorough manner. Data values must be as accurate as possible. Rounding off, or not including enough decimal places in certain parameters, can lead to erroneous results. Influential parameters must not be ignored.

The details of the models of the system components are addressed in Chapter 4. This section will show how the data is organized for input into the load flow program and give some comments on the preparation of the data. The data will be divided into data categories: system data, bus data, generator data, line data, and transformer data as this organization is typical of most load flow analysis software. In order to illustrate the approach required when using a typical load flow analysis software system, the details of one such system will be described.

6.3.1 System Data. As noted in Chapter 4, most load flow programs perform their calculations using a per unit representation of the system rather than working with volts, amperes, and ohms. The per unit system allows for easier, more efficient solution algorithms and also simplifies load flow analysis. Converting the system data to a per unit representation requires the selection of a base kVA and base voltage. Selection of the base kVA and base voltage specifies the base impedance and current.

The system data specifies the base kVA (or MVA) for the entire system. A base kVA of 10 000 kVA (10 MVA) is often used for industrial studies. For utility systems, the accepted convention is a base of 100 MVA.

The base kV is chosen for each voltage level. Selecting the nominal voltage to be the base voltage simplifies the analyses and reduces the chance of errors in interpretation of results.

6.3.2 Bus Data. The bus data describes each bus and the load and shunts connected to that bus. The data includes:

- Bus number
- Bus name
- Bus type
- Load
- Shunt
- Per unit voltage and angle
- Bus base kV

The bus number is normally the primary index to the information about the bus. For example, it is used to define the line connections in the line data and will be used to get output about a bus during program execution. The bus name is normally used only for informational purposes, allowing the user to give a descriptive name to the bus to make program output more easily understood. Some programs allow the use of the bus name as the primary index.

The bus type is a program input code to allow the program to properly organize the buses for load flow solution. Typically, the four types are:

(1) Load buses
(2) Generator buses
(3) Swing buses
(4) Disconnected buses

The terms "load" bus and "generator" bus should not be taken too literally. A "load" bus is any bus that does not have a generator. A "load" bus need not have load; it may simply be an interconnection point for two or more lines. A "generator" bus could also have load connected to it. The "swing" or "slack" bus is a special type of generator bus that is needed by the solution process. The swing generator adjusts its scheduled power to supply the system MW and Mvar losses that are not otherwise accounted for. This is explained in more detail in the section on load flow solution. There is normally only one swing bus. In utility studies, a large generator is picked as the swing bus. In industrial studies, the utility source is usually represented as the swing bus. A "disconnected" bus is a bus that is temporarily deenergized. It is not included in the load flow solution and must not have in-service lines connected to it. The data is retained so the bus could be reenergized (connected to the system) later in the studies.

Bus load is normally entered in MW and Mvar at nominal voltage. Normally, the load is treated as a constant MVA, that is, independent of voltage. In some cases, a constant current or constant impedance component of load will also be entered so that the load is a function of voltage as explained in 4.9. Bus shunts generally are entered in Mvar at nominal voltage. Care must be taken to ensure the proper sign convention is used to distinguish reactors from capacitors as defined for the particular load flow program being used.

The bus data usually contains an initial estimate of the voltage to be used as a starting point for the load flow solution. A voltage of one per unit at an angle of 0° is usually adequate. The bus base kV is often entered to allow output reports to show voltages in kV and currents in amperes.

6.3.3 Generator Data. Generator data is entered for each generator in the system including the system swing generator. The data defines the generator power output and how voltage is controlled by the generator. The data items normally entered are:

- Real power output in MW
- Maximum reactive power output in Mvar (i.e., machine maximum reactive limit)
- Minimum reactive power output in Mvar (i.e., machine minimum reactive limit)
- Scheduled voltage in per unit
- Generator in-service/out-of-service code

Other data items that may be included are the generator MVA base and the generator's internal impedance for use in short-circuit and dynamic studies. The use of the above data items to determine the generator voltage and reactive power output is discussed in the section on load flow solution. Some programs may allow a generator to regulate a remote bus voltage.

6.3.4 Line Data. Data is also entered for each line in the system. Here the term "line" refers to all elements that connect two buses including transmission lines, cables, series reactors, series capacitors, and transformers. The data items include:

- Resistance
- Reactance
- Charging susceptance (shunt capacitance)
- Line ratings
- Line in-service/out-of-service code
- Line-connected shunts

As described in 4.7, lines are represented by a π model with series resistance and reactance and one-half of the charging susceptance placed on each end of the line. The resistance, reactance, and susceptance are usually input in either per unit or percent depending on program convention.

Line rating are normally input in MVA. Current ratings can be converted to MVA with the formula:

$$\text{Rating}_{\text{MVA}} = \frac{\sqrt{3} \times \text{kV}_{\text{BASE}} \times \text{Rating}_{\text{AMPS}}}{1000} \qquad \text{(Eq 93)}$$

A series reactor, series capacitor, or transformer would not have a charging susceptance term.

The modeling of the charging susceptance is often ignored for short overhead lines and industrial plant systems.

6.3.5 Transformer Data. Additional data is required for transformers. This can either be entered as part of the line data or as a separate data category depending on the particular load flow program being used. This additional data usually includes:

- Tap setting in per unit
- Tap angle in degrees
- Maximum tap position
- Minimum tap position
- Scheduled voltage range with tap step size or a fixed scheduled voltage using a continuous tap approximation

The last three data items are needed only for Load Tap Changing (LTC) transformers that automatically vary their tap setting to control voltage on one side of the transformer.

The organization of transformer tap data requires an understanding of the tap convention used by the load flow program to ensure the representation gives the correct boost or buck in voltage. Transformers whose rated primary or secondary voltages do not match the system nominal (base kV) voltages on the terminal buses will require an off-nominal tap representation in the load flow (and possibly require corresponding adjustment of the transformer impedance).

6.4 Load Flow Solution Methods. The computational task of determining power flows and voltages for even a small network for a given system condition is formidable. Solution of large networks for many system conditions as required for analysis of present-day power systems requires sophisticated computational tools.

The first load flow solution device was a special purpose analog computer called the ac network analyzer developed in the late 1920's. Power system networks under study were represented by an equivalent, scaled-down network. The device allowed the analysis of a variety of operating conditions and expansion plans. However, setup time was long. Due to the large amount of hardware involved, only about 50 network analyzers were operational by the mid-1950's.

Digital computers began to emerge in the late 1940's as computational tools. By the mid-1950's, large-scale digital computers of sufficient speed and size to handle the requirements of a power system network calculation were available. Parallel to the hardware development, algorithms to efficiently solve the network equations were developed. Ward and Hale developed a successful load flow program using a modified Newton iterative procedure in 1956 [6].[7] The application of the Gauss-Seidel iteration algorithm followed soon after. Research in algorithms continued and the Newton-Raphson method was introduced in the early 1960's [5]. Considerable research has been performed in the interim years to improve the performance of these algorithms, making them more robust, able to handle additional power system components, and allowing much larger network sizes.

6.4.1 Problem Formulation. The load flow calculation is a network solution problem. The voltages and currents are related by the following equation:

$$[I] = [Y] [V] \hspace{4cm} \text{(Eq 94)}$$

where:

[I] = Vector of total positive sequence currents flowing into the network nodes (buses)

[V] = Vector of positive sequence voltages at the network nodes (buses)

[Y] = The network admittance matrix

Equation 94 is a linear algebraic equation with complex coefficients. If either [I] or [V] were known, the solution for the unknown quantities could be obtained by application of widely used numerical solution techniques for linear equations.

Partly because of tradition and partly because of the physical characteristics of generation and load, the terminal conditions at each bus are normally desribed in terms of active and reactive power (P and Q). The bus current at bus i is related to these quantities as follows:

$$I_i = \frac{(P_i + jQ_i)^*}{V_i^*} \hspace{3cm} \text{(Eq 95)}$$

where * designates the conjugate of a complex quantity. Combining Eqs 94 and 95 yields

---

[7]The numbers in brackets correspond to those in the References at the end of this chapter.

$$\left[\frac{P - jQ}{V^*} \right] = [Y][V] \qquad \text{(Eq 96)}$$

Equation 96 is nonlinear and cannot be readily solved by closed-form matrix techniques. Because of this, load flow solutions are obtained by procedures involving iterative techniques.

6.4.2 Iterative Solution Algorithms. Since the original technical papers describing digital load flow solution algorithms appeared in the mid-1950's, a seemingly endless collection of iterative schemes has been developed and reported. Many of these are variations of one or the other of two basic techniques that are in widespread use by the industry today—the Gauss-Seidel technique and the Newton-Raphson technique.

Both of these solve bus equations in admittance form, as described in the previous section. This system of equations has gained widespread application because of the simplicity of data preparation and the ease with which the bus admittance matrix can be formed and changed in subsequent cases.

In a load flow study, the primary parameters are:

P = Active power into the transmission network
Q = Reactive power into the transmission network
$|V|$ = Magnitude of bus voltage (voltage to ground)
δ = Angle of bus voltage referred to a common reference

In order to define the load flow problem to be solved, it is necessary to specify two of the four quantities at each bus. For generating units, it is reasonable to specify P and $|V|$ because these quantities are controllable through governor and excitation controls, respectively. For the loads, one generally specifies the real power demand P and the reactive power Q. Since there are losses in the transmission system and these losses are not known before the load flow solution is obtained, it is necessary to retain one bus where P is not specified. At this bus, called a swing bus, $|V|$ as well as δ are specified. Since δ is specified (that is, held constant during the load flow solution), it is the reference angle for the system. The swing bus is therefore also called the reference bus. Since the real power, P, and reactive power, Q, are not specified at the swing bus, they are free to adjust to "cover" transmission losses in the system.

Table 10 summarizes the standard electrical specifications for the three bus types. The classifications "generator bus" and "load bus" should not be taken as absolute. There will, for example, be occasions where a pure load bus may be specified by P and $|V|$.

Bus specification is the tool with which the engineer manipulates the load flow solution to obtain the desired information. The objective of the load flow solution is to determine the two quantities at each bus that are not specified.

The generator specification of holding the bus voltage constant and calculating the reactive power output will be overridden in the load flow solution if the generator reactive output reaches its maximum or minimum var limit. In this case, the generator reactive power will be held at the respective limit and the bus voltage will be allowed to vary.

Table 10
Load Flow Bus Specifications
(Quantities checked are the bus boundary conditions.)

BUS TYPE	P	Q	$|V|$	δ	COMMENTS
LOAD	√	√			Usual Load Representation
GENERATOR OR SYNCHRONOUS CONDENSER	√		√ when $Q^- < Q_g < Q^+$		Generator or Synchronous Condenser (P=0) with var Limits
	√	√ when $Q_g < Q^-$ or $Q_g > Q^+$			Q^- = Minimum var Limit Q^+ = Maximum var Limit $|V|$ is held as long as Q_g is within limit.
SWING			√	√	"Swing bus" must adjust net power to hold voltage constant (essential for solution).

NOTE: $[P,\delta]$, $[Q,|V|]$, and $[Q,\delta]$ combinations are generally not used.

6.4.3 Gauss-Seidel Iterative Technique. Descriptions of load flow solution techniques can become rather complicated, due more to the notation required for complex arithmetic rather than the basic concepts of the solution method. In the following sections, therefore, the basic techniques are developed by considering their application to a dc circuit. Applications to ac problems are then a natural extension of the dc problem.

The performance of the Gauss-Seidel technique will be illustrated using the direct current circuit shown in Fig 60.

Bus 3 is a load bus with specified per unit power. Bus 2 is a generator bus with power specified, and bus 1 is the swing bus with voltage specified. The voltages V_2 and V_3 are sought.

The system equations on an admittance basis are from Eq 94.

$$\begin{bmatrix} I_1 \\ I_2 \\ I_3 \end{bmatrix} = \begin{bmatrix} Y_{11} & Y_{12} & Y_{13} \\ Y_{21} & Y_{22} & Y_{23} \\ Y_{31} & Y_{32} & Y_{33} \end{bmatrix} \begin{bmatrix} V_1 \\ V_2 \\ V_3 \end{bmatrix} \qquad \text{(Eq 97)}$$

The terms of the admittance matrix are easily determined from the circuit [2], [3], [4], [7]. The off-diagonal term Y_{ij} is the negative of the line admittance between bus i and bus j.

$$Y_{ij} = \frac{-1}{Z_{ij}}$$

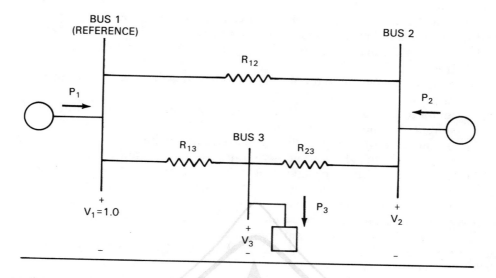

Fig 60
Three-Bus DC Network

The diagonal terms are the sum of the admittances of the lines leaving a bus plus the admittance of the bus shunt plus one-half of the charging admittance for each connected line. The Y matrix is very sparse (few nonzero elements), so special matrix techniques are often used to minimize computer storage requirements. From Eq 97,

$$I_2 = Y_{21}V_1 + Y_{22}V_2 + V_{23}V_3 \qquad \text{(Eq 98)}$$

or

$$V_2 = \frac{1}{Y_{22}} (I_2 - (Y_{21}V_1 + Y_{23}V_3)) \qquad \text{(Eq 99)}$$

Substituting

$$I_2 = P_2/V_2 \qquad \text{(Eq 100)}$$

$$V_2 = \frac{1}{Y_{22}} \left(\frac{P_2}{V_2} - (Y_{21}V_1 + Y_{23}V_3) \right) \qquad \text{(Eq 101)}$$

This is a nonlinear equation in V_2.

For bus 3, a similar procedure yields:

$$V_3 = \frac{1}{Y_{33}} \left(\frac{-P_3}{V_3} - (Y_{31}V_1 + Y_{32}V_2) \right) \qquad \text{(Eq 102)}$$

where the negative sign on P_3 is from the load sign convention.

Equations 101 and 102 are in a form convenient for the application of the Gauss-Seidel iterative solution technique. The steps in this procedure are as follows:

(1) Assign an estimate of V_2 and V_3 (for example, $V_2 = V_3 = 1.0$). Note that V_1 is fixed.

(2) Compute a new value for V_2 using the initial estimates for V_2 and V_3 (see Eq 101).

(3) Compute a new value for V_3 using the initial estimate for V_3 and the just computed value for V_2 (see Eq 102).

(4) Repeat (2) and (3) using the latest computed voltages V_2 and V_3 until the solution is reached. One complete computation of V_2 and V_3 is one iteration.

The computed voltages are said to converge when, for each iteration, they come closer and closer to the actual solution satisfying the network equations. Since the computer time increases linearly with the number of iterations, it is necessary to have the computer program make a check after each iteration and decide whether the last computed voltages are sufficiently close to the true solution or whether further computations are required. The criterion specifying the desired accuracy is called the "convergence criterion."

A reliable convergence criterion is the power mismatch check. Based on the last computed voltage solution, the sum of the power flows (real and reactive) on all lines connected to the bus and to the bus shunt is compared with the specified bus real and reactive power. The difference, which is the power mismatch, is a measure of how close the computed voltages are to the true solution. The power mismatch tolerance is generally specified in the range of 0.01 to 0.0001 pu on the system MVA base.

A different convergence check evaluates the maximum change in any bus voltage from one iteration to the next. A solution with desired accuracy is assumed when the change is less than a specified small value, for example, 0.0001 pu.

A voltage check is dependent on the rate of convergence and is thus less reliable than the power mismatch check. However, the voltage check is much faster (computationally, on a digital computer) than the power mismatch check and since the power mismatch will be large until the voltage change is quite small, one may economically use a procedure where computation of mismatch is avoided until a small amount of voltage change occurs.

The convergence of the Gauss-Seidel iteration algorithm is asymptotic; i.e., a particular bus voltage approaches the final value in smaller and smaller increments and is always greater or always less than the desired solution. In many cases, the rate of convergence for the Gauss-Seidel technique can be increased by applying an acceleration factor to the approximate solution obtained from each iteration. If α is the specified acceleration factor, then

$$V_i^{m\prime} = V_i^{m-1} + \alpha \left(V_i^m - V_i^{m-1} \right) \qquad \text{(Eq 103)}$$

where $V_i^{m\prime}$ is the accelerated estimate of V_i^m at iteration m.

Typical acceleration factors range from about 1.2 to 1.8. In some cases, different acceleration factors are used for the real and imaginary components of voltage. The acceleration factor should never be set greater than 2, and optimum values seldom exceed about 1.8.

Some experimentation may be necessary to select the proper acceleration factor for a specific problem. If the voltage change each iteration, as shown by a load flow convergence monitor, is oscillating in magnitude or sign, the acceleration factor should be decreased. If the voltage change each iteration is decreasing smoothly but slowly, convergence may be improved by increasing acceleration. These concepts are illustrated in Fig 61.

While the solution for a dc circuit was described, solution of an ac circuit would be very similar. For the three-bus example, voltage magnitude and angle at bus 1, generator power and bus voltage at bus 2, and real and reactive load power at bus 3 would be specified. The load flow solution would determine the voltage angle and generator reactive power output of bus 2 and the voltage magnitude and angle at bus 3.

The ac version of Eqs 101 and 102 can be obtained from Eq 96 as follows:

$$V_i^{(m)} = \frac{1}{Y_{ii}} \left(\frac{P_i - jQ_i}{V_i^{*(m-1)}} - \sum_{k=1}^{i-1} Y_{ik} V_k^{(m)} - \sum_{k=i+1}^{N} Y_{ik} V_k^{(m-1)} \right) \quad i = 1,2,...,N-1$$

(Eq 104)

where:

N = The number of buses in the system, and the swing bus is bus N
m = The present iteration number
i and k = Bus indexes
V and Y = Complex voltage and admittance, respectively
V^* = The complex conjugate of V

Fig 61
Dependence of Load Flow Convergence on Acceleration Factors

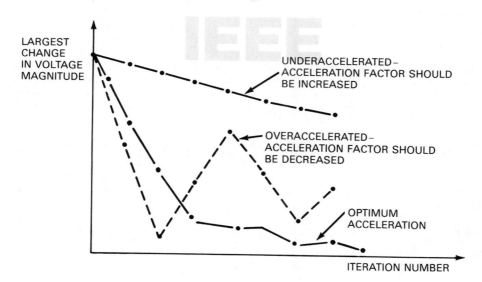

6.4.4 Newton-Raphson Iterative Technique. Not all load flow problems can be solved efficiently using the Gauss-Seidel technique. For some problems, this scheme converges rather slowly. For others, it does not converge at all. Problems that cannot be solved using the Gauss-Seidel technique may often be solved using the Newton-Raphson technique.

This approach utilizes the partial derivatives of the load flow relationships to estimate the changes in the independent variables required to find the solution. In general, the Newton-Raphson technique achieves convergence using fewer iterations than the Gauss-Seidel technique. However, the computational effort per iteration is somewhat greater.

To apply the Newton-Raphson technique to the three-bus example in Fig 60, the bus powers are expressed as nonlinear functions of the bus voltage.

$$
\begin{aligned}
P_1 &= V_1(Y_{11}V_1 + Y_{12}V_2 + Y_{13}V_3) \\
P_2 &= V_2(Y_{21}V_1 + Y_{22}V_2 + Y_{23}V_3) \\
P_3 &= V_3(Y_{31}V_1 + Y_{32}V_2 + Y_{33}V_3)
\end{aligned}
\tag{Eq 105}
$$

Small changes in bus voltages (ΔV) will cause corresponding, small changes in bus powers (ΔP). A linearized approximation to the power change as a function of voltage changes can be obtained as:

$$
\begin{bmatrix} \Delta P_1 \\ \Delta P_2 \\ \Delta P_3 \end{bmatrix} =
\begin{bmatrix}
\dfrac{\partial P_1}{\partial V_1} & \dfrac{\partial P_1}{\partial V_2} & \dfrac{\partial P_1}{\partial V_3} \\
\dfrac{\partial P_2}{\partial V_1} & \dfrac{\partial P_2}{\partial V_2} & \dfrac{\partial P_2}{\partial V_3} \\
\dfrac{\partial P_3}{\partial V_1} & \dfrac{\partial P_3}{\partial V_2} & \dfrac{\partial P_3}{\partial V_3}
\end{bmatrix}
\begin{bmatrix} \Delta V_1 \\ \Delta V_2 \\ \Delta V_2 \end{bmatrix}
\tag{Eq 106}
$$

or symbolically:

$$[\Delta P] = [J] [\Delta V]$$

where [J], the Jacobian matrix, contains the partial derivatives of power with respect to voltages for a particular set of voltages, V_1, V_2, and V_3, that is, the partial derivations of Eq 105. When one or more of the voltages changes substantially, a new Jacobian matrix must be computed.

In the load flow problem, V_1 is specified; that is, $\Delta V_1 = 0$. Also, since ΔP_1 does not enter the computations explicitly, Eq 106 may be reduced to:

$$
\begin{bmatrix} \Delta P_2 \\ \Delta P_3 \end{bmatrix} =
\begin{bmatrix}
\dfrac{\partial P_2}{\partial V_2} & \dfrac{\partial P_2}{\partial V_3} \\
\dfrac{\partial P_3}{\partial V_2} & \dfrac{\partial P_3}{\partial V_3}
\end{bmatrix}
\begin{bmatrix} \Delta V_2 \\ \Delta V_3 \end{bmatrix}
\tag{Eq 107}
$$

Changes in V_2 and V_3 due to changes in P_2 and P_3 are obtained by inverting $[J]$ to obtain:

$$[\Delta V] = [J]^{-1}[\Delta P] \tag{Eq 108}$$

The Newton-Raphson load flow solution method is then as follows:

(1) Assign estimates of V_2 and V_3 (for example, $V_2 = V_3 = 1.0$).
(2) Compute P_2 and P_3 from Eq 105.
(3) Compute the differences (ΔP) between computed and specified powers:

$$\Delta P_2 = P_2 - P_2'$$
$$\Delta P_3 = P_3 - P_3' \tag{Eq 109}$$

where the "prime" indicates specified value.

(4) Since $\Delta P \neq 0$ is caused by errors in the voltages, it seems that the voltages should be incorrect by an amount that is closely approximated by ΔV as evaluated from Eq 108.

Therefore, the new estimate for the bus voltages is:

$$\begin{bmatrix} V_2 \\ V_3 \end{bmatrix}_{New} = \begin{bmatrix} V_2 \\ V_3 \end{bmatrix}_{Old} - [J]^{-1} \begin{bmatrix} \Delta P_2 \\ \Delta P_3 \end{bmatrix} \tag{Eq 110}$$

This is the basic equation in the Newton-Raphson method. The negative sign is due to the way ΔP was defined.

(5) Recompute and "invert" the Jacobian matrix using the last computed voltages and compute the new estimate for the voltages using Eqs 109 and 110. Repeat this procedure until ΔP_2 and ΔP_3 are less than a small value (convergence criterion).

Digital computer programs solving large power system load flows do not explicitly compute the Jacobian inverse. Rather, the voltage correction ΔV is obtained by a numerical technique known as Gaussian elimination. This technique is much faster and requires much less storage than matrix inversion.

The convergence of the Newton-Raphson technique is not asymptotic as was the case with the Gauss-Seidel iterative scheme. The convergence is very rapid for the first few iterations and slows as the solution is neared.

For the ac load flow solution, the Jacobian matrix may be arranged as follows:

$$\begin{bmatrix} \Delta P \\ \hline \Delta Q \end{bmatrix} = \begin{bmatrix} J_1 & J_2 \\ \hline J_3 & J_4 \end{bmatrix} \begin{bmatrix} \Delta \delta \\ \hline \Delta |V| \end{bmatrix} \tag{Eq 111}$$

where the complex bus voltage is written in polar coordinates, $|V| \angle \delta$. The Jacobian matrix can be arranged in many different ways to fit the particular programming techniques selected.

An approximation to the Newton-Raphson formulation can be obtained by observing that, for a small change in the magnitude of bus voltage $\Delta |V|$, the real power, P, does not change appreciably. Similarly, for a small change in bus voltage phase angle $\Delta\delta$, the reactive power, Q, does not change very much. Thus, in Eq 111:

$$[J_2] = [\partial P / \partial |V|] \simeq 0 \qquad \text{(Eq 112)}$$

$$[J_3] = [\partial Q / \partial \delta] \simeq 0 \qquad \text{(Eq 113)}$$

This allows Eq 111 to be "decoupled'" into the following form:

$$[\Delta P] = [J_1] [\Delta\delta] \qquad \text{(Eq 114)}$$

$$[\Delta Q] = [J_4] [\Delta|V|] \qquad \text{(Eq 115)}$$

Note that these two equations can be solved independently and sequentially, thereby reducing the storage and solution time requirements compared to using the full Jacobian. The decoupled Newton-Raphson technique may be used in applications where computational speed is important and the starting solution is close to the actual solution. This situation often occurs where a series of contingencies are being investigated about a previously solved reference case.

Although it is useful to understand how load flow solution techniques work, it is more important to understand the characteristics they exhibit.

6.4.5 Comparison of Load Flow Solution Techniques. Because their convergence characteristics are dependent upon network, load, and generator conditions, each of the iterative techniques discussed has its own strengths and weaknesses. Load flow programs that allow the user to select the technique to be applied tend to be more useful than those that rely on one specific approach. Even more useful are those programs that allow the user to switch from one technique to another during the course of a solution.

All of the methods discussed can fail to converge on the solution of some system conditions. It is rare, however, to find a system condition whose correct voltage solution cannot be found by the application of one or more of the methods. There are some system conditions that are difficult to solve with a single iterative method but which can be solved readily by the successive application of more than one method.

The following guidelines are offered as an aid to determine which technique may be most appropriate for a particular system condition.

(1) The Gauss-Seidel method is generally tolerant of power system operating conditions involving poor voltage distribution and difficulties with generator reactive power allocation, but does not converge well in situations where real power transfers are close to the limits of the system.

(2) The Newton-Raphson method is generally tolerant of power system situations in which there are difficulties in transferring real power, but is prone to failure if there are difficulties in the allocation of generator reactive power output or if the solution has a particularly low voltage magnitude profile.

(3) The Gauss-Seidel method is quite tolerant of poor starting voltage estimates but converges slowly as the voltage estimate gets close to the true solution.

(4) The Newton-Raphson method is prone to failure if given a poor starting voltage estimate, but is usually superior to the Gauss-Seidel method once the voltage solution has been brought close to the true solution.

(5) The decoupled Newton method will not converge when the network contains lines with resistance close to or greater than the reactance. This is often the case in low-voltage systems.

(6) The Gauss-Seidel method will not converge if negative reactance branches are present in the network, such as due to series capacitors or three-winding transformer models. With modifications, the Gauss-Seidel method can be adapted to handle negative reactance within certain constraints.

Some experimentation may be needed to determine the best combination of methods for each particular model. The following approach is suggested for new load flow cases whose specific characteristics have yet to be learned (assuming both solution techniques are available):

(1) Initialize all voltages to either unity magnitude, or to scheduled magnitude if given, and initialize all phase angles to zero. (This step is referred to as a flat start.)

(2) Execute Gauss-Seidel iterations until the adjustments to the voltage estimates decrease to 0.01 or 0.005 pu in both real and imaginary parts.

(3) Switch to Newton-Raphson iterations until either the problem is converged or the reactive power output estimates for generators show signs of failure to converge.

(4) Switch back to Gauss-Seidel iterations if the Newton-Raphson method does not settle down to a smooth convergence within 8 to 10 iterations.

6.5 Load Flow Analysis. A load flow *solution* determines the bus voltages and the flows in all branches for a given set of conditions. A load flow *study* is a series of such calculations made when certain equipment parameters are set at different values, or circuit configuration is changed by opening or closing breakers, adding or removing a line, etc. Load flow studies are performed to check the operation of an existing system under normal or outage conditions, to see if the existing system is capable of supplying planned additional loads, or to check and compare new alternatives for system additions to supply new load or improve system performance.

Generally, the study engineer has a predefined set of criteria that the system must meet. These include:

• Voltage criteria, such as defined in Reference [1].
• Flows on lines and transformers must be within defined thermal ratings
• Generator reactive outputs must be within the limits defined by the generator capability curves

The voltage criteria are usually divided into an acceptable voltage range for normal conditions and a wider range of acceptable voltage under outage conditions. The thermal criteria for lines and transformers may also have such a division, allowing for a temporary overload capability due to the thermal time constant of the equipment or additional forced cooling capabilities of transformers.

A study normally begins with the preparation of base cases to represent the different operating modes of the system or plant. The operating condition normally chosen is maximum load. (Here maximum load refers to the maximum amount of coincident loaded, not the sum of all the loads. See 4.9 for an explanation of load diversity and load modeling.) When maximum load occurs at different times on different parts of the system, several base cases may be needed. The base cases should represent realistic operating conditions. Abnormal conditions and worst-case scenarios will be addressed later in the study.

The base cases are analyzed to determine if voltages and flows are within acceptable ranges. Sample outputs are shown in 6.6.

If voltages or overload problems are noted, system changes can be made to the load flow data and the case resolved to see if the changes are effective in remedying the problem. To remedy low-voltage problems, possible changes include:

- Change in transformer tap positions
- Increase in generator schedule voltage
- Addition of shunt capacitors
- System reconfiguration to shift load to less heavily loaded lines
- Disconnection of shunt reactors
- Addition of lines or transformers

To remedy heavy line or transformer loadings, most of the same remedies apply. In general, the first two of the above remedies will not help heavy loadings due to large real power (watt) flows. Real power flows from the generator to the loads. Real power flow is determined by the phase angle of the supply bus leading the phase angle of the load bus, with voltage magnitudes having a secondary effect. However, reactive power flow is primarily determined by the voltage magnitude with reactive power flowing from the higher voltage bus to the lower voltage bus. Real and reactive power flow, being primarily influenced by different constraints, can flow in different directions on the same line.

Transformer off-nominal taps can change the relative relationship of the voltage on the primary and secondary bus and thus can change the reactive power flow, while the real power flow is largely unaffected by a change in tap position.

When the base case voltages and flows are in the desired range, the system must be examined to check operation under abnormal conditions (contingency analysis). These conditions include:

- Loss of a transmission line or cable
- Loss of a transformer
- Loss of a generator
- Abnormal supply conditions

When the load flow model is changed, for example, to represent a line outage, a new solution is obtained. The voltage and flows are checked against their respective criterion. If necessary, further system changes are made to correct the problems noted. In contingency analysis, it is important to note that several outages may cause system problems; but the different remedies applied may not help equally for all outages. To minimize the number and cost of the remedies, it is necessary to choose those remedies that have the most beneficial effects for the most outages.

The load flow analysis is used to design a system that has a good voltage profile and acceptable line loadings during normal operation and that will continue to operate acceptably when one or more lines become inoperative due to line damage, lightning strokes, failure of transformers, etc. Performing a series of load flow cases and analyzing the results provides operating intelligence in a short time that might take years of actual operating experience to obtain.

In addition to the benefits described above, a study of reactive power flows on the branches can lead to reduced line losses and improved voltage distribution. Reduction in kVA demand due to power factor correction can lead to lower utility bills for an industrial plant. The size and placement of power factor correction capacitors and the setting of generator scheduled voltages and transformer tap positions can be studied with load flows.

Knowledge of branch flows supplies the protection engineer with requirements for proper relay settings. The load flow studies can also provide data for automatic load and demand control, if needed.

The load flow is also used to check the effects of future load growth and the effectiveness of planned additions. These studies are performed in the same way as studies of the present system. The future loads are determined and entered into the model. Base case conditions are studied and additions made, if necessary, to get the system to meet the performance criteria. Then outage conditions are studied and again system changes may be required. Studies of future systems vary in that there are usually more alternative ways of solving the problems encountered. The load flow is the tool that allows the alternatives to be compared in terms of their effectiveness under normal and contingency conditions. Coupled with other studies as well as cost and reliability data, the results lead to the selection of the best alternative.

6.6 Load Flow Study Example. To illustrate the use of a load flow program, a typical industrial plant will be studied. The single-line diagram of the plant electrical system was shown previously in Fig 57.

The impedance diagram in Fig 62 is similar to the single-line diagram except that impedances of interconnecting lines and equipment have been added along with load values.

The first step in performing a load flow study is the preparation of the input data file as explained in 6.3. This data will be input into the load flow program and the network solved. The input file for the sample system is shown in Fig 63. A data listing by the program is shown in Fig 64.

For existing systems, the network configuration, load, and generation are often chosen to match a known operating condition so results can be compared to values known from operating experience to help validate the model. The base case represents the system in the normal operating mode supplying present maximum loads. Most load flow programs have data checking and analysis routines to help find input errors. These include a check of the network topology to see that all in-service buses are connected to the swing bus and range checking of certain data items to flag uncharacteristic values. A fundamental check of the base case is to examine the ability of the load flow solution to converge. As noted in 6.4, convergence should lead to a very small amount of mismatch on every bus, where the mismatch is simply the sum of all the powers or reactive powers entering the bus. The mismatch should ideally equal zero to satisfy Kirchoff's laws; provided its percentage is small in comparison to the total bus load, some amount of mismatch is tolerable without adversely affecting the accuracy of the calculated bus voltages. If the load flow solution cannot approach this point for a known normal operating condition, then

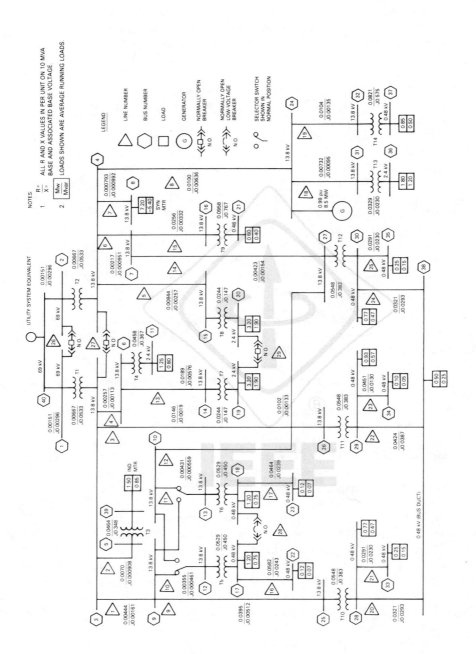

Fig 62

Impedance Diagram of Typical Industrial Power System for Load Flow Study Example

```
0 10
TYPICAL INDUSTRIAL PLANT ELECTRICAL SYSTEM (IEEE BROWN BOOK)     CASE TITLE
BASE CASE LOAD FLOW
 1 1 0 0 0 0 1 1 0 'TIE 1' 69
 2 1 0 0 0 0 1 1 0 'TIE 2' 69
 3 1 0 0 0 0 1 1 0 'MAIN 1' 13.8
 4 1 0 0 0 0 1 1 0 'MAIN 2' 13.8
 5 1 0 0 0 0 1 1 0 'FDR 32' 13.8
 6 1 0 0 0 0 1 1 0 'FDR 34' 13.8
 7 1 0 0 0 0 1 1 0 'FDR 42' 13.8       BUS DATA
 8 1 7.2 -5.4 0 0 1 1 0 'SYNMOTOR' 13.8
 9 1 0 0 0 0 1 1 0 'FDR 31' 13.8
10 1 0 0 0 0 1 1 0 'FDR 271' 13.8      FORMAT:  BUS #, BUS TYPE, LOAD MW, LOAD Mvar,
11 1 1.25 0.8 0 0 1 1 0 'LOAD 2' 2.4            SHUNT MW, SHUNT Mvar, AREA #,
12 1 0 0 0 0 1 1 0 'FDR 92' 13.8               VOLTAGE MAGNITUDE, VOLTAGE ANGLE,
13 1 0 0 0 0 1 1 0 'FDR 101' 13.8              BUS NAME, BASE kV
14 1 0 0 0 0 1 1 0 'FDR 61' 13.8
15 1 0 0 0 0 1 1 0 'FDR 41' 13.8
16 1 0 0 0 0 1 1 0 'FDR 72' 13.8
17 1 1.2 0.75 0 0 1 1 0 'T 5' 0.48
18 1 1.2 0.75 0 0 1 1 0 'T 6' 0.48     BUS TYPES:  1 = LOAD BUS
19 1 3.2 1.9 0 0 1 1 0 'LOAD 6' 2.4                2 = GENERATOR BUS
20 1 3.2 1.9 0 0 1 1 0 'LOAD 7' 2.4                3 = SWING BUS
21 1 0.6 0.4 0 0 1 1 0 'LOAD 8' 0.48
22 1 0.12 0.07 0 0 1 1 0 'LOAD 9' 0.48
23 1 0.12 0.07 0 0 1 1 0 'LOAD 10' 0.48
24 2 0 0 0 0 1 1 0 'GEN 1' 13.8
25 1 0 0 0 0 1 1 0 'FDR 91' 13.8
26 1 0 0 0 0 1 1 0 'FDR 33' 13.8
27 1 0 0 0 0 1 1 0 'FDR 71' 13.8
28 1 0.77 0.47 0 0 1 1 0 'T10' 0.48
29 1 0.93 0.57 0 0 1 1 0 'T11' 0.48
30 1 0.77 0.47 0 0 1 1 0 'T12' 0.48
31 1 0 0 0 0 1 1 0 'FDR 242' 13.8
32 1 0 0 0 0 1 1 0 'FDR 243' 13.8
33 1 0.25 0.15 0 0 1 1 0 'LOAD 11' 0.48
34 1 0.10 0.05 0 0 1 1 0 'LOAD 12' 0.48
35 1 0.25 0.15 0 0 1 1 0 'LOAD 13' 0.48
36 1 1.8 1.2 0 0 1 1 0 'LOAD 14' 2.4
37 1 0.85 0.5 0 0 1 1 0 'LOAD 15' 0.48
38 1 0.5 0.25 0 0 1 1 0 'LOAD 16' 0.48
39 1 1.5 0.85 0 0 1 1 0 'LOAD 1' 4.16
40 3 0 0 0 0 1 1 0 'UTILITY' 69
0
24 1 8.5 0 5 -2 0.99                    GENERATOR DATA
40 1 16 10 25 -10 1.0
0                                       FORMAT:  BUS #, GENERATOR #, GENERATOR
40 1 1 .00151 .00296 .00513 30                  POWER IN MW, REACTIVE POWER IN Mvar,
40 2 1 .00151 .00296 .00513 30                  MAXIMUM REACTIVE POWER LIMIT IN
 1 3 1 .00667 .0533 0.0 15 0 0 1.0              Mvar, MINIMUM REACTIVE POWER LIMIT
 2 4 1 .00667 .0533 0.0 15 0 0 1.0              IN Mvar, SCHEDULED VOLTAGE IN PU.
 3 9 1 .00444 .00161 0.0 3.94
 3 5 1 .00700 0.000908 0.0 1.98
 5 39 1 .0464 .348 0.0 1.725 0 0 1.0
 3 26 1 .0146 .00190 0.0 1.98
 3 6 1 .00257 .00113 0.0 4.66
 4 15 1 .00844 .00257 0.0 3.46
 4 7 1 .00217 .000951 0.0 4.66
 4 8 1 .000793 .000992 0.0 9.80
 4 24 1 .0100 .00636 0.0 5.97
 9 25 1 .0395 .00512 0.0 1.98
 9 12 1 .00355 .000461 0.0 1.98
10 13 1 .00431 .000559 0.0 1.98
10 27 1 .0102 .00133 0.0 1.98
 6 14 1 .0189 .00576 0.0 3.46
 6 11 1 .0458 .367 0.0 1.5 0 0 1.0
 7 27 1 .00423 .00154 0.0 3.94          LINE DATA
 7 16 1 .0256 .00332 0.0 1.98
12 17 1 .0529 .450 0.0 1.5 0 0 1.0      FORMAT:  FROM BUS #, TO BUS #, CIRCUIT #,
17 22 1 .0582 .0243 0.0 .178                     RESISTANCE IN PU, REACTANCE IN PU,
13 18 1 .0529 .450 0.0 1.5 0 0 1.0               TOTAL LINE CHARGING IN PU, LINE
18 23 1 .0464 .0239 0.0 .210                     RATING A IN MVA, RATING B, RATING C,
14 19 1 .0244 .147 0.0 3.75 0 0 1.0              TAP POSITION IN PU.
15 20 1 .0244 .147 0.0 3.75 0 0 1.0
16 21 1 .0958 .767 0.0 .75 0 0 1.0      TAP POSITION ENTERED ONLY FOR
25 28 1 .0548 .383 0.0 1.5 0 0 1.0      TRANSFORMERS.
28 38 1 .0321 .0293 0.0 .6
28 33 1 .0291 .0230 0.0 .35
26 29 1 .0548 .383 0.0 1.5 0 0 1.0
29 38 1 .0424 .0387 0.0 .6
29 34 1 .0451 .0130 0.0 .135
27 30 1 .0548 .383 0.0 1.5 0 0 1.0
30 38 1 .0321 .0293 0.0 .6
30 35 1 .0291 .0230 0.0 .35
24 31 1 .00732 .000950 0.0 1.98
31 36 1 .0329 .230 0.0 2.5 0 0 1.0
24 32 1 .0104 .00135 0.0 1.98
32 37 1 .0821 .575 0.0 1.0 0 0 1.0
0
0
0
0
0
0
0
0
```

Fig 63
Input Data File for Sample System

```
         PTI INTERACTIVE POWER SYSTEM SIMULATOR--PSS/E      FRI. DEC 15 1989  09:15
TYPICAL INDUSTRIAL PLANT ELECTRICAL SYSTEM (IEEE BROWN BOOK)               BUS DATA
BASE CASE LOAD FLOW
BUS#   NAME  BSKV CODE    VOLT  ANGLE   PLOAD   QLOAD      S H U N T    AREA ZONE
   1 TIE 1   69.0   1   0.9965  -0.1    0.0     0.0      0.0     0.0     1    1
   2 TIE 2   69.0   1   0.9983  -0.1    0.0     0.0      0.0     0.0     1    1
   3 MAIN 1  13.8   1   0.9545  -2.9    0.0     0.0      0.0     0.0     1    1
   4 MAIN 2  13.8   1   0.9853  -2.6    0.0     0.0      0.0     0.0     1    1
   5 FDR 32  13.8   1   0.9533  -2.9    0.0     0.0      0.0     0.0     1    1
   6 FDR 34  13.8   1   0.9530  -2.9    0.0     0.0      0.0     0.0     1    1
   7 FDR 42  13.8   1   0.9843  -2.6    0.0     0.0      0.0     0.0     1    1
   8 SYNMOTOR13.8   1   0.9852  -2.6    7.2    -5.4      0.0     0.0     1    1
   9 FDR 31  13.8   1   0.9532  -2.9    0.0     0.0      0.0     0.0     1    1
  10 FDR 271 13.8   1   0.9812  -2.5    0.0     0.0      0.0     0.0     1    1
  11 LOAD 2   2.40  1   0.9134  -5.7    1.3     0.8      0.0     0.0     1    1
  12 FDR 92  13.8   1   0.9526  -2.9    0.0     0.0      0.0     0.0     1    1
  13 FDR 101 13.8   1   0.9806  -2.5    0.0     0.0      0.0     0.0     1    1
  14 FDR 61  13.8   1   0.9452  -2.8    0.0     0.0      0.0     0.0     1    1
  15 FDR 41  13.8   1   0.9819  -2.5    0.0     0.0      0.0     0.0     1    1
  16 FDR 72  13.8   1   0.9826  -2.5    0.0     0.0      0.0     0.0     1    1
  17 T 5      0.48  1   0.9020  -6.6    1.2     0.8      0.0     0.0     1    1
  18 T 6      0.48  1   0.9317  -5.9    1.2     0.8      0.0     0.0     1    1
  19 LOAD 6   2.40  1   0.9045  -5.6    3.2     1.9      0.0     0.0     1    1
  20 LOAD 7   2.40  1   0.9430  -5.1    3.2     1.9      0.0     0.0     1    1
  21 LOAD 8   0.48  1   0.9429  -5.1    0.6     0.4      0.0     0.0     1    1
  22 LOAD 9   0.48  1   0.9010  -6.6    0.1     0.1      0.0     0.0     1    1
  23 LOAD 10  0.48  1   0.9309  -5.9    0.1     0.1      0.0     0.0     1    1
  24 GEN 1   13.8   2   0.9900  -2.3    0.0     0.0      0.0     0.0     1    1
  25 FDR 91  13.8   1   0.9483  -2.8    0.0     0.0      0.0     0.0     1    1
  26 FDR 33  13.8   1   0.9527  -2.9    0.0     0.0      0.0     0.0     1    1
  27 FDR 71  13.8   1   0.9827  -2.5    0.0     0.0      0.0     0.0     1    1
  28 T10      0.48  1   0.9228  -5.4    0.8     0.5      0.0     0.0     1    1
  29 T11      0.48  1   0.9231  -5.4    0.9     0.6      0.0     0.0     1    1
  30 T12      0.48  1   0.9261  -5.5    0.8     0.5      0.0     0.0     1    1
  31 FDR 242 13.8   1   0.9885  -2.2    0.0     0.0      0.0     0.0     1    1
  32 FDR 243 13.8   1   0.9890  -2.2    0.0     0.0      0.0     0.0     1    1
  33 LOAD 11  0.48  1   0.9216  -5.4    0.3     0.2      0.0     0.0     1    1
  34 LOAD 12  0.48  1   0.9225  -5.4    0.1     0.1      0.0     0.0     1    1
  35 LOAD 13  0.48  1   0.9249  -5.5    0.3     0.2      0.0     0.0     1    1
  36 LOAD 14  2.40  1   0.9526  -4.5    1.8     1.2      0.0     0.0     1    1
  37 LOAD 15  0.48  1   0.9503  -5.0    0.9     0.5      0.0     0.0     1    1
  38 LOAD 16  0.48  1   0.9231  -5.5    0.5     0.3      0.0     0.0     1    1
  39 LOAD 1   4.16  1   0.9118  -6.1    1.5     0.9      0.0     0.0     1    1
  40 UTILITY 69.0   3   1.0000   0.0    0.0     0.0      0.0     0.0     1    1
```

```
         PTI INTERACTIVE POWER SYSTEM SIMULATOR--PSS/E      FRI. DEC 15 1989  09:15
TYPICAL INDUSTRIAL PLANT ELECTRICAL SYSTEM (IEEE BROWN BOOK)            GENERATING
BASE CASE LOAD FLOW                                                    PLANT DATA
BUS#   NAME  BSKV COD MCNS PGEN  QGEN QMAX  QMIN VSCHED   VACT. REMOT PCT Q
  24 GEN 1   13.8   2   1    8.5   0.1   5.0   -2.0 0.9900 0.9900
  40 UTILITY 69.0   3   1   17.8   8.4  25.0  -10.0 1.0000 1.0000
```

```
         PTI INTERACTIVE POWER SYSTEM SIMULATOR--PSS/E      FRI. DEC 15 1989  09:15
TYPICAL INDUSTRIAL PLANT ELECTRICAL SYSTEM (IEEE BROWN BOOK)            GENERATING
BASE CASE LOAD FLOW                                                     UNIT DATA
BUS#   NAME  BSKV COD ID ST PGEN QGEN QMAX QMIN PMAX PMIN MBASE  Z S O R C E   X T R A N   GENTAP
  24 GEN 1   13.8   2   1  1   9    0    5   -2 9999-9999  10  0.0000 1.0000
  40 UTILITY 69.0   3   1  1  18    8   25  -10 9999-9999  10  0.0000 1.0000
```

```
         PTI INTERACTIVE POWER SYSTEM SIMULATOR--PSS/E      FRI. DEC 15 1989  09:15
TYPICAL INDUSTRIAL PLANT ELECTRICAL SYSTEM (IEEE BROWN BOOK)            BRANCH DATA
BASE CASE LOAD FLOW
FROM  TO CKT NAME       NAME     LINE R  LINE X  CHRGING TP ST RATA RATB RATC
   1   3  1  TIE 1      MAIN 1   0.0067  0.0533  0.0000  F  1   15   0    0
   1  40  1  TIE 1      UTILITY  0.0015  0.0030  0.0051     1   30   0    0
   2   4  1  TIE 2      MAIN 2   0.0067  0.0533  0.0000  F  1   15   0    0
   2  40  1  TIE 2      UTILITY  0.0015  0.0030  0.0051     1   30   0    0
   3   5  1  MAIN 1     FDR 32   0.0070  0.0009  0.0000     1    2   0    0
   3   6  1  MAIN 1     FDR 34   0.0026  0.0011  0.0000     1    5   0    0
   3   9  1  MAIN 1     FDR 31   0.0044  0.0016  0.0000     1    4   0    0
   3  26  1  MAIN 1     FDR 33   0.0146  0.0019  0.0000     1    2   0    0
   4   7  1  MAIN 2     FDR 42   0.0022  0.0010  0.0000     1    2   0    0
   4   8  1  MAIN 2     SYNMOTOR 0.0008  0.0010  0.0000     1   10   0    0
   4  15  1  MAIN 2     FDR 41   0.0084  0.0026  0.0000     1    3   0    0
   4  24  1  MAIN 2     GEN 1    0.0100  0.0064  0.0000     1    6   0    0
   5  39  1  FDR 32     LOAD 1   0.0464  0.3480  0.0000  F  1    2   0    0
   6  11  1  FDR 34     LOAD 2   0.0458  0.3670  0.0000  F  1    2   0    0
   6  14  1  FDR 34     FDR 61   0.0189  0.0058  0.0000     1    3   0    0
   7  16  1  FDR 42     FDR 72   0.0256  0.0033  0.0000     1    2   0    0
   7  27  1  FDR 42     FDR 71   0.0042  0.0015  0.0000     1    4   0    0
   9  12  1  FDR 31     FDR 92   0.0036  0.0005  0.0000     1    2   0    0
   9  25  1  FDR 31     FDR 91   0.0395  0.0051  0.0000     1    2   0    0
  10  13  1  FDR 271    FDR 101  0.0043  0.0006  0.0000     1    2   0    0
  10  27  1  FDR 271    FDR 71   0.0102  0.0013  0.0000     1    2   0    0
  12  17  1  FDR 92     T 5      0.0529  0.4500  0.0000  F  1    2   0    0
  13  18  1  FDR 101    T 6      0.0529  0.4500  0.0000  F  1    2   0    0
  14  19  1  FDR 61     LOAD 6   0.0244  0.1470  0.0000  F  1    4   0    0
  15  20  1  FDR 41     LOAD 7   0.0244  0.1470  0.0000  F  1    2   0    0
  16  21  1  FDR 72     LOAD 8   0.0958  0.7670  0.0000  F  1    1   0    0
  17  22  1  T 5        LOAD 9   0.0582  0.0243  0.0000     1    0   0    0
  18  23  1  T 6        LOAD 10  0.0464  0.0239  0.0000     1    0   0    0
  24  31  1  GEN 1      FDR 242  0.0073  0.0010  0.0000     1    2   0    0
  24  32  1  GEN 1      FDR 243  0.0104  0.0014  0.0000     1    2   0    0
  25  28  1  FDR 91     T10      0.0548  0.3830  0.0000  F  1    2   0    0
  26  29  1  FDR 33     T11      0.0548  0.3830  0.0000  F  1    2   0    0
  27  30  1  FDR 71     T12      0.0548  0.3830  0.0000  F  1    2   0    0
  28  33  1  T10        LOAD 11  0.0291  0.0230  0.0000     1    0   0    0
  28  38  1  T10        LOAD 16  0.0321  0.0293  0.0000     1    1   0    0
  29  34  1  T11        LOAD 12  0.0451  0.0130  0.0000     1    0   0    0
  29  38  1  T11        LOAD 16  0.0424  0.0387  0.0000     1    1   0    0
  30  35  1  T12        LOAD 13  0.0291  0.0230  0.0000     1    0   0    0
  30  38  1  T12        LOAD 16  0.0321  0.0293  0.0000     1    1   0    0
  31  36  1  FDR 242    LOAD 14  0.0329  0.2300  0.0000  F  1    3   0    0
  32  37  1  FDR 243    LOAD 15  0.0821  0.5750  0.0000  F  1    1   0    0
```

Fig 64
Data Listing for Sample System

a problem in the model is indicated. Scanning of the load flow output to see buses with large values of mismatch or abnormal voltages will often help find the problem area. The problem could be incomplete or inconsistent data. Short low-impedance lines in close proximity with long lines may make convergence difficult. Very low-impedance lines will likely cause convergence problems unless the load flow program contains special logic in the solution techniques to handle them. Engineering judgment is needed to determine whether it is more appropriate to model these elements explicitly or to lump them with adjacent elements.

Most load flow programs will have the ability to take a solved load flow base case and store all the necessary data including the solution in a file on the computer disk. This will allow easy retrieval of the base case to incorporate future changes or to perform outage condition studies.

One page of the solved load flow output for the sample base case is shown in Fig 65 in tabular form. For each bus, the bus voltage magnitude and angle are given. The voltage magnitude may be given as per unit or kV (in this case, both are listed). Each line going from that bus to another bus is listed, giving the MW and Mvar flow (or kW and kvar) on the line out of the "from" bus. A negative flow means the flow is coming into the "from" bus. For transformers, the tap is also listed. If there is significant mismatch on the bus, it will also be listed. Different programs will use somewhat different formats; but they will present basically the same information.

A more concise and usually more informative method of presenting load flow results is to display them graphically on the system single-line diagram. System flows can be quickly analyzed from this visual presentation that relates system configuration, operating conditions, and equipment parameters.

Figure 66 displays the base case load flow results in graphical form. This figure shows the voltages on all buses and flows on all lines. As well as this output data, the system configuration is shown clearly: which buses are supplied by each feeder, loads being modeled, generator output, transformer tap ratios, and shunt capacitor values.

Many commercial load flow programs will generate such drawings. As noted before, output format will vary. In Fig 66, the line flows are shown near one of the buses and a positive value indicates flow going from that bus to the other bus.

Load flow output as shown in Fig 66 provides an excellent opportunity to document study results. At a convenient location on the drawing, or on a facing page in the analytical report, comments can be entered as to good or poor conditions, or both, that exist in circuit parameters or configuration. It is desirable to list corrective action taken for the next load flow run, which will hopefully improve the operation.

6.6.1 Analysis of Sample System. Now that a load flow has been run for the base condition, what has been learned about the system, and what can be done to improve its operation? Analysis of load flow output shows the following:

(1) Voltage on bus 3 (MAIN 1) is 0.955 pu, which is low. Voltage on bus 4 (MAIN 2) is 0.985 pu.
(2) Voltages on all buses supplied from bus 3 are relatively low.
(3) Voltage on many of the 2400 V and 480 V buses are too low for proper operation. This is due partially to the low voltage on bus 3.

```
1   PTI INTERACTIVE POWER SYSTEM SIMULATOR--PSS/E         FRI. DEC 15 1989   09:03
    TYPICAL INDUSTRIAL PLANT ELECTRICAL SYSTEM (IEEE BROWN BOOK)              RATING
    BASE CASE LOAD FLOW                                                       SET A

BUS   1 TIE 1      69.0  AREA CKT         KW      KVAR        KVA   %I  0.9965PU   -0.11   1
                                                                       68.759KV
TO    3 MAIN 1     13.8    1    1      9702.8    6849.5   11876.8    79  1.0000LK
TO   40 UTILITY    69.0    1    1     -9702.8   -6849.6   11876.9    40

BUS   2 TIE 2      69.0  AREA CKT         KW      KVAR        KVA   %I  0.9983PU   -0.12   2
                                                                       68.883KV
TO    4 MAIN 1     13.8    1    1      8094.1    1595.3    8249.8    55  1.0000LK
TO   40 UTILITY    69.0    1    1     -8094.2   -1595.1    8249.8    28

BUS   3 MAIN 1     13.8  AREA CKT         KW      KVAR        KVA   %I  0.9545PU   -2.95   3
                                                                       13.173KV
TO    1 TIE 1      69.0    1    1     -9608.0   -6092.3   11376.7    79  1.0000UN
TO    5 FDR 32     13.8    1    1      1519.1     974.7    1804.9    95
TO    6 FDR 34     13.8    1    1      4544.1    3058.9    5477.8   123
TO    9 FDR 31     13.8    1    1      2444.1    1454.3    2844.1    76
TO   26 FDR 33     13.8    1    1      1100.1     603.9    1254.9    66

BUS   4 MAIN 2     13.8  AREA CKT         KW      KVAR        KVA   %I  0.9853PU   -2.58   4
                                                                       13.597KV
TO    2 TIE 2      69.0    1    1     -8048.6   -1251.3    8142.2    55  1.0000UN
TO    7 FDR 42     13.8    1    1      3373.9    2678.0    4307.5    94
TO    8 SYNMOTOR   13.8    1    1      7206.6   -5391.2    9000.0    93
TO   15 FDR 41     13.8    1    1      3251.1   -2133.0    3888.4   114
TO   24 GEN 1      13.8    1    1     -5782.1    1811.2    6059.1   103

BUS   5 FDR 32     13.8  AREA CKT         KW      KVAR        KVA   %I  0.9533PU   -2.91   5
                                                                       13.156KV
TO    3 MAIN 1     13.8    1    1     -1516.0    -974.4    1802.7    95  1.0000LK
TO   39 LOAD 1      4.16   1    1      1516.6     974.4    1802.6   110

BUS   6 FDR 34     13.8  AREA CKT         KW      KVAR        KVA   %I  0.9530PU   -2.93   6
                                                                       13.151KV
TO    3 MAIN 1     13.8    1    1     -4535.7   -3055.2    5468.7   123  1.0000LK
TO   11 LOAD 2      2.40   1    1      1262.1     896.9    1548.3   108
TO   14 FDR 61     13.8    1    1      3273.4    2158.6    3921.0   119

BUS   7 FDR 42     13.8  AREA CKT         KW      KVAR        KVA   %I  0.9843PU   -2.56   7
                                                                       13.583KV
TO    4 MAIN 2     13.8    1    1     -3369.7   -2676.2    4303.1    94
TO   16 FDR 72     13.8    1    1       607.0     445.1     752.7    39
TO   27 FDR 71     13.8    1    1      2761.5    2231.0    3550.0    92

BUS   8 SYNMOTOR13.8  13.8 AREA CKT        KW     KVAR        KVA   %I  0.9852PU   -2.64   8
                                                                       13.596KV
TO    4 MAIN 2     13.8    1    1      7200.0   -5400.0    9000.0    94
   LOAD-PQ                            -7200.0    5399.5    8999.7    93

BUS   9 FDR 31     13.8  AREA CKT         KW      KVAR        KVA   %I  0.9532PU   -2.93   9
                                                                       13.154KV
TO    3 MAIN 1     13.8    1    1     -2440.1   -1452.9    2839.9    76
TO   12 FDR 92     13.8    1    1      1336.8     953.8    1642.2    87
TO   25 FDR 91     13.8    1    1      1103.2     499.1    1210.8    64
```

FORMAT:

BUS #, NAME, BASE KV, AREA #, CIRCUIT #, kW FLOW, kvar FLOW, kVA FLOW, LINE FLOW IN % OF RATING

BUS VOLTAGE IN PER UNIT, ANGLE IN DEGREES
BUS VOLTAGE IN kV

TAP POSITION: (LK+TAP ON FROM BUS, UN+TAP ON TO BUS)

Fig 65
Sample Load Flow Output

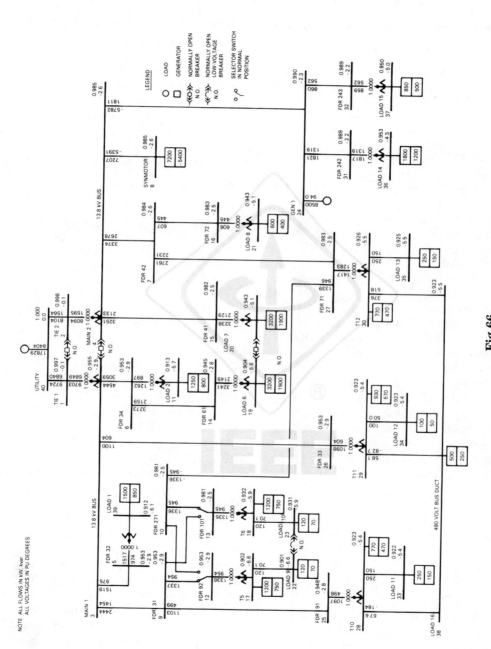

Fig 66

Example System Base Case Load Flow Output

(4) The reactive power output of generator 1 could be increased to supply additional vars to the system.

In addition to the above voltage-related problem, the following branches are overloaded:

BRANCH LOADINGS ABOVE 100.0% OF RATING

X-----FROM BUS-------X   X-----TO BUS--------X

BUS	NAME	BASE kV	BUS	NAME	BASE kV	CKT	LOADING	RATING	%
3	MAIN 1	13.8	6	FDR 34	13.8	1	5738.6	4660.0	123.1
4	MAIN 2	13.8	15	FDR 41	13.8	1	3946.5	3460.0	114.1
4	MAIN 2	13.8	24	GEN 1	13.8	1	6149.6	5970.0	103.0
5	FDR 32	13.8	39	LOAD 1	4.16	1	1890.9	1725.0	109.6
6	FDR 34	13.8	11	LOAD 2	2.40	1	1624.7	1500.0	108.3
6	FDR 34	13.8	14	FDR 61	13.8	1	4114.6	3460.0	118.9
12	FDR 92	13.8	17	T5	0.48	1	1722.9	1500.0	114.9
13	FDR 101	13.8	18	T6	0.48	1	1668.0	1500.0	111.2
14	FDR 61	13.8	19	LOAD 6	2.40	1	4114.5	3750.0	109.7
15	FDR 41	13.8	20	LOAD 7	2.40	1	3946.4	3750.0	105.2
16	FDR 72	13.8	21	LOAD 8	0.48	1	764.8	750.0	102.0
24	GEN 1	13.8	31	FDR 242	13.8	1	2271.0	1980.0	114.7
27	FDR 71	13.8	30	T12	0.48	1	1945.3	1500.0	129.7
30	T12	0.48	38	LOAD 16	0.48	1	691.2	600.0	115.2
32	FDR 243	13.8	37	LOAD 15	0.48	1	1037.7	1000.0	103.8

Many of these overloaded branches are transformers. These transformer overloads can be corrected by reduction of the reactive power flows through the transformers. The load power factors are generally around 0.85 in the sample system. Shunt capacitors were added to bring the power factor to about 0.95. The added shunt capacitors are listed below. Actual analysis of the proper size and placement of capacitors for power factor improvement requires more in-depth analysis such as explained in the chapter on power factor improvement in Reference [1].

CAPACITORS ADDED TO IMPROVE POWER FACTOR

BUS	NAME	VOLTAGE VOLTS	LOAD kW	LOAD kvar	INSTALLED CAPACITOR kvar
11	LOAD 2	2400	1250	800	300
17	T5	480	1200	750	300
18	T6	480	1200	750	300
19	LOAD 6	2400	3200	1900	750
20	LOAD 7	2400	3200	1900	750
21	LOAD 8	480	600	400	150
28	T10	480	770	470	150
29	T11	480	930	570	300
30	T12	480	770	470	150
36	LOAD 14	2400	1800	1200	600
37	LOAD 15	480	850	500	150
39	LOAD 14	4160	1500	850	300

Solution of the load flow after the addition of the capacitors showed that voltage was still low on the load buses supplied from bus 3. This suggests that the tap should be changed on transformer T1 (from bus 1 to bus 3). The transformer can be set on a 67.3 to 13.8 kV tap. This is represented by a 0.975 pu tap (67.3/69.0) on the primary. Since several of the load bus voltages are quite low, changing the taps on the 13.8 to 4.16 kV and 13.8 to 480 V transformers would also be beneficial.

The following tap changes were made:

X-----FROM BUS-----X			X--------TO BUS--------X				TAP	TAP
BUS	NAME	BASE kV	BUS	NAME	BASE kV	CKT	(kV)	(pu)
1	TIE 1	13.8	3	MAIN 1	13.8	1	67.3/13.8	0.975
5	FDR 32	13.8	39	LOAD 1	4.16	1	13.46/4.16	0.975
6	FDR 34	13.8	11	LOAD 2	2.40	1	13.46/2.4	0.975
12	FDR 92	13.8	17	T5	0.48	1	13.46/0.48	0.975
13	FDR 101	13.8	18	T6	0.48	1	13.46/0.48	0.975
14	FDR 61	13.8	19	LOAD 6	2.40	1	13.46/2.4	0.975
25	FDR 91	13.8	28	T10	0.48	1	13.46/0.48	0.975
26	FDR 33	13.8	29	T11	0.48	1	13.46/0.48	0.975
27	FDR 71	13.8	30	T12	0.48	1	13.46/0.48	0.975

In addition, the scheduled voltage of the generator on bus 24 was increased to 1.01 pu to cause the generator to supply additional reactive power to the system.

The load flow solution after the above changes is shown in Fig 67. The voltages are now all in the acceptable range. There are still a couple of heavily loaded lines that would require further analysis. In addition, the above changes must be checked to see if they result in acceptable conditions for different operating conditions, such as light load in the plant or the utility system supply voltage being higher or lower than the nominal voltage modeled in the base case.

When the base case condition is satisfactory, it is necessary to analyze outage conditions. A couple of examples would be outage of one of the main supply transformers with the closing of the breaker between buses 3 and 4 and the alternate supply of buses 12, 17, and 22 from bus 10. The load flow allows the analysis of outages to see if the system will operate in an acceptable (although possibly degraded) condition or if curtailment of the plant's operations would be necessary and allows alternatives for system improvement to be tested.

Analysis of the system load flow outputs after each set of changes results in the system being gradually tuned to obtain the most efficient and reliable operation. Experience with load flows improves the engineer's ability to make corrections with a minimum number of load flow solutions. However, it is stressed that any change affects the whole system, and a cure at one spot can create unexpected problems at another location in the system. For this reason, it is better not to make too many changes in a single run as the effects on the system may be difficult to understand. Each change case should be documented showing changes made and results obtained in order to keep future changes consistent with improving the system. A good place to record changes and their results is directly on the single-line diagram.

6.7 Load Flow Programs. Many load flow programs are presently available, both in the public domain and as commercial products. Many are available for purchase or on a timeshare basis. These programs may be designed to run on large mainframe computers, medium-sized minicomputers, or on microcomputers, such as personal computers. The various load flow programs differ in the ease of use, program documentation, program sophistication and, of course, cost.

Most load flow programs now operate in an interactive mode, prompting the user for input and direction in basically a question and answer mode. This approach is much more efficient from an engineering standpoint than the previous batch-job-oriented programs.

There is a wide range in the level of sophistication in the available programs. There is a corresponding range in the level of user need for this sophistication. The more sophisticated programs often contain several load flow solution techniques allowing for easier solution of a wide range of problems, more data checking activities to help in debugging data input errors, more data handling activities to ease changes to system data or configuration, graphic display of load flow results, ability to handle much larger networks, the modeling of additional power system components, and the incorporation of additional control functions into the solution techniques. For example, load flows are available that will handle 12 000 buses, incorporate such power system elements as HVDC lines and phase shifting transformers, and perform utility area interchange control as part of the solution proc-

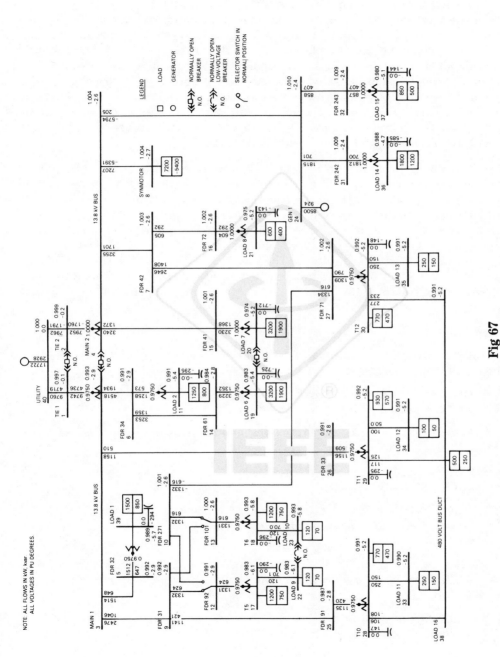

Fig 67

Example System Load Flow Output After Corrective Changes

ess. Although essential for large utility systems, a high level of sophistication may not be needed by industrial engineers to analyze their systems.

Since the same network data is used for load flow, short-circuit, transient stability, motor starting, and harmonic studies, it is useful to select a program package that integrates these calculations so that they all use the same load flow network data. Of course, all of the above study capabilities will not be needed by all users.

As an alternative to in-house use of a load flow program, there are consultants who can do the analysis and present the user with a complete report including a technical analysis of the computer output and suggestions and advice on system improvement. Even in this study mode, it is to the user's advantage to understand the data requirements and how a study is performed.

6.8 Conclusions. It should be evident to designers and operators of industrial plant electrical systems, as well as to utility system engineers, that a tool which predicts the performance of their electrical systems under various operating conditions that can actually be encountered *before* these conditions occur is of value. Such a tool can be used to prevent expensive outages, damaged equipment, and possibly even loss of life. The load flow solution provides a means to study systems under real or hypothetical conditions. The solution results should be evaluated and analyzed with respect to optimum present and future operation. This leads to a diagnosis of the system as it exists. The analysis can also point the way to improved operation and provide a meaningful basis for future system planning.

6.9 References

[1] IEEE Std 141-1986, Recommended Practice for Electric Power Distribution for Industrial Plants (ANSI).[8]

[2] BROWN, H. E. *Solution of Large Networks by Matrix Methods*, New York, NY: John Wiley and Sons, 1975.

[3] STAGG, G. W. and EL-ABIAD, A. H. *Computer Methods in Power System Analysis*, New York, NY: McGraw-Hill, 1968.

[4] STEVENSON, W. D. *Elements of Power System Analysis*, New York, NY: McGraw-Hill, 1975.

[5] TINNEY, W. F. and HART, C. E. "Power Flow Solution by Newton's Method," *IEEE Transactions on Power Apparatus and Systems*, vol. PAS-86, pp. 1444–1460, Nov. 1967.

[6] WARD, J. B. and HALE, H. W. "Digital Computer Solution of Power Flow Problems," *AIEE Transactions*, Part III–Power Apparatus and Systems, vol. 75, pp. 398–404, June 1956.

[7] WOOD, A. J. and WOLLENBERG, B. F. *Power Generation, Operation and Control*, New York, NY: John Wiley and Sons, 1984.

---

[8]IEEE publications are available from the Institute of Electrical and Electronics Engineers, IEEE Service Center, 445 Hoes Lane, Piscataway, NJ 08855-1331.

Chapter 7
Short-Circuit Studies

7.1 Introduction. Even the most carefully designed power systems may be subject to damaging arc-blast or overheating and the explosive magnetic forces associated with high magnitude currents flowing during a short-circuit condition. To ensure that circuit protective equipment can isolate faults quickly and minimize system component damage, personnel hazard and outage severity, it is essential that a short-circuit study be included in the electrical design of new industrial and commercial power systems, and for modifications to existing systems.

A short-circuit study can be used to determine any or all of the following:

(1) Calculated system fault current duties that can be compared with the first cycle momentary or *close-and-latch* rating and with the interrupting rating of circuit-interrupting devices, such as circuit breakers and fuses.

(2) Calculated system fault current duties to compare with short-time or withstand ratings of system components, such as busway, cables, transformers, reactors, disconnect switches, etc.

(3) The selection and rating or setting of short-circuit protective devices, such as molded case breakers, solid-state trip units, fuses, relays.

(4) Evaluation of short-circuit current flow and voltage levels in the overall system for short circuits in specific areas.

Details of fundamental concepts involved in the rigorous calculation of short-circuit currents are included in Chapters 3 and 4. Simplifying techniques and their limitations, and step-by-step procedures to follow for manual short-circuit calculations are given in Chapter 6 of IEEE Std 141-1986 [4] [9] and Chapter 2 of IEEE Std 242-1986 [5]. An understanding of the information contained in these sources is an essential prerequisite to attempting a detailed short-circuit study.

Short-circuit studies are an excellent example of system analysis for which digital computers can be applied effectively. The intent of this chapter, therefore, is:

(1) To supplement Chapter 6 of IEEE Std 141-1986 [4] and Chapter 2 of IEEE Std 242-1986 [5] by providing a summary of the steps required to perform a short-circuit study in an industrial or commercial power system, with particular emphasis on the use of the digital computer as a *tool* in the process.

---

[9]The numbers in brackets correspond to those in the References at the end of this chapter. IEEE publications are available from the Institute of Electrical and Electronics Engineers, IEEE Service Center, 445 Hoes Lane, Piscataway, NJ 08855-1331.

(2) To discuss some specific ways in which computers can be used effectively in performing most of the steps of a short-circuit study.

(3) To provide an illustrative example of a short-circuit study.

7.2 Short-Circuit Study Procedure

7.2.1 Preparation of a Study Single-Line Diagram.

The starting point in performing a short-circuit study is the preparation of a basic single-line diagram that represents the system to be studied. The accuracy and usefulness of the study results depend to a large extent on the reliability of this diagram.

Usually, for existing plants, a general purpose single-line diagram containing most, if not all, of the required information will be available. Such a drawing can be used as the basis for the preparation of a single-line diagram that is specifically for use in the short-circuit study. It is wise, however, to check the general purpose drawing carefully to ensure that it actually represents the system under study, and if not, it will be necessary to obtain additional data required for the study. For proposed new plants, extensions to existing plants, and/or existing plants where a single-line diagram is not available, it will be necessary to prepare a suitable diagram.

As indicated in the foregoing paragraph, it is usually worthwhile to use the "raw" data obtained from the general purpose single-line diagram to prepare a basic "study-oriented" single-line diagram. All system components that are relevant to the short-circuit study should be shown on this diagram. Typically such components are:

(1) Power company supply(ies)
(2) In-plant generators
(3) Transformers
(4) Reactors
(5) Feeder cables and duct systems
(6) Overhead lines
(7) Individual or grouped synchronous motors
(8) Individual or grouped induction motors
(9) Switching equipment
(10) Static loads (e.g., heaters, inverters)

It is desirable, when practical, to show on the diagram the relevant switching equipment short-circuit ratings, for example:

(1) Low-voltage motor control and switchgear bus bracing and "interrupting" ratings
(2) Medium-voltage motor control and switchgear "momentary" or "close-and-latch" ratings and "interrupting" ratings
(3) Disconnect switch momentary bracing and "close-and-latch" ratings

7.2.2 Determination of Study Requirements.

The next step is to determine, from the basic single-line diagram, how extensive the short-circuit study must be, and what degree of accuracy is required.

There are five possibilities for a short circuit in a three-phase system:

(1) Three-phase short circuit. Three phase conductors shorted together.
(2) Line-to-line short circuit. Any two phase conductors shorted together.

(3) Single line-to-ground short circuit. One phase conductor shorted to neutral or ground.

(4) Double line-to-ground short circuit. Any two phase conductors shorted together and simultaneously to neutral or ground.

(5) Three line-to-ground short circuit. All three phase conductors shorted together and to neutral or ground.

If a short circuit of one type is not interrupted promptly, it often progresses to another type, which generally results in more severe damage. For example, in a solidly grounded low-voltage system, a single line-to-ground short circuit, if not interrupted, can quickly escalate to a double line-to-ground or three line-to-ground short circuit. The choice of study (or studies) that is required for a particular system is a matter of engineering judgment based on an analysis of the basic single-line diagram and a determination of the specific purpose of the study.

Generally, it is useful to think in terms of conducting a study that achieves the specified purpose with sufficient accuracy to be reliable using methodology that is practical.

For three-phase industrial and commercial power systems, the most common study is the calculation of three-phase (balanced) short-circuit current specifically for comparison with switching equipment capability. The short-circuit current determined from this type of study generally represents the highest value at a particular location in the system. If information regarding lesser balanced short-circuit values, or unbalanced short circuits (line-to-line or line-to-ground) is not required, the balanced three-phase study is sufficient.

It is important to realize that single line-to-ground or double line-to-ground short-circuit current magnitudes can exceed three-phase short-circuit current under certain conditions. This can occur near:

(1) Solidly grounded synchronous machines
(2) Solidly grounded wye (Y) connection of a delta-wye (Δ-Y) transformer of the three-phase core (three-leg) design
(3) Grounded Y-Δ tertiary autotransformers
(4) Grounded Y-grounded Y-Δ tertiary three-winding transformers

In systems where any of these machine and/or transformer connections exist, it may be necessary to conduct a single line-to-ground short-circuit study. Medium- and high-voltage circuit breakers have 15% higher interrupting capability for single line-to-ground short circuits than for phase-to-phase or three-phase short circuits. This difference must be taken into account when comparing short-circuit duty with equipment ratings.

In resistance-grounded systems, unless it is essential to determine specific "less-than-maximum" short-circuit values, it is not necessary to calculate the phase-to-ground short-circuit current since neutral grounding resistors limit maximum short-circuit current to a known value.

Where several resistance-grounded zero sequence sources are connected in parallel, feeding a loop system, the ground fault current flowing in each segment of the loop must be calculated.

7.2.3 Determination and Use of System Impedances. The next step is to determine the appropriate values of impedance for all passive and rotating (active) system components that will affect the study.

At this point, it is necessary to decide how to model each of the system components. Chapter 4 provides detailed modeling information that is helpful in deciding on the most suitable representation of system components for any specific study. Refer also to Chapter 6 of IEEE Std 141-1986 [4] and Chapter 2 of IEEE Std 242-1986 [5] for further detailed information on the various types of component representation and the assumptions and limitations that must be recognized when simplification is made to rigorous modeling.

For short-circuit studies in industrial and commercial power systems, the impedance of passive system elements can be considered to be constant with respect to time (over the range of times considered in short-circuit studies). For three-phase short-circuit studies in balanced systems, only positive sequence impedance Z_1 or $R_1 + jX_1$ is required. For phase-to-phase short-circuit studies, positive and negative sequence impedances Z_1 and Z_2 or $R_2 + jX_2$ are required. For resistance-grounded systems, the zero sequence impedance must include the grounding resistor impedance value. For single and double line-to-ground short-circuit studies, positive, negative, and zero sequence impedance Z_1, Z_2, and Z_0 or $R_0 + jX_0$ are required. Each rotating (active) component is represented for short-circuit purposes as a constant voltage source behind an impedance. This impedance varies with time after a short circuit, and accounts for the ac component decay. In addition to the different impedances listed for the passive elements, different impedance values are required for the active elements. Basic impedance values required for three-phase short-circuit studies are listed in Table 11. For synchronous and induction rotating machines, this is accomplished by specifying certain impedance multiplying factors larger than 1 depending on whether ½ cycle momentary duties, 3 to 5 cycles interrupting duties, or 30 cycles extended-time short-circuit duties are considered. (Refer to Table 12.)

In addition to impedance values listed for three-phase studies, machine negative sequence reactance X_2 is required for line-to-line studies and both X_2 and the zero sequence reactance X_0 are required for line-to-ground studies. Although positive sequence machine reactances vary with time proceeding from X_d'' through X_d' to X_d, negative and zero sequence reactances are constant.

Actual impedance values should be obtained for specific equipment wherever possible. However, many other excellent of typical data are available to assist when actual values are not known. References IEEE Std 141-1986 [4] and IEEE Std 242-1986 [5] contain such data. Table 13 also provides a summary of typical data.

Two established forms of expressing impedances are in ohms and in per unit (or percent that differs from per unit by a factor of 100). For short-circuit study purposes, the per unit (or percent) system is generally the simplest to use, particularly when the system under study has more than one voltage level.

Details on the development and use of the per unit system are given in Chapter 4, 4.8.1.

When the system component *raw* impedances have been determined, the next step is to convert all impedances to a common base. Impedances expressed as per unit on a common base can be combined directly, in series or parallel, regardless of how many voltage levels exist and so can be ultimately reduced to a single equiva-

Table 11
Basic Impedance Values Required for Three-Phase Short-Circuit Studies

	½ Cycle	Interrupting (3 to 5 Cycles)	Approximately 30 Cycles
Induction motors	X_d'', R	X_d'', R	—
Synchronous motors	X_d'', R	X_d'', R	See Note 3
Synchronous generators and condensers	X_d'', R	X_d'', R	X_d' or X_d, R
Electric power company systems	X_s, R_s	X_s, R_s	X_s, R_s
Passive system components, i.e., transformers, cables, reactors	X, R	X, R	X, R

where:

X_d'' = Subtransient reactance (for induction motors, X_d'' is approximately equal to locked-rotor reactance)
X_d' = Transient reactance
X_d = Synchronous reactance
X = Equivalent reactance
R = Equivalent resistance
X_s, R_s = Power company system equivalent reactance and resistance

NOTES: (1) X_d'' of synchronous-rotating machines is the rated-voltage (saturated) direct-axis sub-transient reactance.
(2) X_d' of synchronous-rotating machines is the rated-voltage (saturated) direct-axis transient reactance.
(3) For calculations of minimum short-circuit values, contribution to be neglected. For calculations of maximum short-circuit values, use X_d' and R values.

lent impedance to the point of short circuit.

7.2.4 Preparation of an Impedance Diagram. It is useful at this point to develop an *impedance* diagram that is a companion document to the study single-line diagram. This diagram should show:

(1) All impedances separated into X and R components converted to a common base, with rotating equipment separated into appropriate groups as shown in Table 12 for any specific study.

For rotating (active) equipment, the multipliers applied to X are also applied to R for ½ cycle momentary, 3 to 5 cycle interrupting, and extended time (30 cycles) duty calculations, in order to preserve the equipment X/R ratio.

(2) Identification of each system bus and each system component.

(3) Power company system equivalent impedance converted to the common base.

(4) System pre-fault voltage levels that are to be used for the study, provided the computer program has this optional input facility.

7.2.5 Calculations. The next step in any short-circuit study is to use the information from the impedance diagram to calculate short-circuit current values at each location in the system as required to meet the objectives of the study. Basically, this is accomplished by combining all system impedances using appropriate

Table 12
Reactance Values for First Cycle and Interrupting Duty Calculations
(See Note 1)

Duty Calculation	System Component	Reactance Value for Medium- and High-Voltage Calculations per IEEE C37.010-1979 [1] and IEEE C37.5-1979 [3]	Reactance Value for Low-Voltage Calculations (See Note 2)
First Cycle (Momentary Calculations)	**Power Company Supply**	X_s	X_s
	All turbine generators; all hydrogenerators with amortisseur windings; all condensers	$1.0\ X_d''$	$1.0\ X_d''$
	Hydrogenerators without amortisseur windings	$0.75\ X_d'$	$0.75\ X_d'$
	All synchronous motors	$1.0\ X_d''$	$1.0\ X_d''$
	Induction Motors		
	Above 1000 hp at 1800 rev/min or less	$1.0\ X_d''$	$1.0\ X_d''$
	Above 250 hp at 3600 rev/min	$1.0\ X_d''$	$1.0\ X_d''$
	All others, 50 hp and above	$1.2\ X_d''$	$1.2\ X_d''$
	All smaller than 50 hp	Neglect (See Note 6)	$1.67\ X_d''$
Interrupting Calculations	**Power Company Supply**	X_s	N/A
	All turbine generators; all hydrogenerators with amortisseur windings; all condensers	$1.0\ X_d''$	N/A
	Hydrogenerators without amortisseur windings	$0.75\ X_d'$	N/A
	All synchronous motors	$1.5\ X_d''$	N/A
	Induction Motors		
	Above 1000 hp at 1800 rev/min or less	$1.5\ X_d''$	N/A
	Above 250 hp at 3600 rev/min	$1.5\ X_d''$	N/A
	All others 50 hp and above	$3.0\ X_d''$	N/A
	All smaller than 50 hp	Neglect (See Note 6)	N/A

NOTES: (1) First-cycle duty is the momentary (or close-and-latch) duty for medium-/high-voltage equipment and is the interrupting duty for low-voltage equipment.

(2) Reactance (X) values to be used for low-voltage breaker duty calculations. (See IEEE C37.13-1981 [2] and IEEE Std 242-1986 [5].)

(3) X_d'' of synchronous-rotating machines is the rated-voltage (saturated) direct-axis subtransient reactance.

(4) X_d' of synchronous-rotating machines is the rated-voltage (saturated) direct-axis transient reactance.

(5) X_d'' of induction motors equals 1 divided by per unit locked-rotor current at rated voltage.

(6) For comprehensive multivoltage system calculations, motors less than 50 hp are represented in medium- and high-voltage short-circuit calculations. (See IEEE Std 141-1986, Chapter 6 [4].)

**Table 13
Typical Values of Motor Impedances and kVA Ratings
to Use When Exact Values Are Not Known**

Induction motor	1 hp = 1 kVA
Synchronous motor, 0.8 pf	1 hp = 1 kVA
Synchronous motor, 1.0 pf	1 hp = 0.8 kVA

Motor Type	X_d'' (See note)
Synchronous motors	
2–6 poles	0.15
8–14 poles	0.20
16 poles or more	0.28
Individual large induction motors, usually medium voltage.	0.167
All others, 50 hp and above.	0.167
All smaller than 50 hp, usually low voltage.	0.167

NOTE: Motor impedances are in per unit on motor kVA rating. X_d'' for induction motors is approximately equal to the locked-rotor reactance. For induction motors, the locked-rotor reactance is the reciprocal value of the locked-rotor current. Reactances and motor base kVA ratings listed were taken from data and assumptions in IEEE Std 141-1986 [4].

techniques (see Chapter 6 of IEEE Std 141-1986 [4] and Chapter 2 of IEEE Std 242-1986 [5]).

The end result is to determine a single equivalent value of Z and X/R for the entire system to the point of short circuit, and to determine to what extent the sources of short-circuit current are to be considered *remote* or *local*. Short-circuit current from sources located *remote* from the point of fault generally has a slower ac current decay compared to *local* sources. This must be taken into account where the calculated short-circuit current is adjusted to reflect the system X/R ratio at the point of fault. This adjustment can be performed and processed by the computer to calculate the actual short-circuit current at each specific point of interest in the system.

7.2.6 Interpretation and Application of Study Results. The final step involves the interpretation and application of results to provide recommendations for system changes (in the case of existing plants) or for initial design (in the case of new plants).

General guidelines for interpretation and application of study results are usually not very useful since sound engineering judgment can only be made on the basis of treating each case individually. However, some important questions to ask are:

(1) Is circuit-interrupting equipment adequately rated for the maximum short-circuit momentary and interrupting duties imposed? If not, what is the most economical method of making system changes while still maintaining a satisfactory degree of system flexibility?

(2) Is there any short-circuit capability margin for future expansion? If not, is it necessary? If it is necessary, what is the most suitable method of effecting changes to the system?

(3) Is noninterrupting equipment, i.e., reactors, cables, bus systems, bus duct, overhead lines, transformers, adequately rated to withstand short-circuit current until cleared by circuit-interrupting equipment?

(4) Do load-break or disconnect switches in the system have sufficient momentary bracing and/or close-and-latch capability?

(5) What will be the effect on short-circuit levels in the plant system if the power company short-circuit level increases significantly? Economically, what can be done now to anticipate such an event in the future?

(6) Is special protective equipment or circuitry necessary to provide protective device selectivity for both maximum and minimum values of short-circuit current?

(7) Does the voltage of unfaulted buses in the system drop to values that will cause motor-starter contactor drop-out or unnecessary operation of undervoltage relays? If so, can practical measures be taken using additional equipment, or can modifications be made to the system configuration, etc., to minimize the extent of the system outage?

7.3 Use of the Computer. Chapter 5 provides detailed information on the types of computing systems available for system analysis including short-circuit studies.

Recent trends have been to more extensive use of personal computers rather than large mainframe computers for industrial and commercial power system analysis. With this trend, software is now available that makes it possible to perform virtually all of the short-circuit study steps using a desktop computer.

To illustrate, programs are available that can be used for:

(1) Preparation of the system single-line diagram described in 7.2.1.

(2) Converting *raw* system impedance data to impedances on a common base.

(3) Reduction of a number of system impedances in series or parallel, or both, to a single equivalent impedance value.

(4) Calculation of equivalent wye and delta impedances to assist in impedance reduction.

(5) Determination of short-circuit sources to be considered *remote* and *local* per the specifications in IEEE C37.010-1979 [1].

(6) Performing a wide range of short-circuit calculations, such as:

    (a) Complex short-circuit studies using fundamental circuit analysis.

    (b) Calculations of three-phase, line-to-line, line-to-ground, and double line-to-ground short circuits in either simple or complex systems. Where the system neutral is *brought out* and wired in three phase-four wire systems, line-to-neutral faults must be considered separately. Load flow data is required for input to establish initial conditions. Output gives positive, negative, and zero sequence line currents and system voltages.

    (c) Simple calculation of three-phase faults. Output gives total fault E/X current, line currents to faulted buses, and system voltages. No load flow data required to establish initial conditions for input to the short-circuit program.

    (d) Calculation of three-phase short-circuit duties for use in comparison with interrupting-device ratings. Output gives total faults E/X or E/Z

current, line current flows from other selected buses, X/R ratio values, and applies appropriate multipliers, including remote and local source adjustment factors to fault values to allow direct comparison with interrupting-device ratings based on applicable American National Standards. No load flow data required.

(7) Production of short-circuit diagrams directly using (in some cases) the data produced in the short-circuit calculations.

(8) Production of a short-circuit summary sheet that provides a comparison between calculated short-circuit currents and equipment ratings.

Most computer programs come with user manuals that assist in transferring the *raw* data to a form that can be accepted by the computer. Program requirements, limitations, and output format are also generally explained.

Usually, the programs are menu driven and so are designed to be interactive with the user. That is, questions will be asked of the user that appear on the computer screen and that call for a typed response.

7.4 Short-Circuit Study Example. To illustrate the use of a computer in performing the various tasks associated with a short-circuit study, an example system will be analyzed. The basic steps previously described will be followed. The example system represents a typical industrial plant where a three-phase short-circuit study is necessary to ensure that switchgear and other equipment is adequately rated to withstand and interrupt the maximum short-circuit current. Conditions of maximum load, maximum power company short-circuit duty, and maximum generation are considered since this is the most severe fault condition.

7.4.1 The Computer Program Capability. The computer is a PC using the MS-DOS system. The program used is contained on a hard disk. The program has been specifically developed for calculating momentary or close-and-latch (first ½ cycle), and interrupting short-circuit values using techniques in accordance with IEEE C37.13-1980 [2] for low-voltage breakers and IEEE C37.010-1979 [1] and IEEE C37.5-1979 [3] for medium- and high-voltage breakers.

IEEE C37.010-1979 [1] is used to compare short-circuit values with circuit breakers rated on a symmetrical current basis (generally applicable for all new North American medium- and high-voltage breakers).

IEEE C37.5-1979 [3] for breakers rated on a total current basis is generally applicable for older (i.e., pre-1964) medium- and high-voltage North American breakers in existing plants.

The program uses the matrix inversion technique to calculate an equivalent fault point impedance at each bus. It then derives total current levels at the bus and breaks this down into current flows to the fault from other connected buses.

The program calculates both separate R and X networks and a complex Z network. The complex Z network is used to calculate the total bus short-circuit current and connected bus contributions.

The R and X network calculations are used to determine the X/R ratio at the point of fault.

The computer program will modify the impedance input data as required to perform the calculation of first cycle and interrupting fault duties. Table 14 identi-

Table 14
Modified Data File Impedance Values Used in the Short-Circuit Computer Program

Types of Calculations	Line or Transformer	Utility Source	Generator	Synchronous Motor	Large Induction Motor*	Induction Motor ≥ 50 hp (Not Code 4)	Induction Motor <50 hp
Group Classification Code Number	0	1	2	3	4	5	6
FCYHV – Calculation of first cycle fault duties for evaluation of high- and medium-voltage breakers in accordance with IEEE C37.010-1979 [1] and IEEE C37.5-1979 [3]. — Calculation of first cycle fault duties for evaluation of medium-voltage fused contactors.	$R+jX$	$R+jX$	$R+jX''_d$	$R+jX''_d$	$R+jX''$	$1.2(R+jX'')$	—
INTHV – Calculation of interrupting fault duties for evaluation of high- and medium-voltage breakers in accordance with IEEE C37.010-1979 [1] and IEEE C37.5-1979 [3].	$R+jX$	$R+jX$	$R+jX''_d$	$1.5(R+jX''_d)$	$1.5(R+jX'')$	$3(R+jX'')$	—
FCYLV – Calculation of first cycle fault duties for evaluation of low-voltage breakers in accordance with IEEE C37.13-1981 [2] or for calculation of first cycle fault duties on multivoltage systems in accordance with IEEE Std 141-1986 [4], Chapter 6.	$R+jX$	$R+jX$	$R+jX''_d$	$R+jX''_d$	$R+jX''$	$1.2(R+jX'')$	$1.67(R+jX'')$

NOTES: (1) X''_d is the subtransient reactance for synchronous motors.
(2) X'' is the locked-rotor reactance for induction motors.
(3) A large induction motor is one that is > 1000 hp @ ≤1800 rev/min or
> 250 hp @ 3600 rev/min.

fies the modifications made to the impedance input data based on the calculation type being used and the group classification assigned for each type of equipment when creating the input data file.

Frequently for systems with both in-plant generation and a power company tie, in performing interrupting calculations it is necessary to determine a special composite adjustment factor that is a combination of the multiplier used for the *remote* short-circuit contribution and the *local* short-circuit contribution. In this context, *remote* is defined in IEEE C37.010-1979 [1] and IEEE C37.5-1979 [3] as a source of short-circuit current fed from generators through:

(1) Two or more transformations
(2) A per unit reactance external to the generator that is equal to or exceeds 1.5 times the generator per unit subtransient reactance on a common system megavolt-ampere base.

Chapter 6 of IEEE Std 141-1986 [4], 6.6.11 and 6.6.12, provide details on the methods that can be used to determine the appropriate composite adjustment factor to account for remote and local short-circuit contributions.

The computer program in this example utilizes the American National Standard curves and the *weighted interpolation* method and provides a composite adjustment factor directly. The breaker duty, along with X/R ratio and breaker speed, is printed with remote/local-plus-remote interpolated multipliers applied to symmetrical current.

The short-circuit program prints enough information so other interpretations of applicable IEEE/ANSI standards can be used.

7.4.2 Input Data Requirements. In the computer, the electrical power system is represented in matrix form. Each of the power system components, e.g., power company sources, generators, motors, transformers, cables, etc., is represented by a resistance value and a reactance value. In this example, system X/R values are required, so both the reactance and the resistance of each component is required. In studies where X/R values are not required and when the reactance of a component is very large compared to its resistance, it is acceptable to neglect resistance and so its value for computer input becomes zero.

The computer program places an assumed three-phase fault at the desired bus location in the system, and a set of short-circuit currents are calculated for comparison with published short-circuit ratings of the power system equipment.

Figure 68 is the single-line diagram for the system to be studied. It is the same diagram that is used in 6.6 and elsewhere in this text, with minor modifications for specific use in the three-phase short-circuit study example.

Figure 69 is the impedance diagram for the system. It has been developed from Fig 68. All buses are numbered, and all impedances are in a form that can be used as input to the computer. Major equipment impedances have been converted from values shown on the single-line diagram to per unit on a common 100 MVA base. Bus 38 load, which will not contribute short-circuit current, is not shown. Selected fault locations are shown as F3, F19, and F30.

The power company is represented by a per unit impedance, which is equivalent to a maximum short-circuit duty of 1000 MVA with $X/R = 22$. The impedance for computer input is computed as follows:

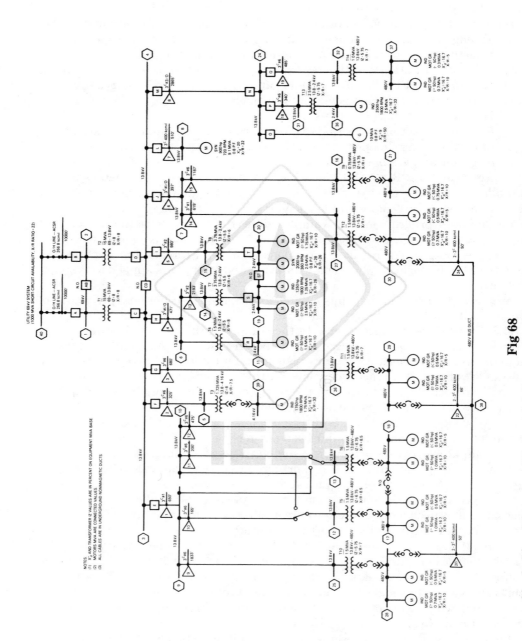

Fig 68

Single-Line Diagram of Industrial System for Three-Phase Short-Circuit Study Example

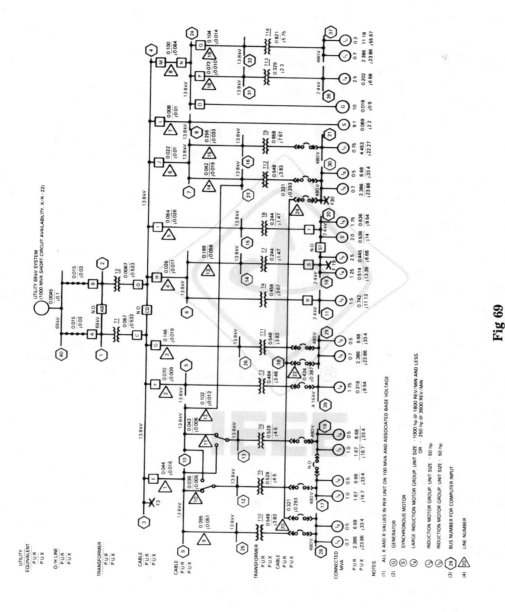

Fig 69

Impedance Diagram for Three-Phase Short-Circuit Study Example

$$X_{pu} = \frac{\text{base MVA}}{\text{fault availability in MVA}} = \frac{100}{1000} = 0.1 \qquad \text{(Eq 116)}$$

$$R_{pu} = \frac{0.1}{22} = 0.0045$$

System cable impedances are values taken from Fig 62 in Chapter 6. The impedances for computer input are derived as follows:

$$X_{pu} \text{ or } R_{pu} = \frac{\text{actual ohms} \times \text{base MVA}}{(\text{base kV})^2} \qquad \text{(Eq 117)}$$

All motors are grouped in accordance with the categories shown in Table 14 and then the computer combines several impedances into one equivalent impedance for each substation. Where specific motor kVA ratings and impedances were not known, the typical values shown in Table 13 were used.

Modified subtransient (X''_d) and transient (X'_d) reactance values are used in first cycle (momentary) and interrupting duty calculations. These values are in accordance with applicable circuit breaker application standards (see IEEE C37.010-1979 [1] and IEEE C37.5-1979 [3]).

American National Standards for calculating short-circuit duties require that actual motor or generator reactances be modified under certain conditions with modification factors as listed in Table 14 for both momentary (close-and-latch) and interrupting duty calculations.

In this example, the calculation of medium-voltage short-circuit duties does not include the contribution from low-voltage motors less than 50 hp.

The X/R ratios of synchronous and induction motors and transformers were determined by using medium typical curves from IEEE C37.010-1979 [1]. Typical X/R ratios for various electrical equipment are also given in Chapter 11. When exact values can be obtained, they should be used. For groups of induction motors smaller than 50 hp, the X/R ratio was estimated to be 5. For motors larger than 50 hp, the X/R ratio was estimated to be 10.

Examples of the preparation of impedance data for input to the computer are as follows:
For all rotating machines:

$$X = \frac{\text{Machine } X''_d \text{ in pu} \times \text{base MVA}}{\text{Machine-rated MVA}} \text{ per unit} \qquad \text{(Eq 118)}$$

$$R = \frac{X \text{ in pu}}{\text{Machine } X/R} \text{ per unit}$$

Examples:
  (1) 13.8 kV, 9000 hp synchronous motor for medium- and low-voltage calculations (see Figs 68 and 69, 13.8 kV Bus 8)
     Group Classification: 3

$$X''_d = 0.2 \text{ pu}; \quad X_{pu} = \frac{0.2 \times 100}{9.1} = 2.2$$

$$\frac{X}{R} = 32; \quad R_{pu} = \frac{X}{X/R} = \frac{2.2}{32} = 0.069$$

(2) 2.4 kV motors for medium- and low-voltage calculations (see Figs 68 and 69, 2.4 kV Bus 20)

    (a) 2000 hp synchronous motor
        Group Classification: 3

$$X''_d = 0.28 \text{ pu}; \quad X_{pu} = \frac{0.28 \times 100}{2.0} = 14.0$$

$$\frac{X}{R} = 26; \quad R_{pu} = \frac{X}{X/R} = \frac{14}{26} = 0.54$$

    (b) 1.75 MVA of induction motors between 50 hp and 1000 hp
        Group Classification: 5

$$X''_d = 0.167 \text{ pu}; \quad X_{pu} = \frac{0.167 \times 100}{1.75} = 9.54$$

$$\frac{X}{R} = 15; \quad R_{pu} = \frac{X}{X/R} = \frac{9.54}{15} = 0.63$$

(3) 480 V motors for medium-voltage calculations (see Figs 68 and 69, 480 V bus 17 and bus 18)

    (a) 1.0 MVA of induction motors 50 hp to 150 hp
        Group Classification: 5

$$X''_d = 0.167 \text{ pu}; \quad X_{pu} = \frac{0.167 \times 100}{1.0} = 16.7$$

$$\frac{X}{R} = 10.0; \quad R_{pu} = \frac{X}{X/R} = \frac{16.7}{10} = 1.67$$

    (b) 0.5 MVA of induction motors less than 50 hp, neglect.

(4) 480 V motors for low-voltage calculations (see Figs 68 and 69, 480 V bus 17 and bus 18)

    (a) 1.0 MVA of induction motors 50 hp to 150 hp
        X and R values as for example 3(a).
    (b) 0.5 MVA of induction motors less than 50 hp.
        Group Classification: 6

$$X''_d = 0.167 \text{ pu}; \quad X_{pu} = \frac{0.167 \times 100}{0.5} = 33.4$$

$$\frac{X}{R} = 5; \quad R_{pu} = \frac{X}{X/R} = \frac{33.4}{5} = 6.68$$

For the example system, all motors and the generator are assumed to be operating. This creates the highest possible short-circuit currents that the equipment may be subjected to, since the total short-circuit currents from all system motors, generators, and power company connections are present. The 13.8 kV bus tie is open with both power company transformers in service. If the tie breaker were closed for normal operation, the fault duty would be more severe and the study would be based on this operating mode.

The example study does not include pre-fault steady-state load currents. The effect of system load currents is usually negligible in short-circuit current studies for industrial and commercial power distribution systems.

7.4.3 Computer Program Input and Output Records.

To arrange the system data contained on the impedance diagram so it can be accepted by the computer program, it is necessary to make up an input data file for the high- and medium-voltage momentary calculation, the high- and medium-voltage interrupting calculation, and the low-voltage momentary calculation. Figure 70 shows the data on the input file for medium-voltage momentary and interrupting calculations. Data is usually listed in this way so it can be checked for errors before proceeding with short-circuit calculations.

Change cases can easily be run by modifying data lines in the file. The data file must be given a name. In this sample study, the data file name for medium-voltage interrupting calculation is IEEE01.SC.

Figure 71 is a sample of the computer output for medium-voltage momentary faults giving modified input data, using multiplying factors as shown in Table 14 and the result of the calculations for a fault on 13.8 kV bus 3 (F3) and on 2.4 kV bus 19 (F19).

Figure 72 is a sample of the computer output for the medium-voltage interrupting faults similar to the format for Fig 71.

Also shown are X/R ratios at the faulted bus, adjustment factors taken from the standards to apply to the E/Z values for 8 cycle, 5 cycle, and 3 cycle breakers, and the fault duty for direct comparison with the circuit breaker interrupting rating. Remote (in this case, the power company system) and local (in this case, the in-plant generator) sources of short-circuit current are listed separately. As indicated previously, it is possible to produce such diagrams using a computer program that takes for its input the short-circuit study output data.

Figure 73 shows the input data necessary for the first cycle (momentary) low-voltage fault duty calculations. The format is similar to that of Fig 70. The sample computer output for this data is shown in Fig 74 and is in a format similar to Figs 71 and 72.

Short-circuit diagrams showing short-circuit flows will provide the best means of presenting study results as depicted on Figs 75 and 76. The calculated fault currents are shown directly on these diagrams at each point of interest.

Using the complete computer output information, study results can be analyzed and applied to the specific equipment shown on the single-line diagram.

Table 15 shows a sample summary of the short-circuit study results and a comparison with switchgear ratings. Not all computer output information is shown since it is not possible to clearly present this information in one table. Several tables might be necessary to present complete study results.

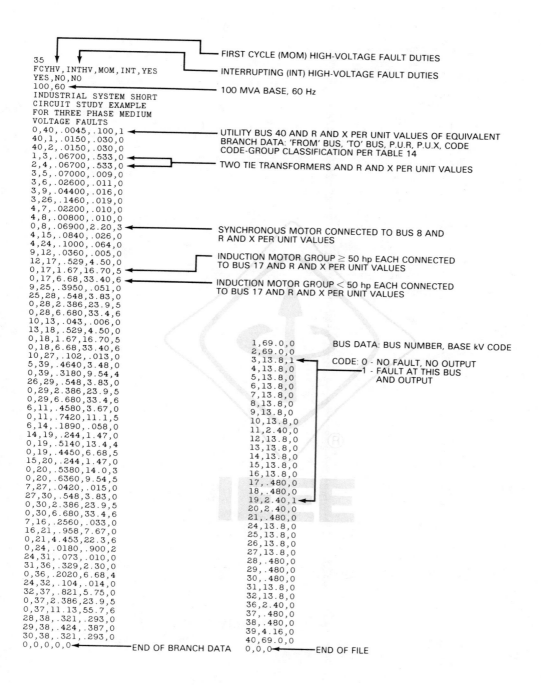

```
35
FCYHV,INTHV,MOM,INT,YES
YES,NO,NO
100,60
INDUSTRIAL SYSTEM SHORT
CIRCUIT STUDY EXAMPLE
FOR THREE PHASE MEDIUM
VOLTAGE FAULTS
0,40,.0045,.100,1
40,1,.0150,.030,0
40,2,.0150,.030,0
1,3,.06700,.533,0
2,4,.06700,.533,0
3,5,.07000,.009,0
3,6,.02600,.011,0
3,9,.04400,.016,0
3,26,.1460,.019,0
4,7,.02200,.010,0
4,8,.00800,.010,0
0,8,.06900,2.20,3
4,15,.0840,.026,0
4,24,.1000,.064,0
9,12,.0360,.005,0
12,17,.529,4.50,0
0,17,1.67,16.70,5
0,17,6.68,33.40,6
9,25,.3950,.051,0
25,28,.548,3.83,0
0,28,2.386,23.9,5
0,28,6.680,33.4,6
10,13,.043,.006,0
13,18,.529,4.50,0
0,18,1.67,16.70,5
0,18,6.68,33.40,6
10,27,.102,.013,0
5,39,.4640,3.48,0
0,39,.3180,9.54,4
26,29,.548,3.83,0
0,29,2.386,23.9,5
0,29,6.680,33.4,6
6,11,.4580,3.67,0
0,11,.7420,11.1,5
6,14,.1890,.058,0
14,19,.244,1.47,0
0,19,.5140,13.4,4
0,19,.4450,6.68,5
15,20,.244,1.47,0
0,20,.5380,14.0,3
0,20,.6360,9.54,5
7,27,.0420,.015,0
27,30,.548,3.83,0
0,30,2.386,23.9,5
0,30,6.680,33.4,6
7,16,.2560,.033,0
16,21,.958,7.67,0
0,21,4.453,22.3,6
0,24,.0180,.900,2
24,31,.073,.010,0
31,36,.329,2.30,0
0,36,.2020,6.68,4
24,32,.104,.014,0
32,37,.821,5.75,0
0,37,2.386,23.9,5
0,37,11.13,55.7,6
28,38,.321,.293,0
29,38,.424,.387,0
30,38,.321,.293,0
0,0,0,0,0
```

```
1,69.0,0
2,69.0,0
3,13.8,1
4,13.8,0
5,13.8,0
6,13.8,0
7,13.8,0
8,13.8,0
9,13.8,0
10,13.8,0
11,2.40,0
12,13.8,0
13,13.8,0
14,13.8,0
15,13.8,0
16,13.8,0
17,.480,0
18,.480,0
19,2.40,1
20,2.40,0
21,.480,0
24,13.8,0
25,13.8,0
26,13.8,0
27,13.8,0
28,.480,0
29,.480,0
30,.480,0
31,13.8,0
32,13.8,0
36,2.40,0
37,.480,0
38,.480,0
39,4.16,0
40,69.0,0
0,0,0
```

FIRST CYCLE (MOM) HIGH-VOLTAGE FAULT DUTIES

INTERRUPTING (INT) HIGH-VOLTAGE FAULT DUTIES

100 MVA BASE, 60 Hz

UTILITY BUS 40 AND R AND X PER UNIT VALUES OF EQUIVALENT
BRANCH DATA: 'FROM' BUS, 'TO' BUS, P.U.R, P.U.X, CODE
CODE-GROUP CLASSIFICATION PER TABLE 14

TWO TIE TRANSFORMERS AND R AND X PER UNIT VALUES

SYNCHRONOUS MOTOR CONNECTED TO BUS 8 AND
R AND X PER UNIT VALUES

INDUCTION MOTOR GROUP ≥ 50 hp EACH CONNECTED
TO BUS 17 AND R AND X PER UNIT VALUES

INDUCTION MOTOR GROUP < 50 hp EACH CONNECTED
TO BUS 17 AND R AND X PER UNIT VALUES

BUS DATA: BUS NUMBER, BASE kV CODE

CODE: 0 - NO FAULT, NO OUTPUT
1 - FAULT AT THIS BUS
AND OUTPUT

END OF BRANCH DATA

END OF FILE

**Fig 70
Computer Input File for Medium-Voltage Faults**

```
INDUSTRIAL SYSTEM SHORT
CIRCUIT STUDY EXAMPLE

CASE: MOM(ENTARY)
FOR THREE PHASE MEDIUM
VOLTAGE FAULTS

INPUT DATA
BUS   TO   BUS        R P.U.      X P.U.      CODE
 0         40        0.00450     0.10000       1
40          1        0.01500     0.03000       0
40          2        0.01500     0.03000       0
 1          3        0.06700     0.53300       0
 2          4        0.06700     0.53300       0
 3          5        0.07000     0.00900       0
 3          6        0.02600     0.01100       0
 3          9        0.04400     0.01600       0
 3         26        0.14600     0.01900       0
 4          7        0.02200     0.01000       0
 4          8        0.00800     0.01000       0
 0          8        0.06900     2.20000       3
 4         15        0.08400     0.02600       0
 4         24        0.10000     0.06400       0
 9         12        0.03600     0.00500       0
12         17        0.52900     4.50000       0
 0         17        2.00400    20.04000       5
 9         25        0.39500     0.05100       0
25         28        0.54800     3.83000       0
 0         28        2.86320    28.68000       5
10         13        0.04300     0.00600       0
13         18        0.52900     4.50000       0
 0         18        2.00400    20.04000       5
10         27        0.10200     0.01300       0
 5         39        0.46400     3.48000       0
 0         39        0.31800     9.54000       4
26         29        0.54800     3.83000       0
 0         29        2.86320    28.68000       5
 6         11        0.45800     3.67000       0
 0         11        0.89040    13.32000       5
 6         14        0.18900     0.05800       0
14         19        0.24400     1.47000       0
 0         19        0.51400    13.40000       4
 0         19        0.53400     8.01600       5
15         20        0.24400     1.47000       0
 0         20        0.53800    14.00000       3
 0         20        0.76320    11.44800       5
 7         27        0.04200     0.01500       0
27         30        0.54800     3.83000       0
 0         30        2.86320    28.68000       5
 7         16        0.25600     0.03300       0
16         21        0.95800     7.67000       0
 0         24        0.01800     0.90000       2
24         31        0.07300     0.01000       0
31         36        0.32900     2.30000       0
 0         36        0.20200     6.68000       4
24         32        0.10400     0.01400       0
32         37        0.82100     5.75000       0
 0         37        2.86320    28.68000       5
28         38        0.32100     0.29300       0
29         38        0.42400     0.38700       0
30         38        0.32100     0.29300       0
```

```
*BUS 3 (F3)    E/Z=  8.423 KA( 201.32MVA)AT-82.43DEG.,X/R=  7.95, 13.800 KV
               Z=  0.065415 +J  0.492387
               1.6*ISYM= 13.48    IASYM BASED ON X/R=  11.63

    CONTRIBUTIONS   IN KA
    BUS  TO  BUS      MAG      ANG      BUS  TO  BUS      MAG      ANG
     1        3      6.300   -82.523     5        3      0.320   -86.259
     6        3      0.878   -83.946     9        3      0.545   -78.716
    26        3      0.381   -79.559


*BUS 19 (F19)  E/Z= 15.981 KA(  66.43MVA)AT-78.99DEG.,X/R=  8.39,  2.400 KV
               Z= 0.287544 +J  1.477602
               1.6*ISYM= 25.57    IASYM BASED ON X/R=  22.29

    CONTRIBUTIONS IN KA
    BUS  TO  BUS      MAG      ANG      BUS  TO  BUS      MAG      ANG
    14        19     11.256  -75.676    INDMOT   19      1.794   -87.802
    INDMOT    19      2.994  -86.187
```

Fig 71
Computer Output for Medium-Voltage Faults (F3 and F19)
Momentary Duties

```
INDUSTRIAL SYSTEM SHORT
CIRCUIT STUDY EXAMPLE

CASE: INT(ERRUPTING)
FOR THREE PHASE MEDIUM
VOLTAGE FAULTS

INPUT DATA
BUS   TO   BUS      R P.U.        X P.U.      CODE
0          40      0.00450       0.10000       1
40         1       0.01500       0.03000       0
40         2       0.01500       0.03000       0
1          3       0.06700       0.53300       0
2          4       0.06700       0.53300       0
3          5       0.07000       0.00900       0
3          6       0.02600       0.01100       0
3          9       0.04400       0.01600       0
3          26      0.14600       0.01900       0
4          7       0.02200       0.01000       0
4          8       0.00800       0.01000       0
0          8       0.10350       3.30000       3
4          15      0.08400       0.02600       0
4          24      0.10000       0.06400       0
9          12      0.03600       0.00500       0
12         17      0.52900       4.50000       0
0          17      5.01000      50.10000       5
9          25      0.39500       0.05100       0
25         28      0.54800       3.83000       0
0          28      7.15800      71.70000       5
10         13      0.04300       0.00600       0
13         18      0.52900       4.50000       0
0          18      5.01000      50.10000       5
10         27      0.10200       0.01300       0
5          39      0.46400       3.48000       0
0          39      0.47700      14.31000       4
26         29      0.54800       3.83000       0
0          29      7.15800      71.70000       5
6          11      0.45800       3.67000       0
0          11      2.22600      33.30000       5
6          14      0.18900       0.05800       0
14         19      0.24400       1.47000       0
0          19      0.77100      20.10000       4
0          19      1.33500      20.04000       5
15         20      0.24400       1.47000       0
0          20      0.80700      21.00000       3
0          20      1.90800      28.62000       5
7          27      0.04200       0.01500       0
27         30      0.54800       3.83000       0
0          30      7.15800      71.70000       5
7          16      0.25600       0.03300       0
16         21      0.95800       7.67000       0
0          24      0.01800       0.90000       2
24         31      0.07300       0.01000       0
31         36      0.32900       2.30000       0
0          36      0.30300      10.02000       4
24         32      0.10400       0.01400       0
32         37      0.82100       5.75000       0
0          37      7.15800      71.70000       5
28         38      0.32100       0.29300       0
29         38      0.42400       0.38700       0
30         38      0.32100       0.29300       0
```

```
*BUS 3 (F3)    E/Z=7.718 KA( 184.47MVA)AT-82.43DEG.,X/R=  8.01, 13.800 KV
               Z= 0.071449 +J   0.537361

        CIRCUIT BREAKER TYPE        8TOT,SYM      5SYM       5TOT      3SYM
        MAX DUTY LEVEL                7.72         7.72       7.73      7.72
        MULT. FACTOR                 1.000        1.000      1.002     1.000

   CONTRIBUTIONS IN KA
   BUS  TO  BUS     MAG      ANG     BUS  TO  BUS      MAG      ANG
    1        3     6.288  -82.538     5        3      0.235   -86.750
    6        3     0.473  -85.235     9        3      0.400   -77.626
   26        3     0.326  -78.974


*BUS 19 (F19)  E/Z= 13.446 KA(  55.90MVA)AT-77.73DEG.,X/R=  6.84,  2.400 KV
               Z= 0.380239 +J   1.748179
        CIRCUIT BREAKER TYPE        8TOT,SYM      5SYM       5TOT      3SYM
        MAX DUTY LEVEL               13.45        13.45      13.45     13.45
        MULT. FACTOR                 1.000        1.000      1.000     1.000

   CONTRIBUTIONS IN KA
   BUS  TO  BUS     MAG      ANG     BUS  TO  BUS      MAG      ANG
   14       19     11.091 -75.738    INDMOT   19     1.196   -87.802
   INDMOT   19      1.198 -86.187
```

Fig 72
Computer Output for Medium-Voltage Faults (F3 and F19)
Interrupting Duties

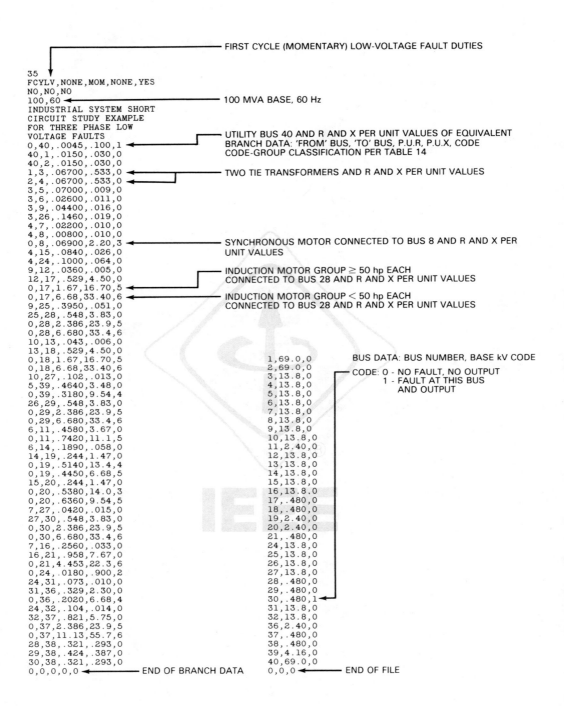

FIRST CYCLE (MOMENTARY) LOW-VOLTAGE FAULT DUTIES

```
35
FCYLV,NONE,MOM,NONE,YES
NO,NO,NO
100,60                                  ◄──── 100 MVA BASE, 60 Hz
INDUSTRIAL SYSTEM SHORT
CIRCUIT STUDY EXAMPLE
FOR THREE PHASE LOW
VOLTAGE FAULTS                          UTILITY BUS 40 AND R AND X PER UNIT VALUES OF EQUIVALENT
0,40,.0045,.100,1    ◄────────          BRANCH DATA: 'FROM' BUS, 'TO' BUS, P.U.R, P.U.X, CODE
40,1,.0150,.030,0                       CODE-GROUP CLASSIFICATION PER TABLE 14
40,2,.0150,.030,0
1,3,.06700,.533,0    ◄────────          TWO TIE TRANSFORMERS AND R AND X PER UNIT VALUES
2,4,.06700,.533,0    ◄────────
3,5,.07000,.009,0
3,6,.02600,.011,0
3,9,.04400,.016,0
3,26,.1460,.019,0
4,7,.02200,.010,0
4,8,.00800,.010,0
0,8,.06900,2.20,3    ◄────────          SYNCHRONOUS MOTOR CONNECTED TO BUS 8 AND R AND X PER
4,15,.0840,.026,0                       UNIT VALUES
4,24,.1000,.064,0
9,12,.0360,.005,0                       INDUCTION MOTOR GROUP ≥ 50 hp EACH
12,17,.529,4.50,0                        CONNECTED TO BUS 28 AND R AND X PER UNIT VALUES
0,17,1.67,16.70,5    ◄────────
0,17,6.68,33.40,6    ◄────────          INDUCTION MOTOR GROUP < 50 hp EACH
9,25,.3950,.051,0                       CONNECTED TO BUS 28 AND R AND X PER UNIT VALUES
25,28,.548,3.83,0
0,28,2.386,23.9,5
0,28,6.680,33.4,6
10,13,.043,.006,0
13,18,.529,4.50,0
0,18,1.67,16.70,5                    1,69.0,0        BUS DATA: BUS NUMBER, BASE kV CODE
0,18,6.68,33.40,6                    2,69.0,0        CODE: 0 - NO FAULT, NO OUTPUT
10,27,.102,.013,0                    3,13.8,0              1 - FAULT AT THIS BUS
5,39,.4640,3.48,0                    4,13.8,0                     AND OUTPUT
0,39,.3180,9.54,4                    5,13.8,0
26,29,.548,3.83,0                    6,13.8,0
0,29,2.386,23.9,5                    7,13.8,0
0,29,6.680,33.4,6                    8,13.8,0
6,11,.4580,3.67,0                    9,13.8,0
0,11,.7420,11.1,5                   10,13.8,0
6,14,.1890,.058,0                   11,2.40,0
14,19,.244,1.47,0                   12,13.8,0
0,19,.5140,13.4,4                   13,13.8,0
0,19,.4450,6.68,5                   14,13.8,0
15,20,.244,1.47,0                   15,13.8,0
0,20,.5380,14.0,3                   16,13.8,0
0,20,.6360,9.54,5                   17,.480,0
7,27,.0420,.015,0                   18,.480,0
27,30,.548,3.83,0                   19,2.40,0
0,30,2.386,23.9,5                   20,2.40,0
0,30,6.680,33.4,6                   21,.480,0
7,16,.2560,.033,0                   24,13.8,0
16,21,.958,7.67,0                   25,13.8,0
0,21,4.453,22.3,6                   26,13.8,0
0,24,.0180,.900,2                   27,13.8,0
24,31,.073,.010,0                   28,.480,0
31,36,.329,2.30,0                   29,.480,0
0,36,.2020,6.68,4                   30,.480,1  ◄────
24,32,.104,.014,0                   31,13.8,0
32,37,.821,5.75,0                   32,13.8,0
0,37,2.386,23.9,5                   36,2.40,0
0,37,11.13,55.7,6                   37,.480,0
28,38,.321,.293,0                   38,.480,0
29,38,.424,.387,0                   39,4.16,0
30,38,.321,.293,0                   40,69.0,0
0,0,0,0,0  ◄──── END OF BRANCH DATA   0,0,0  ◄──── END OF FILE
```

Fig 73
Computer Input File for Low-Voltage Faults

```
INDUSTRIAL SYSTEM SHORT
CIRCUIT STUDY EXAMPLE

CASE: MOM(ENTARY)
FOR THREE PHASE LOW
VOLTAGE FAULTS

INPUT DATA
BUS   TO   BUS        R P.U.        X P.U.       CODE
 0         40        0.00450       0.10000         1
40          1        0.01500       0.03000         0
40          2        0.01500       0.03000         0
 1          3        0.06700       0.53300         0
 2          4        0.06700       0.53300         0
 3          5        0.07000       0.00900         0
 3          6        0.02600       0.01100         0
 3          9        0.04400       0.01600         0
 3         26        0.14600       0.01900         0
 4          7        0.02200       0.01000         0
 4          8        0.00800       0.01000         0
 0          8        0.06900       2.20000         3
 4         15        0.08400       0.02600         0
 4         24        0.10000       0.06400         0
 9         12        0.03600       0.00500         0
12         17        0.52900       4.50000         0
 0         17        2.00400      20.04000         5
 0         17       11.15560      55.77800         6
 9         25        0.39500       0.05100         0
25         28        0.54800       3.83000         0
 0         28        2.86320      28.68000         5
 0         28       11.15560      55.77800         6
10         13        0.04300       0.00600         0
13         18        0.52900       4.50000         0
 0         18        2.00400      20.04000         5
 0         18       11.15560      55.77800         6
10         27        0.10200       0.01300         0
 5         39        0.46400       3.48000         0
 0         39        0.31800       9.54000         4
26         29        0.54800       3.83000         0
 0         29        2.86320      28.68000         5
 0         29       11.15560      55.77800         6
 6         11        0.45800       3.67000         0
 0         11        0.89040      13.32000         5
 6         14        0.18900       0.05800         0
14         19        0.24400       1.47000         0
 0         19        0.51400      13.40000         4
 0         19        0.53400       8.01600         5
15         20        0.24400       1.47000         0
 0         20        0.53800      14.00000         3
 0         20        0.76320      11.44800         5
 7         27        0.04200       0.01500         0
27         30        0.54800       3.83000         0
 0         30        2.86320      28.68000         5
 0         30       11.15560      55.77800         6
 7         16        0.25600       0.03300         0
16         21        0.95800       7.67000         0
 0         21        7.43651      37.24100         6
 0         24        0.01800       0.90000         2
24         31        0.07300       0.01000         0
31         36        0.32900       2.30000         0
 0         36        0.20200       6.68000         4
24         32        0.10400       0.01400         0
32         37        0.82100       5.75000         0
 0         37        2.86320      28.68000         5
 0         37       18.58710      93.01900         6
28         38        0.32100       0.29300         0
29         38        0.42400       0.38700         0
30         38        0.32100       0.29300         0
```

```
*BUS 30      E/Z= 80.574 KA( 66.99MVA)AT-75.15DEG.,X/R= 4.55,  0.480 kV
 (F30)       Z=  0.382489 +J   1.442973
             MAX. LOW VOLTAGE POWER CIRCUIT BREAKER   DUTY LEVEL=  80.57
             MAX. LV MCCB OR ICCB (RATED >20KA INT.)  DUTY LEVEL=  80.57
             MAX. LV MCCB OR ICCB (RATED 10-20KA INT.)DUTY LEVEL=  88.16

     CONTRIBUTIONS IN KA
     BUS  TO  BUS       MAG      ANG      BUS  TO  BUS      MAG      ANG
      27      30      28.303  -81.379     INDMOT   30      4.173  -84.297
      INDMOT  30       2.115  -78.689       38     30     46.368  -70.376
```

Fig 74
Computer Output for Low-Voltage Faults (F30) Momentary Duty

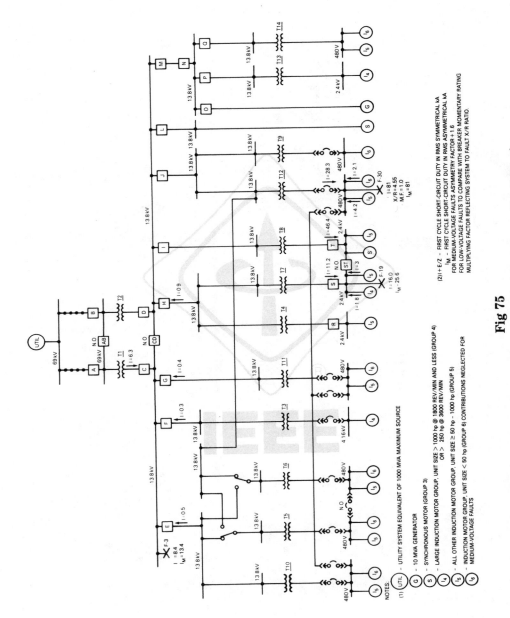

Fig 75

Short-Circuit Diagram Three-Phase Momentary Fault Duties

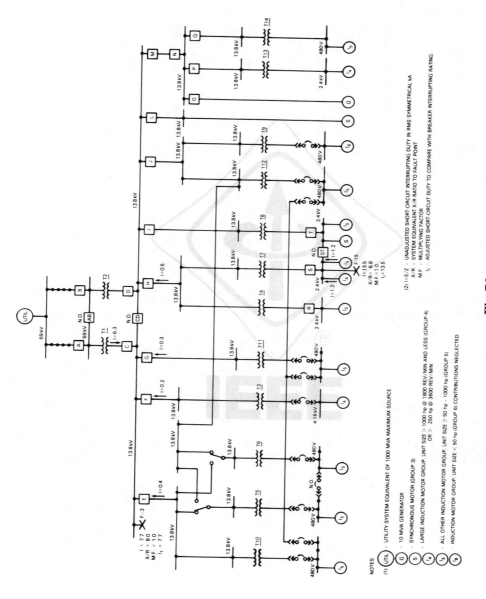

Fig 76

Short-Circuit Diagram Three-Phase Interrupting Fault Duties

Table 15
Sample Summary of Results for Example Short-Circuit Study

Short-Circuit Location	Short-Circuit Duty to Compare Directly with Equipment Rating		Switchgear Short-Circuit Rating	
	Momentary (First ½ Cycle) Asymmetrical kA	Interrupting Symmetrical kA	First ½ Cycle Close-and-Latch Asymmetrical kA	Interrupting Symmetrical kA
F3 13.8 kV Bus 3	(See note) 13.4	7.7	40.0	21.0
F19 2.4 kV Bus 19	(See note) 25.6	13.5	80.0	50.0
F30 480 kV Bus 30	81	—	—	200.0

NOTE: Asymmetrical factor equals 1.6.

7.5 References. The following references were used in the preparation of Chapter 7:

[1] IEEE C37.010-1979, IEEE Standard Application Guide for AC High-Voltage Circuit Breakers Rated on a Symmetrical Current Basis (includes Supplement C37.010d) (ANSI).

[2] IEEE C37.13-1981, IEEE Standard for Low-Voltage AC Power Circuit Breakers Used in Enclosures (ANSI).

[3] IEEE C37.5-1979, IEEE Guide for Calculation of Fault Currents for Application of AC High-Voltage Circuit Breakers Rated on a Total Current Basis (ANSI).

NOTE: IEEE C37.5-1979 has been withdrawn. Copies can be obtained from the IEEE Standards Department at 908-562-3821.

[4] IEEE Std 141-1986, IEEE Recommended Practice for Electric Power Distribution for Industrial Plants (ANSI).

[5] IEEE Std 242-1986, IEEE Recommended Practice for Protection and Coordination of Industrial and Commercial Power Systems (ANSI).

Chapter 8
Stability Studies

8.1 Introduction. For years, system stability has been almost exclusively a problem to electric utility engineers. Within the past decade, however, increasing numbers of industrial and commercial facilities have installed local generation, large synchronous motors, or both. This means that system stability is of concern to a growing number of industrial plant electrical engineers and consultants.

8.2 Stability Fundamentals

8.2.1 Definition of Stability. Fundamentally, stability is a property of a power system containing two or more synchronous machines. The system is stable, *under a specified set of conditions*, if all of its synchronous machines remain in step with one another (or having pulled out of step, regain synchronism soon afterward). The emphasis on specified conditions in this definition is intended to stress the fact that a system which is stable under one set of conditions can be unstable under some other set of conditions.

8.2.2 Steady-State Stability. Although the discussion in the rest of this section revolves around stability under transient conditions such as faults, switching operations, etc., there should also be an awareness that a power system can become unstable under steady-state conditions.

The simplest power system to which stability considerations apply consists of a pair of synchronous machines, one acting as a generator and the other acting as a motor, connected together through a reactance. (See Fig 77.) (In this model, the reactance is the sum of the transient reactances of the two machines and the reactance of the connecting circuit. Losses in the machines and the resistance of the line are neglected for simplicity.)

If the internal voltages of the two machines are E_G and E_M, and the phase angle between them is θ, it can easily be demonstrated (see References [1][10] and [3])

---

[10]The numbers in brackets correspond to those in the References at the end of this chapter.

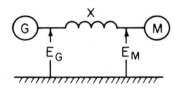

**Fig 77
Simplified Two-Machine Power System**

that the real power transmitted from the generator to the motor is

$$P = \frac{E_G E_M}{X} \sin \theta \qquad \text{(Eq 119)}$$

The maximum value of P obviously occurs when $\theta = 90°$. Thus

$$P_{\max} = \frac{E_G E_M}{X} \qquad \text{(Eq 120)}$$

This is the steady-state stability limit for the simplified two-machine system. Any attempt to transmit more power than $P_{\max}$ will cause the two machines to pull out of step with the given values of internal voltages.

This model of a simple power system shows that at least three electrical characteristics of the system affect stability. They are:

(1) Internal voltage of the generator
(2) Reactances of the machines and transmission system
(3) Internal voltage of the motor

The higher the internal voltages, and the lower the system and machine reactances, the greater power can be transmitted under steady-state conditions.

8.2.3 Transient and Dynamic Stability. The preceding look at steady-state stability serves as a background for an examination of the more complicated problem of transient stability. This is true because the same three electrical characteristics that determine steady-state stability limits affect transient stability in the same way. However, a system that is stable under steady-state conditions is not necessarily stable when subjected to a transient disturbance.

Transient stability means the ability of a power system to survive a sudden change in generation, load, or system characteristics without a prolonged loss of synchronism. To see how a disturbance affects a synchronous machine, first look at the steady-state characteristics described by the steady-state torque equation (see Reference [2]).

$$T = \frac{\pi P^2}{8} \phi_{SR} F_R \sin \delta_R \qquad \text{(Eq 121)}$$

where:

T = Mechanical shaft torque
P = Number of poles of machine
ϕ_{SR} = Air-gap flux
F_R = Rotor field MMF
δ_R = Mechanical angle between rotor and stator field lobes

The air-gap flux ϕ_{SR} stays constant as long as the voltage at the machine does not change, if the effects of saturation of the iron are neglected. Therefore, if the field excitation remains unchanged, a change in shaft torque T will cause a corresponding change in rotor angle δ_R. (This is the angle by which, for a motor, the peaks of the rotating stator field lead the corresponding peaks of the rotor field. For a generator, the relation is reversed.) Figure 78 graphically illustrates the variation of rotor angle with shaft torque. With the machine operating as a motor (when rotor angle and torque are positive), torque increases with rotor angle until δ_R reaches 90 electrical degrees. Beyond 90°, torque decreases with increasing rotor angle. As a result, if we increase the required torque output of a synchronous motor beyond the level corresponding to 90° rotor angle, it will *slip a pole*. Unless the load torque is reduced below the 90° level (the pullout torque), the motor will continue slipping poles indefinitely. The problems that can follow from extended operation in this out-of-step condition will be discussed further in this section.

A generator operates similarly. Increasing torque input until the rotor angle exceeds 90° results in pole slipping and loss of synchronism with the power system, assuming constant electrical load.

Similar relations apply to the other parameters of the torque equation. For example, air-gap flux ϕ_{SR} is a function of voltage at the machine. Thus, if the other factors remain constant, a change in system voltage will cause a change in rotor angle. Likewise, changing the field excitation will cause a change in rotor angle, if constant torque and voltage are maintained.

Fig 78
Torque versus Rotor Angle Relationship
for Synchronous Machines in Steady State

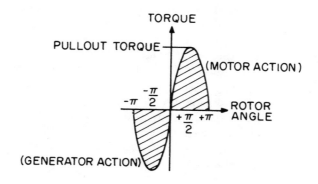

The preceding discussion refers to rather gradual changes in the conditions affecting the torque angle, so that approximate steady-state conditions always exist. The coupling between the stator and rotor fields of a synchronous machine, however, is somewhat elastic. This means that, if an abrupt rather than a gradual change occurs in one or more of the parameters of the torque equation, the rotor angle will tend to overshoot the final value determined by the changed conditions. This disturbance can be severe enough to carry the ultimate steady-state rotor angle past 90° or the transient *swing* rotor angle past 180°. Either event results in the slipping of a pole. If the conditions that caused the original disturbance are not corrected, the machine will then continue to slip poles; in short, pull out of step with the power system to which it is connected.

Of course, if the rotor angle overshoot does not transitorily exceed 180°, or if the disturbance causing the rotor swing is promptly removed, the machine may remain in synchronism with the system. The rotor angle then oscillates in decreasing swings until it settles to its final value (less than 90°). The oscillations are damped by mechanical load and losses in the system, especially in the damper windings of the machine.

A change in rotor angle of a machine generally requires a change in speed of the rotor. For example, if we assume that the stator field frequency is constant, it is necessary to at least momentarily slow down the rotor of a synchronous motor to permit the rotor field to fall farther behind the stator field and thus increases δ_R. The rate at which rotor speed can change is determined by the moment of inertia of the rotor, plus whatever is mechanically coupled to it, prime mover, load, reduction gears, etc. This means a machine with high inertia is less likely to pull out under a disturbance of brief duration than a low-inertia machine, all other characteristics being equal.

Traditionally, *transient* stability is determined by considering only the inherent mechanical and electromagnetic characteristics of the synchronous machines and the impedances of the circuits connecting them. The responses of the excitation or governor systems to the changes in generator speed or electrical output induced by a system disturbance are neglected. On the other hand, *dynamic* stability takes automatic excitation and governor system responses into account.

The traditional definition of transient stability is equivalent to assuming that excitation and governor-prime mover time constants are much longer than the duration of the instability-inducing disturbance. This assumption was usually accurate when power system stability was first defined back in the 1920's, because both the generator excitation voltage and the prime mover throttle were controlled either manually or by very slow feedback mechanisms, and brief short circuits were normally the worst-case disturbances considered. In any case, the computational capacity required to study automatic excitation and governor system responses did not exist.

However, technological advances have rendered the unstated assumption underlying the conventional definition of transient stability obsolete in most cases. These include the advent of fast electronic excitation systems and governors, the recognition of the value of stability analysis for investigating conditions of widely varying severities and durations, and the virtual elimination of computational power as a constraint on system modeling complexity. Most transient stability studies performed today consider at least the generator excitation system, and are therefore actually dynamic studies under the conventional definition. Nevertheless, in deference to traditional usage, the term *transient stability* will continue to be applied to them in this guide.

8.2.4 Two-Machine Systems. The previous discussion of transient behavior of synchronous machines is based on a single machine connected to a good approximation of an infinite bus. An example is the typical industrial situation where a synchronous motor of at most a few thousand horsepower is connected to a utility company system with a capacity of thousands of megawatts. Under these conditions, we can safely neglect the effect of the machine on the power system.

A system consisting of only two machines of comparable size connected through a transmission link, however, becomes more complicated, because the two machines can affect one another's performance. The medium through which this occurs is the air-gap flux. This is a function of machine terminal voltage, which is affected by the characteristics of the transmission system, the amount of power being transmitted, and the power factor, etc.

In the steady state, the rotor angles of the two machines are determined by the simultaneous solution of their respective torque equations. Under a transient disturbance, as in the single-machine system, the rotor angles move toward values corresponding to the changed system conditions. Even if these new values are within the steady-state stability limits of the system, an overshoot can result in loss of synchronism. If not, both rotors will undergo a damped oscillation and ultimately settle to their new steady-state values.

An important concept here is synchronizing power. The more real power transmitted over the transmission link between the two machines, the more likely they are to remain in synchronism in the face of a transient disturbance. Synchronous machines separated by a sufficiently low impedance behave as one composite machine, since they tend to remain in step with one another regardless of external disturbances.

8.2.5 Multimachine Systems. At first glance, it appears that a power system incorporating many synchronous machines would be extremely complex to analyze. This is true if a detailed, precise analysis is needed; a large digital computer and a sophisticated program are required for a complete transient stability study of a multimachine system. However, many of the multimachine systems encountered in industrial practice contain only synchronous motors that are similar in characteristics, closely coupled electrically, and connected to a high-capacity utility system. Under most types of disturbance, motors will remain synchronous with each other, although they can all lose synchronism with the utility. Thus, the problem is similar to a single synchronous motor connected through an impedance to an infinite bus. The simplification should be apparent. More complex systems, where machines are of comparable sizes and are separated by substantial impedances, will usually involve a full-scale computer stability study.

8.3 Problems Caused by Instability. The most immediate hazards of asynchronous operation of a power system are the high transient mechanical torques and currents that usually occur. To prevent these transients from causing mechanical and thermal damage, synchronous motors and generators are almost universally equipped with pullout protection. For motors of small to moderate sizes, this protection is usually provided by a damper protection of pullout relay that operates on the low power factor occurring during asynchronous operation. The same function is usually provided for large motors, generators, and synchronous condensers by loss-of-field relaying. In any case, the pullout relay trips the machine breaker or contactor. Whatever load is being served by the machine is naturally interrupted. Consequently, the primary disadvantage of a system that tends to be unstable is the probability of frequent process interruptions.

Out-of-step operation also causes large oscillatory flows of real and reactive power over the circuits connecting the out-of-step machines. Impedance or distance-type relaying that protects these lines can falsely interpret power surges as a line fault, tripping the line breakers and breaking up the system. Although this is primarily a utility problem, large industrial systems or those where local generation operates in parallel with the utility can be susceptible.

8.4 System Disturbances That Can Cause Instability. The most common disturbances that produce instability in industrial power systems are (not necessarily in order of probability):

(1) Short circuits
(2) Loss of a tie circuit to a public utility
(3) Loss of a portion of on-site generation
(4) Starting a motor that is large relative to a system generating capacity
(5) Switching operations
(6) Impact loading on motors
(7) Abrupt decrease in electrical load on generators

The effect of each of these disturbances should be apparent from the previous discussion of stability fundamentals. Items (1) through (5) tend to reduce voltage levels, ultimately requiring an increase in machine angles to maintain a given load. Items (6) and (7) directly increase the rotor angles of affected machines.

8.5 Solutions to Stability Problems. Generally, anything that decreases the severity or duration of a transient disturbance will make the power system less likely to become unstable under that disturbance. In addition, increasing the moment of inertia per rated kVA of the synchronous machines in the system will raise stability limits by resisting changes in rotor speeds required to change rotor angles.

8.5.1 System Design. System design primarily affects the amount of synchronizing power that can be transferred between machines. Two machines connected by a low-impedance circuit, such as a short cable or bus run, will probably stay synchronized with each other under all conditions except a fault on the connecting circuit, a loss of field excitation, or an overload. The greater the impedance between machines, the less severe a disturbance will be required to drive them out of step. This means that, from the standpoint of maximum stability, all synchronous machines should be closely connected to a common bus. Limitations on short-circuit duties, economics, and the requirements of physical plant layout usually combine to render this radical solution impractical.

8.5.2 Design and Selection of Rotating Equipment. Design and selection of rotating equipment can be a major contributor to improving system stability. Most obviously, use of induction instead of synchronous motors eliminates the potential stability problems associated with the latter. (Under rare circumstances, an induction motor/synchronous generator system can experience instability, in the sense that undamped rotor oscillations occur in both machines, but the possibility is too remote to be of serious concern.) However, economic considerations often preclude this solution.

Where synchronous machines are used, stability can be enhanced by increasing the inertia of the mechanical system. Since the H constant (stored energy per rated kVA) is proportional to the square of the speed, fairly small increases in synchronous speed can pay significant dividends in higher inertia. If carried too far, this can become self-defeating because higher speed machines have smaller diameter rotors. Wk^2 varies with the square of the rotor radius, so the increase in H due to a higher speed may be offset by a decrease due to the lower Wk^2 of a smaller diameter rotor.

A further possibility is to use synchronous machines with low transient reactances that permit the maximum flow of synchronizing power. Applicability of this solution is limited mostly by short-circuit considerations and machine design problems.

8.5.3 System Protection. System protection often offers the best prospects for improving the stability of a power system. The most severe disturbance that an industrial power system is likely to experience is a short circuit. To prevent loss of synchronism, as well as to limit personnel hazards and equipment damage, short circuits should be isolated as rapidly as possible. A system that tends to be unstable should be equipped with instantaneous overcurrent protection on all of its primary feeders, which are the most exposed section of the primary system. As a general rule, instantaneous relaying should be used throughout the system wherever selectivity permits.

8.5.4 Voltage Regulator and Exciter Characteristics. Voltage regulator and exciter characteristics affect stability because, all other things being equal, higher field excitation requires a smaller rotor angle. Consequently, stability is enhanced by a properly applied regulator and exciter that respond rapidly to transient effects and furnish a high degree of field forcing. In this respect, modern solid-state voltage regulators and static exciters can contribute markedly to improved stability. (On the other hand, a mismatch in exciter and regulator characteristics can make an existing stability problem even worse.)

8.6 Transient Stability Studies. Knowing how to correct an unstable power system is not very valuable if, in order to test our proposed recommendations, we have to stage stability tests on the actual system. This is especially true if the system in question is still in the design stage. Consequently, we need a fast, simple, and inexpensive way to simulate transient performance of a power system under a variety of normal and abnormal conditions.

8.6.1 History. The first transient stability studies responding to this need were done by semieducated guesses. When this technique proved insufficiently accurate, and as rotating machine theory was developed, simple power systems consisting of only a few machines were analyzed by manual calculations or mechanical models. These methods were neither precise nor applicable to complex systems. In the 30's and 40's, further development of modeling techniques led to the use of the ac network analyzer in stability studies. This analog device permitted the simulation of much larger systems than previously possible, but still suffered from the disadvantages of imprecise representation of synchronous machines, proneness to human error, limited capacity, and high cost per study run. Finally, in the 50's, the digital computer came on the scene. Its enormous arithmetic capability, precision, and ability to store and retrieve huge amounts of information made it a natural for stability studies. Transient stability programs were written and used by major electrical suppliers, utility companies, consulting firms, and universities, and have now almost totally replaced the older methods.

8.6.2 How Stability Programs Work. Mathematical methods of stability analysis depend on a repeated solution of the swing equation for each machine,

$$P_a = \frac{\text{MVA}\,(H)}{180f} \frac{d^2\delta_R}{dt^2} \qquad \text{(Eq 122)}$$

where:

$$\begin{aligned}
P_a &= \text{Accelerating power (input power minus output power), MW} \\
\text{MVA} &= \text{Rated MVA of machine} \\
H &= \text{Inertia constant of machine, MW} \cdot \text{seconds/MVA} \\
f &= \text{System frequency, Hz} \\
\delta_R &= \text{Rotor angle, degrees} \\
t &= \text{Time, seconds}
\end{aligned}$$

The program begins with the results of a load flow study to establish initial power and voltage levels in all machines and interconnecting circuits. The specified disturbance is applied at a time defined as zero, and the resulting changes in power levels are calculated by a load flow routine. Using the calculated accelerating power values, the swing equation is solved for a new value of δ_R for each machine at an incremental time (for example, 0.01 second) after the disturbance. Voltage and power levels corresponding to this new angular position of the synchronous machines are then used as base information for another iteration. In this way, performance of the system is calculated for every integration interval out to as much as several minutes.

8.6.3 Simulation of the System. A modern transient stability study computer program can simulate virtually any set of power system components in sufficient detail to give accurate results. Simulation of rotating machines and related equipment is of special importance in stability studies. The simplest possible representation for a synchronous motor or generator involves only a constant internal voltage, a constant transient reactance, and the rotating inertia (H) constant. This approximation neglects saturation of core iron, voltage regulator action, the influence of construction of the machine on transient reactances for the direct and quadrature axes, and most of the characteristics of the prime mover or load. Nevertheless, this *classical* representation is often accurate enough to give reliable results, especially when the time period being studied is rather short. (Limiting the study to a short period — say, ½ second or less, means that neither the voltage regulator nor the governor, if any, has time to exert a significant effect.) The classical representation is generally used for the smaller and less influential machines in a system, or where the more detailed information required for better simulations is not available.

As additional data on the machines becomes available, better approximations can be used. This permits more accurate results that remain reliable for longer time periods. Modern large-scale stability programs can simulate all of the following characteristics of a rotating machine:

(1) Voltage regulator and exciter
(2) Steam system or other prime mover, including governor
(3) Mechanical load
(4) Damper windings
(5) Salient poles
(6) Saturation

Induction motors can also be simulated in detail, together with speed-torque characteristics of their connected loads.

In addition to rotating equipment, the stability program can include in its simulation practically any other major system component, including transmission lines, transformers, capacitor banks, and voltage-regulating transformers and dc transmission links in some cases.

8.6.4 Simulation of Disturbances. The versatility of the modern stability study is apparent in the range of system disturbances that can be represented. The most severe disturbance that can occur on a power system is usually a three-phase bolted short circuit. Consequently, this type of fault is most often used to test system stability. Stability programs can simulate a three-phase fault at any location, with provisions for clearing the fault by opening breakers either after a specified time delay, or by the action of overcurrent, underfrequency, overpower, or impedance relays. This feature permits the adequacy of proposed protective relaying to be evaluated from the stability standpoint.

Short circuits other than the bolted three-phase fault cause less disturbance to the power system. Although most stability programs cannot directly simulate line-to-line or ground faults, the effects of these faults on synchronizing power flow can be duplicated by applying a three-phase fault with a properly chosen fault impedance. This means the effects of any type of fault on stability can be studied.

In addition to faults, stability programs can simulate switching of lines and generators. This is particularly valuable in the load-shedding type of study, which will be covered in the following section.

Finally, the starting of large motors on relatively weak power systems and impact loading of running machines can be analyzed.

8.6.5 Data Requirements for Stability Studies. The data required to perform a transient stability study, and the recommended format for organizing and presenting the information for most convenient use are covered in detail in the application guides for particular stability programs. The following is a summary of the generic classes of data needed. Note that some of the more esoteric information is not essential; omitting it merely limits the accuracy of the results, especially at times exceeding five times the duration of the disturbance being studied. The more essential items are marked by an asterisk (*).

(1) System data
  (a) Impedances ($R + jX$) of all significant transmission lines, cables, reactors, and other series components*
  (b) For all signficant transformers and autotransformers
    (i)   kVA rating*
    (ii) Impedance*
    (iii) Voltage ratio*
    (iv) Winding connection*
    (v)   Available taps and tap in use*
    (vi) For regulators and load tap-changing transformers: regulation range, tap step size, type of tap changer control*
  (c) Short-circuit capacity (steady-state basis) of utility supply, if any*
  (d) kvar of all significant capacitor banks*
  (e) Description of normal and alternate switching arrangements*
(2) Load data: real and reactive electrical loads on all significant load buses in the system*
(3) Rotating machine data
  (a) For major synchronous machines (or groups of identical machines on a common bus)
    (i)   Mechanical and/or electrical power ratings (kVA, hp, kW, etc.)*
    (ii)   Inertia constant H or inertia Wk^2 of rotating machine and connected load or prime mover*
    (iii) Speed*
    (iv)   Real and reactive loading, if base-loaded generator*
    (v)   Speed-torque curve or other description of load torque, if motor*
    (vi)   Direct-axis subtransient,* transient,* and synchronous reactances*
    (vii) Quadrature-axis subtransient, transient,* and synchronous reactances
    (viii) Direct-axis and quadrature-axis subtransient and transient* time constants
    (ix)   Saturation information
    (x)   Potier reactance
    (xi)   Damping data
    (xii) Excitation system-type time constants and limits
    (xiii) Governor and steam system or other prime mover type, time constants, and limits
  (b) For minor synchronous machines (or groups of machines)
    (i)   Mechanical and/or electrical power ratings*
    (ii) Inertia*
    (iii) Speed*
    (iv) Direct-axis synchronous reactance*
  (c) For major induction machines or groups of machines

   (i)   Mechanical and/or electrical power ratings*
   (ii) Inertia*
   (iii) Speed*
   (iv) Positive sequence equivalent circuit data (e.g., R_1, X_1, X_M)*
   (v)   Load speed-torque curve*
   (vi) Negative sequence equivalent circuit data (e.g., R_2, X_2)*
   (vii) Description of reduced-voltage or other starting arrangements, if used*
   (d) For major induction machines: detailed dynamic representation not needed, represent as a static load
(4) Disturbance data
   (a) General description of disturbance to be studied, including (as applicable) initial switching status; fault type, location, and duration; switching operations and timing; manufacturer, type, and setting of protective relays; and clearing time of associated breakers*
   (b) Limits on acceptable voltage, current, or power swings*
(5) Study parameters
   (a) Duration of study*
   (b) Integrating interval*
   (c) Output printing interval*
   (d) Data output required*

8.6.6 Stability Program Output. Most stability programs give the user a wide choice of results to be printed out. The program can calculate and print any of the following information as a function of time:

(1) Rotor angles, torques, and speeds of synchronous machines
(2) Real and reactive power flows throughout the system
(3) Voltages and voltage angles at all buses
(4) System frequency
(5) Torques and slips of all induction machines

The combination of these results selected by the user can be printed out for each printing interval (also user selected) during the course of the study period.

The value of the study is strongly affected by the selection of the proper printing interval and the total duration of the simulation. Normally, a printing interval of 0.01 or 0.02 second is used; longer intervals reduce the computer costs slightly, but increase the risk of missing fast swings of rotor angle. The computer time cost is nearly proportional to the total study time, so this parameter should be closely controlled for the sake of economy.

This is especially important if the system and machines have been represented approximately or incompletely, because the errors will accumulate and render the results meaningless after some point. A time limit of five times the duration of the major disturbance being studied is generally long enough to show whether the system is stable or not, while keeping costs to a reasonable level.

8.6.7 Interpreting Results—Swing Curves. The results of a computer transient stability study are fairly easy to understand once the user learns the basic principles underlying stability problems. The most direct way to determine from the study results whether a system is stable is to look at a set o swing curves for the machines in the system. Swing curves are simply plots of rotor angles against time; if the curves of all the machines involved are plotted on common axes, we can easily see whether they diverge (indicating instability) or settle to new steady-state values.

For example, Fig 79 shows swing curves for a system that is stable under the disturbance applied. This is a reproduction of an actual computer printout. A simplified single-line diagram of the system appears as Fig 80. Note that while the three-phase bolted fault on a synchronous bus feeder, Case I, cleared by instantaneous tripping of the feeder breaker, causes all five generators to experience swings of varying magnitude, the oscillations in the rotor angles are obviously damped and can be expected to die out.

By contrast, in Case II, the fault is applied to the tie between the synchronizing bus and one of the generator buses and is cleared by tripping the tie circuit breaker. The swing curves for this condition are shown in Fig 81. Generator No. 1 is disconnected from the system and suffers a severe overload, causing it to decelerate, as shown by a unidirectional negative change in rotor angle. The other machines stay in synchronism.

8.7 Stability Studies on a Typical System. Probably the best way to examine some of the typical applications of stability analysis to industrial power systems is to look at the stability studies that would go into the design of a typical large

Fig 79
Computer Printout of Swing Curves for Case I Fault on System in Fig 80

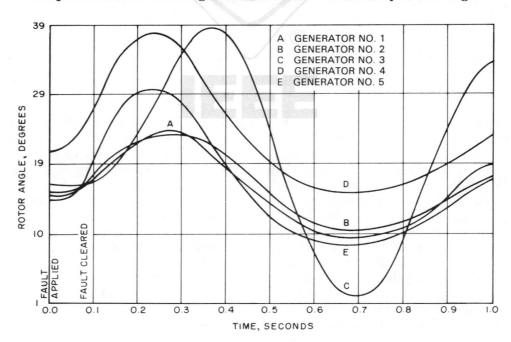

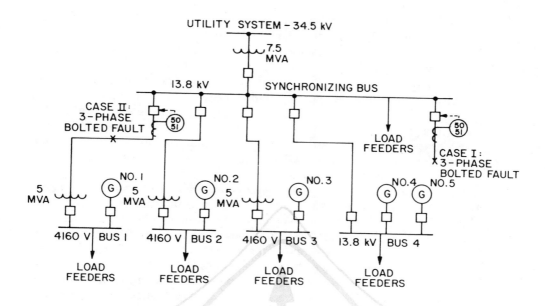

**Fig 80
Single-Line Diagram of System Whose Swing Curves Appear in Figs 79 and 81**

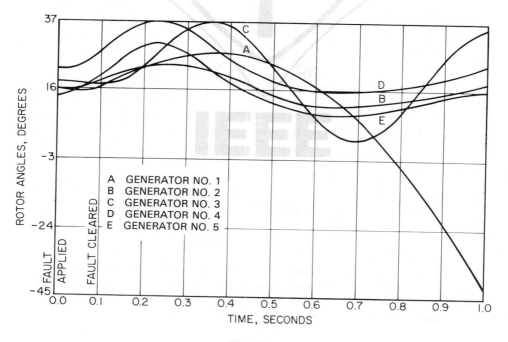

**Fig 81
Computer Printout of Swing Curves for Case II Fault in System Shown in Fig 80**

industrial system including 20 MVA of local generation and 40 MVA of purchased power capacity. The stability studies that might be applied to this system are:

(1) The basic layout of the primary system can be affected by stability considerations. For example, an initial design choice might be to connect the generated and purchased power buses through only one tie circuit. However, stability studies could show that inadequate synchronizing power is available to prevent the generators and the utility from losing synchronism during primary system faults unless two ties are provided. The same sort of considerations might dictate that the 4160 V bus ties be operated closed, to ensure the lowest possible impedance between the synchronous motors and the power sources to enhance stability.

(2) Related to the design of the basic layout is the problem of protective relaying. The system can be designed for maximum inherent stability by closely coupling all machines. Or the same objective can oten be obtained by designing the protective scheme for the fastest possible clearing of faults. Since the former choice may involve economic sacrifices in the form of higher capacity switchgear, often the latter choice represents the best solution. Extra-fast relaying can conflict with the requirements of a selectively coordinated system, however, unless expensive zone protection schemes (bus differential, pilot wire, etc.) are used. Balancing all of these factors, probably the best procedure is to design the system layout around process requirements, provide the fastest relaying possible within the constraints of selectivity and economics, and then check the proposed layout and relaying by a series of stability studies simulating the more probable fault conditions. In the system shown in Fig 82, three-phase faults are applied on one 138 kV utility line ahead of the plant transformers, on a feeder from each of the 13.8 kV buses, and on a feeder from each 4160 V bus. Of course, the simulation would include clearing of the fault via the proposed relaying. If any of these studies show an unstable condition, further stability studies might be required to test the effectiveness of various proposed solutions.

(3) In the system shown in Fig 82, some considerations should certainly be given to automatic load shedding. If the power company suffers an outage on the 138 kV lines while the plant is running at nearly full load, the 20 MVA of local generation will abruptly be subjected to an overload approaching 300%. This overload will promptly cause the generators to trip off, leaving the plant with no power at all, even though 20 MVA of perfectly sound generation is available to maintain service to the most critical loads. Obviously, a method of automatically interrupting noncritical loads commensurately with the loss of system capacity would be valuable.

One such possibility would be to trip noncritical feeders whenever the utility tie beakers are opened. However, this wired-in scheme lacks flexibility. To permit shedding only the amount of load required to prevent system collapse, many industrial plants with local generation use underfrequency relaying. This scheme depends on the fact that an overloaded generator slows down, dropping the system frequency.

A two-stage load-shedding scheme might operate as follows: The first-

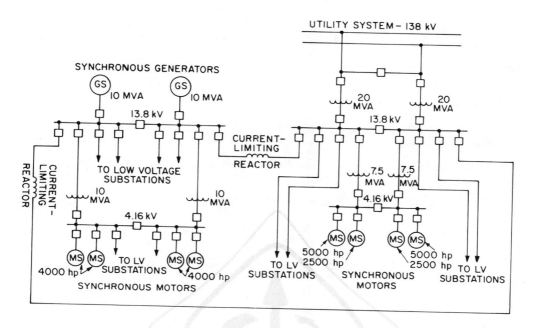

Fig 82
Single-Line Diagram of a Typical Large Industrial Power System
with On-Site Generation

stage relay operates at 59 Hz and a time delay of 6 cycles, tripping 10 or 15 MVA of noncritical load; the second stage operates at 58 Hz and a delay of 30 cycles, tripping an additional 20 MVA of somewhat more critical load.

In designing the load-shedding scheme for the system in Fig 82, first run a stability study to calculate the decay of system frequency when the utility tie is lost, without any load shedding. Using this frequency decrement curve, estimates can be made of the amounts of load to be shed and the frequency and time delay settings for the underfrequency relays. Then these data can be used in the stability study program to calculate the system frequency versus time curve with the proposed load shedding. If sufficient load is shed fast enough to prevent system collapse, the validity of the proposed relay scheme and settings is confirmed. Usually several runs are made with different system conditions in each load-shedding analysis.

(4) In the system shown in Fig 82, the effect of starting one of the large synchronous motors could be substantial, especially under abnormal conditions when one or more power sources are out of service. This effect can be evaluated by a series of stability studies simulating motor starting under various conditions of system capacity, prestart motor terminal voltage, etc. The study results yield motor accelerating times, real and reactive power flows, and bus voltages at all critical points in the system.

As this discussion indicates, transient stability analysis should be an integral part of the design or expansion of any industrial power system containing several large synchronous machines, and should be considered even if only one machine is being applied.

8.8 References

[1] *Electrical Transmission and Distribution Reference Book*, Westinghouse Electric Corporation, East Pittsburgh, PA, 1964, chapter 13.

[2] FITZGERALD, A. E. and KINGSLEY, CHARLES Jr. Electric Machinery, New York, McGraw-Hill, 1961, chapter 5.

[3] KIMBARK, E. W. Power System Stability, vol. 1, New York, John Wiley, 1948.

Chapter 9
Motor Starting Studies

9.1 Introduction. This section discusses benefits obtained from motor starting studies and examines various types of computer-aided studies normally performed. Data or information required to permit these studies along with expected results of a motor starting study effort are also reviewed.

9.2 Need for Motor Starting Studies

9.2.1 Problems Revealed. Motors on modern industrial systems are becoming increasingly larger. Some are considered large even in comparison to the total capacity of large industrial power systems. Starting large motors, especially across-the-line, can cause severe disturbances to the motor and any locally connected load, and also to buses electrically remote from the point of motor starting.

Ideally, a motor starting study should be made before a large motor is purchased. A starting voltage requirement and preferred locked-rotor current should be stated as part of the motor specification. A motor starting study should be made if the motor horsepower exceeds approximately 30% of the supply transformer(s) base kVA rating, if no generators are present. It may be necessary to make a study for smaller horsepower sizes depending on the daily fluctuation of nominal voltage, voltage level, size and length of the motor feeder cable, amount of load, regulation of the supply voltage, the impedance and tap ratio of the supply transformer(s), load torque versus motor torque, and the allowable starting time. If generation is present, and no other sources are involved, a study should be considered whenever the motor horsepower exceeds 10–15% of the generator kVA rating, depending on actual generator characteristics. The study should also recognize contingent condition(s), i.e., the loss of a source (if applicable).

A brief discussion of major problems associated with starting large motors, and therefore, of significance in power system design and evaluation follows.

9.2.2 Voltage Dips. Probably the most widely recognized and studied effect of motor starting is the voltage dip experienced throughout an industrial power system as a direct result of starting large motors. Available accelerating torque drops appreciably at the motor bus as voltage dips to a lower value, extending the starting interval and affecting, sometimes adversely, overall motor starting performance. Acceptable voltage for motor starting depends on motor and load torque characteristics. Requirements for minimum starting voltage can vary over a wide range, depending on the application. (Voltages can range from 80% or lower to 95% or higher.)

During motor starting, the voltage level at the motor terminals should, as a minimum, be maintained at approximately 80% of rated voltage or above for a standard National Electrical Manufacturers Association (NEMA) type B motor having a standard 150% starting torque and with a constant torque load applied. This value results from examination of speed-torque characteristics of this type motor (150% starting torque at full voltage) and the desire to successfully accelerate a fully loaded motor at reduced voltage (that is, torque varies with the square of the voltage $T = 0.8^2 \times 150\% \simeq 100\%$). When other motors are affected, or when lower shaft loadings are involved, the minimum permissible voltage may be either higher or lower, respectively. The speed-torque characteristics of the starting motor along with any other affected motors and all related loads should be examined to specifically determine minimum acceptable voltage. Assuming reduced voltage permits adequate accelerating torque, it should also be verified that the longer starting interval required at reduced torque caused by a voltage dip does not result in the I^2t damage limit of the motor being exceeded.

Several other problems may arise on the electrical power system due to the voltage dips caused by motor starting. Motors that are running normally on the system, for example, will slow down in response to the voltage dip occurring when a large motor is started. The running machines must be able to reaccelerate once the machine being started reaches operating speed. When the voltage depression caused by the starting motor is severe, the loading on the running machines may exceed their breakdown torque (at the reduced voltage), and they may decelerate significantly or even stall before the starting interval is concluded. The decelerating machines all impose heavy current demands that only compound the original distress caused by the machine that was started. The result is a "dominoing" voltage depression that can lead to the loss of all load.

In general, if the motors on the system are standard NEMA design B, the speed-torque characteristics (200% breakdown torque at full voltage) should prevent a stall, provided the motor terminal voltage does not drop below about 71% of motor nameplate voltage. This is a valid guideline to follow anytime the shaft load does not exceed 100% rated since the developed starting torque is again proportional to the terminal voltage squared (V^2), and the available torque at 71% voltage would thus be slightly above 100%. If other than NEMA design B motors are used on the system, a similar criterion can be established to evaluate reacceleration following a motor starting voltage dip based on the exact speed-torque characteristics of each particular motor.

Other types of loads, such as electronic devices and sensitive control equipment, may be adversely effected during motor starting. There is a wide range of variation in the amount of voltage drop that can be tolerated by static drives and computers. Voltage fluctuations may also cause objectionable fluctuations in lighting. Tolerable voltage limits should be obtained from the specific equipment manufacturers.

By industry standards (see Reference [5][11]), ac control devices are not required

---

[11]The numbers in brackets correspond to those in the References at the end of this chapter.

to pick-up at voltages below 85% of rated nameplate voltage, whereas dc control devices must operate dependably (i.e., pick-up) at voltages above 80% of their rating. Critical control operations may, therefore, encounter difficulty during motor starting periods where voltage dips are excessive. A motor starting study might be required to determine if this is a problem with thoughts to using devices rated at 110 V rather than the normal 115 V nominal devices. Contactors are required to hold-in with line voltage as low as 80% of their rating (see Reference [5]). The actual dropout voltages of contactors used in industrial applications commonly range between 60–70% of rated voltage, depending on the manufacturer. Voltages in this range, therefore, may be appropriate and are sometimes used as the criteria for the lower limit that contactors can tolerate. Depending on where lighting buses are located, with respect to large starting motors, this may be a factor requiring a motor starting study. Table 16 summarizes some critical system voltage levels of interest when performing a motor starting study for the purpose of evaluating the effects of voltage dips.

9.2.3 Weak Source Generation. Smaller power systems are usually served by limited capacity sources, which generally magnify voltage drop problems on motor starting, especially when large motors are involved. Small systems can also have on-site generation, which causes an additional voltage drop due to the relatively higher impedance of the local generators during the (transient) motor starting interval. The type of voltage regulator system applied with the generators can dramatically influence motor starting as illustrated in Fig 83. A motor starting

Table 16
Summary of Representative Critical System
Voltage Levels When Starting Motors

Voltage Drop Location or Problem	Minimum Allowable Voltage (% Rated)
At Terminals of Starting Motor	80%[1]
All Terminals of Other Motors That Must Reaccelerate	71%[1]
AC Contactor Pick-up (By Standard) (see Reference [5])	85%
DC Contactor Pick-up (By Standard) (See Reference [5])	80%
Contactor Hold-In (Average of Those in Use)	60–70%[2]
Solid-State Control Devices	90%[3]
Noticeable Light Flicker	3% Change

[1] Typical for NEMA design B motors only. Value may be higher (or lower) depending on actual motor and load characteristics.
[2] Value may be as high as 80% for certain conditions during prolonged starting intervals.
[3] May typically vary by ± 5% depending on available tap settings of power supply transformer when provided.

More detailed information is provided in Table 51 of Reference [4].

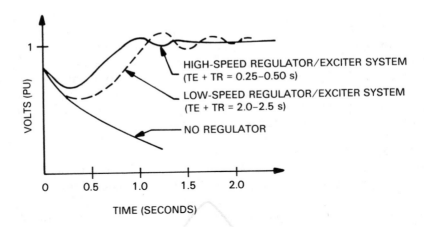

Fig 83
Typical Generator Terminal Voltage Characteristics
for Various Exciter/Regulator Systems

study can be useful, even for analyzing the performance of small systems. Certain digital computer programs can accurately model generator transient behavior and exciter/regulator response under motor starting conditions, providing meaningful results and conclusions.

9.2.4 Special Torque Requirements. Sometimes special loads must be accelerated under carefully controlled conditions without exceeding specified torque limitations of the equipment. An example of this is starting a motor connected to a load through gearing. This application requires a special period of low torque *cushioned* acceleration to allow slack in the gears and couplings to be picked up without damage to the equipment. Certain computer-aided motor starting studies allow an instant-by-instant shaft output torque tabulation for comparison to allowable torque limits of the equipment. This study can be used for selecting a motor or a starting method, or both, with optimum speed-torque characteristics for the application. The results of a detailed study are used for sizing the starting resistors for a wound rotor motor or in analyzing rheostatic control for a starting wound rotor motor that might be used in a *cushioned* starting application involving mechanical gearing or a coupling system that has torque transmitting limitations. High inertia loads increase motor starting time, and heating in the motor due to high currents drawn during starting can be intolerable. A computer-aided motor starting study allows accurate values of motor current and time during acceleration to be calculated. This makes it possible to determine if thermal limits of standard motors will be exceeded for longer than normal starting intervals.

Other loads have special starting torque requirements or accelerating time limits that require special *high* starting torque (and inrush) motors. Additionally, the starting torque of the load or process may not permit low inrush motors in situations where these motors might reduce the voltage dip caused by starting a motor having standard inrush characteristics. A simple inspection of the motor and load speed-torque curves is not sufficient to determine whether such problems exist. This is another area where the motor torque and accelerating time study can be useful.

9.3 Recommendations

9.3.1 Voltage Dips.
A motor starting study can expose and identify the extent of a voltage drop problem. The voltage at each bus in the system can, for example, be readily determined by a digital computer study. Equipment locations likely to experience difficulty during motor starting can be immediately determined.

In situations where a variety of equipment voltage ratings are available, the correct rating for the application can be selected. Circuit changes, such as off-nominal tap settings for distribution transformers and larger than standard conductor size cable, can also be readily evaluated. On a complex power system, this type of detailed analysis is very difficult to accomplish with time-consuming hand solution methods.

Several methods of minimizing voltage dip on starting motors are based on the fact that during starting time, a motor draws an inrush current directly proportional to terminal voltage, and therefore a lower voltage causes the motor to require less current, thereby reducing the voltage dip. Auto-transformer starters are a very effective means of obtaining a reduced voltage during starting with standard taps ranging from 50% to 80% of normal rated voltage. A motor starting study is used to select the proper voltage tap and the lower line current inrush for the electrical power system during motor start. Other special reduced voltage starting methods include resistor or reactor starting, part-winding starting, and wye (Y) - start delta (Δ) - run motors. All are examined by an appropriate motor starting study and the best method for the particular application involved can be selected. In all reduced voltage starting methods, torque available for accelerating the load is a very critical consideration once bus voltage levels are judged otherwise acceptable. Only 25% torque is available, for example, with 50% of rated voltage applied at the motor terminals. Any problems associated with reduced starting torque imposed by special starting methods are automatically uncovered by a motor starting study.

Another method of reducing high inrush currents when starting large motors is a capacitor starting system (see Reference [10]). This maintains acceptable voltage levels throughout the system. With this method, the high inductive component of normal reactive starting current is offset by the addition, during the starting period only, of capacitors to the motor bus. This differs from the practice of applying capacitors for running motor power factor correction. A motor starting study can provide information to allow optimum sizing of the starting capacitors and determination of the length of time the capacitor must be energized. The study can also establish whether the capacitor and motor can be switched together, or because of an excessive voltage drop that might result from the impact of capacitor transient charging current when added to the motor inrush current, the capacitor must be energized momentarily ahead of the motor. The switching procedure can appreciably affect the cost of final installation.

Use of special starters or capacitors to minimize voltage dips can be an expensive method of maintaining voltage at acceptable levels (see Reference [10]). Where possible, off-nominal tap settings for distribution transformers are an effective, economical solution for voltage dips. By raising no-load voltage in areas of the system experiencing difficulties during motor starting, the effect of the voltage dip can often be minimized. In combination with a load flow study, a motor starting study can provide information to assist in selecting proper taps and ensure that light load voltages are not excessively high.

The motor starting study can be used to prove the effectiveness of several other solutions to the voltage dip problem as well. With a wound rotor motor, differing values of resistance are inserted into the motor circuit at various times during the starting interval to reduce maximum inrush (and accordingly starting torque) to some desired value. Figure 84 shows typical speed-torque characteristic curves for a wound rotor motor. With appropriate switching times (dependent on motor speed) of resistance values, practically any desired speed-torque (starting) characteristic can be obtained. A motor starting study aids in choosing optimum current and torque values for a wound rotor motor application whether resistances are switched in steps by timing relays or continuously adjusted values obtained through a liquid rheostat feedback starting control.

For small loads, voltage stabilizers are sometimes used. These devices provide essentially instantaneous response to voltage fluctuations by "stabilizing" line voltage

Fig 84
Typical Wound Rotor Motor Speed-Torque Characteristics

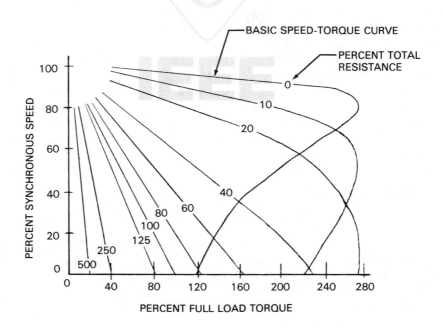

variations of as great as $\pm 15\%$ to within $\pm 1\%$ at the load. The cost and limited loading capability of these devices, however, have restricted their use mostly to controlling circuit power supply applications.

Special inrush motors can be purchased for a relatively small price increase over standard motors. These motors maintain nearly the same speed-torque characteristics as standard machines, but the inrush current is limited (usually to about 4.6 times full load current compared with 6 times full load current for a standard motor).

9.3.2 Analyzing Starting Requirements. A speed-torque and accelerating time study often in conjunction with the previously discussed voltage dip study permits a means of exploring a variety of possible motor speed-torque characteristics. This type of motor starting study also confirms that starting times are within acceptable limits. The accelerating time study assists in establishing the necessary thermal damage characteristics of motors or verifies that machines with locked-rotor protection supervised by speed switches will not experience nuisance tripping on starting.

The speed-torque/accelerating time motor starting study is also used to verify that special motor torque and/or inrush characteristics specified for motors to be applied on the system will produce the desired results. Mechanical equipment requirements and special ratings necessary for motor starting auxiliary equipment are based on information developed from a motor starting study.

9.4 Types of Studies. From the above discussion, it is apparent that, depending on the factors of concern in any specific motor starting situation, more than one type of motor starting study can be required.

9.4.1 The Voltage Drop Snapshot. One method of examining the effect of voltage dip during motor starting is to ensure the maximum instantaneous drop that occurs leaves bus voltages at acceptable levels throughout the system. This is done by examining the power system that corresponds to the worst-case voltage. Through appropriate system modeling, this study can be performed by various calculating methods using the digital computer.

The *snapshot* voltage drop study is useful only for finding system voltages. Except for the recognition of generator transient impedances when appropriate, machine inertias, load characteristics, and other transient effects are usually ignored. This type of study, while certainly an approximation, is often sufficient for many applications.

9.4.2 The Detailed Voltage Profile. This type of study allows a more exact examination of the voltage drop situation. Regulator response, exciter operation, and sometimes governor action are modeled to accurately represent transient behavior of local generators. This type of study is similar to a simplified transient stability analysis and can be considered a series of voltage *snapshots* throughout the motor starting interval including the moment of *minimum* or *worst-case* voltage.

9.4.3 The Speed-Torque and Acceleration Time Analysis. Perhaps the most exacting analysis for motor starting conditions is the detailed speed-torque analysis. Similar to a transient stability study (some can also be used to accurately investigate motor starting), speed-torque analysis provides electrical and accelerating torque calculations for specified time intervals during the motor starting period. Motor slip, load and motor torques, terminal voltage magnitude and angle, and the complex value of motor current drawn are values to be examined at time zero and at the end of each time interval.

Under certain circumstances, even across-the-line starting, the motor may not be able to *break away* from standstill or it may stall at some speed before acceleration is complete. A speed-torque analysis, especially when performed using a computer program, and possibly in combination with one or more previously discussed studies, can predict these problem areas and allow corrections to be made before difficulties arise. When special starting techniques are necessary, such as auto-transformer reduced voltage starting, speed-torque analysis can account for the auto-transformer magnetizing current and it can determine the optimum time to switch the transformer out of the circuit. The starting performance of wound rotor motors is examined through this type of study.

9.4.4 Adaptations. A particular application can require a slight modification of any of the above studies to be of greatest usefulness. Often combinations of several types of studies described are required to adequately evaluate system motor starting problems.

9.5 Data Requirements

9.5.1 Basic Information. Since other loads on the system during motor starting affect voltage available at the motor terminals, essentially the same information necessary for a load flow or short-circuit study is also required for a motor starting study. Although this information is summarized below, details are available elsewhere in this standard (see Chapters 6 and 7 as well as References [8] and [12]).

(1) *Utility and Generator Impedances.* These values are extremely significant and should be as accurate as possible. Generally, they are obtained from local utility representatives and generator manufacturers. When representing the utility impedance, it should be based on the *minimum* capacity of the utility system in order to yield the most pessimistic results insofar as voltage drop problems are concerned. This is in direct opposition to the approach normally used for a short-circuit analysis discussed in Chapter 7 of this standard. Where exact generator data cannot be obtained, typical impedance values are available from References [8] and [12].

(2) *Transformers.* Manufacturers' impedance information should be obtained where possible, especially for large units (that is, 5000 kVA and larger). Standard impedances can usually be used with little error for smaller units, and typical X/R ratios are available in IEEE C37.010-1979 [2].

(3) *Other Components.* System elements (such as cables) should be specified as to the number and size of conductor, conductor material, and whether magnetic duct or armor is used. All system elements should be supplied with R and X values so an equivalent system impedance can be calculated.

(4) *Load Characteristics.* System loads should be detailed including type (constant current, constant impedance, or constant kVA), power factor, and load

factor, if any. Exact inrush (starting) characteristics should also be given for the motor to be started.

(5) *Machine and Load Data.* Along with the aforementioned basic information, which is also required for a voltage drop type of motor starting analysis, several other items are also required for the detailed speed-torque and accelerating time analysis. These include the Wk^2 of the motor and load (with the Wk^2 of the mechanical coupling or any gearing included), and speed-torque characteristics of both the motor and load. Typical speed-torque curves are shown in Fig 85.

For additional accuracy, speed versus current and speed versus power factor characteristics should be given for as exact a model as possible for the motor during starting. For some programs, constants for the motor equivalent circuit given in Fig 86 can be either required input information or typical default values. This data must be obtained from the manufacturer since values are critical. Exciter/regulator data should also be obtained from the manufacturer for studies involving locally connected generators.

9.5.2 Simplifying Assumptions. Besides using standard impedance values for transformers and cables, it is often necessary to use typical or assumed values for other variables when making motor starting voltage drop calculations. This is particularly true when calculations are for evaluating a preliminary design and exact motor and load characteristics are unknown. Some common assumptions used in the absence of more precise data follow:

(1) *Horsepower to kVA Conversion.* A reasonable assumption is 1 hp equals 1 kVA. For induction motors and synchronous motors with 0.8 leading, running power factor, it can easily be seen from the equation:

$$\text{hp} = \frac{\text{kVA (PF) (EFF)}}{0.746} \qquad\qquad \text{(Eq 123)}$$

Fig 85
Typical Motor and Load Speed-Torque Characteristics

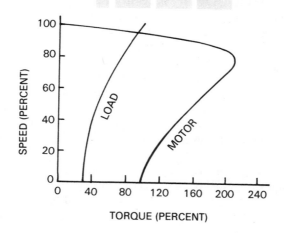

219

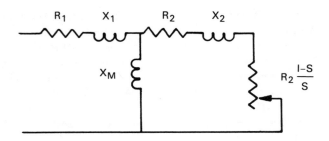

Fig 86
Simplified Equivalent Circuit for a Motor on Starting

The ratio of 0.746 to efficiency times the power factor approaches unity for most motors given the 1 hp/kVA approximation. Therefore, for synchronous motors operating at 1.0 PF, a reasonable assumption is 1 hp equals 0.8 kVA.

(2) *Inrush Current.* Usually, a conservative multiplier for motor starting inrush currents is obtained by assuming the motor to have a code G characteristic with locked-rotor current equal to approximately 6 times the full load current with full voltage applied at motor terminals (see Reference [1]).

(3) *Starting Power Factor.* The power factor of a motor during starting determines the amount of reactive current that is drawn from the system, and thus to a large extent the maximum voltage drop. Typical data (see Reference [12]) suggest the following:

(a) Motors under 1000 hp, PF = 0.20

(b) Motors 1000 hp and over, PF = 0.15

9.6 Solution Procedures and Examples. Regardless of the type of study required, a basic voltage drop calculation is always involved. When voltage drop is the only concern, the end product is this calculation when all system impedances are at maximum value and all voltage sources are at minimum expected level. In a more complex motor speed-torque analysis and accelerating time study, voltage drop calculations are required. These are performed at regular time intervals following the initial impact of the motor starting event and take into account variations in system impedances and voltage sources. Results of each *iterative* voltage drop calculation are used to calculate output torque, which is dependent on the voltage at machine terminals and motor speed. Since the interval of motor starting usually ranges from a few seconds to 10 or more seconds, effects of generator voltage regulator and governor action are evident, sometimes along with transformer tap control depending on control settings. Certain types of motor starting studies account for generator voltage regulator action while a transient stability study is usually required in cases where other transient effects are considered important. A summary of fundamental equations used in various types of motor starting studies follows, along with examples illustrating applications of fundamental equations to actual problems, which illustrate typical computer program outputs.

9.6.1 The Mathematical Relationships. There are basically three ways to solve for bus voltages realized throughout the system on motor starting. These are presented in this section and then examined in detail by examples in the next section.

(1) *Impedance Method.* This method involves reduction of the system to a simple voltage divider network (see Reference [14]) where voltage at any point (bus) in a circuit is found by taking known voltage (source bus) times the ratio of impedance to the point in question over total circuit impedance. For the circuit of Fig 87,

$$V = E \ \frac{X_1}{X_1 + X_2} \tag{Eq 124}$$

or, more generally,

$$V = E \ \frac{Z_1}{Z_1 + Z_2} \tag{Eq 125}$$

The effect of adding a large capacitor bank at the motor bus is seen by the above expression for V. The addition of negative vars causes X_1, or Z_1, to become larger in both numerator and denominator so bus voltage V is increased and approaches 1.0 per unit as the limiting improvement. Locked-rotor impedance for three-phase motor is simply

$$Z_{LR} = \ \frac{\text{rated voltage L-L}}{(\sqrt{3} \ \text{LRA})} \ \text{in } \Omega \tag{Eq 126}$$

where:

LRA = Locked-rotor current at rated voltage

This value in per unit is equal to the inverse of the inrush multiplier on the motor rated kVA base:

$$Z_{LR} = \ \frac{1}{\text{LRA/FLA}} \ \text{in per unit} \tag{Eq 127}$$

**Fig 87
Simplified Impedance Diagram**

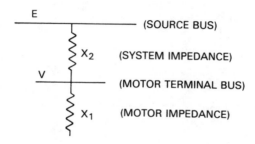

Since a starting motor is accurately represented as a constant impedance, the impedance method is a very convenient and acceptable means of calculating system bus voltages during motor starting. Validity of the impedance method can be seen and is usually used for working hand calculations. Where other than simple radial systems are involved, the digital computer greatly aids in obtaining necessary network reduction. To obtain results with reasonable accuracy, however, various system impedance elements must be represented as complex quantities rather than as simple reactances.

(2) *Current Method.* For any bus in the system represented in Figs 88 and 89, the basic equations for the current method are as follows:

$$I_{\text{per unit}} = \frac{\text{MVA}_{\text{load}}}{\text{MVA}_{\text{base}}} \text{ at 1.0 per unit voltage} \qquad \text{(Eq 128)}$$

$$V_{\text{drop}} = I_{\text{per unit}} \cdot Z_{\text{per unit}} \qquad \text{(Eq 129)}$$

$$V_{\text{bus}} = V_{\text{source}} - V_{\text{drop}} \qquad \text{(Eq 130)}$$

The quantities involved should be expressed in complex form for greatest accuracy, although reasonable results can be obtained by using magnitudes only for first-order approximations.

The disadvantage to this method is that, since all loads are not of constant current type, the current to each load varies as voltage changes. An iterative type solution procedure is therefore necessary to solve for the ultimate voltage at every bus, and such tedious computations are readily handled by a digital computer.

**Fig 88
Typical Single-Line Diagram**

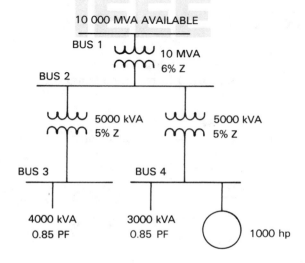

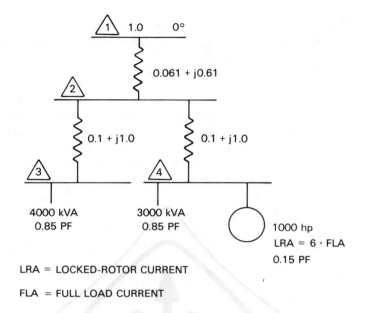

LRA = LOCKED-ROTOR CURRENT

FLA = FULL LOAD CURRENT

Fig 89
Impedance Diagram for System in Fig 88

(3) *Load Flow Solution Method.* From the way loads and other system elements are portrayed in Figs 88 and 89, it appears that bus voltages and the voltage dip could be determined by a conventional load flow program. This is true. By modeling the starting motor as a constant impedance load, the load flow calculations yield the bus voltages during starting. The basic equations involved in this process are repeated here (see References [15] and [17]).

$$I_k = \frac{P_k - jQ_k}{V_k^*} - Y_k V_k \qquad\qquad \text{(Eq 131)}$$

$$V_k = V_{\text{ref}} + \sum_{i=1}^{n} Z_{ki} \left(\frac{P_i - jQ_i}{V_i} - Y_i V_i \right) \qquad\qquad \text{(Eq 132)}$$

where:

I_k = Current in the kth branch of any network

P_k, Q_k = Real and reactive powers representative of the loads at the kth bus

V_k = The voltage at the kth bus

Y_k = The admittance to ground of bus k

V_{ref} = Voltage of the swing or slack bus

n = Number of buses in the network

Z_{ki} = The system impedance between the kth and ith buses

The load flow solution to the motor starting problem is very precise for finding bus voltages at the instant of maximum voltage drop. It is apparent from the expressions for I_k and V_k that this solution method is ideally suited for the digital computer any time the system involves more than two or three buses.

9.6.2 Other Factors. Unless steady-state conditions exist, all of the above solution methods are valid for one particular instant and provide the single *snapshot* of system bus voltages as mentioned earlier. For steady-state conditions, it is assumed that generator voltage regulators have had time to increase field excitation sufficiently to maintain the desired generator terminal voltage. Accordingly, the presence of the internal impedance of any local generators connected to the system is ignored. During motor starting, however, the influence of machine transient behavior becomes important. To model the effect of a close-connected generator on the maximum voltage drop during motor starting requires inclusion of generator transient reactance in series with other source reactances. In general, use of the transient reactance as the representation for the machine results in calculated bus voltages and, accordingly, voltage drops that are reasonably accurate and conservative, even for exceptionally slow-speed regulator systems.

Assuming, for example, that bus 1 in the system shown in Fig 89 is at the line terminals of a 12 MVA generator rather than being an infinite source ahead of a constant impedance utility system, the transient impedance of the generator would be added to the system. The resulting impedance diagram is shown in Fig 90. A new bus 99 is created. Voltage at this new bus is frequently referred to as *voltage behind the transient reactance*. It is actually the internal machine transient driving voltage (see Chapter 3).

When the steady-state operating voltage is 1.0 per unit, the internal machine transient driving voltage can be considered the voltage that must be present ahead of the generator transient reactance with the terminal voltage maintained at 1.0 per unit (within exciter tolerances) during steady-state conditions while supplying power to the other loads on the system. The transient driving voltage V is calculated as follows:

$$\begin{aligned} V &= V_{\text{terminal}} + (jX'_d)(I_{\text{load}}) \\ &= 1 + (jX'_d)(I_{\text{load}}) \end{aligned} \qquad \text{(Eq 133)}$$

where:

$$V_{\text{terminal}} = 1.0 \text{ per unit}$$

$$I_{\text{load}} = \frac{\text{MVA}_{\text{load}}}{\text{MVA}_{\text{base}}} \text{ per unit} \qquad \text{(Eq 134)}$$

Treatment of a locally connected generator is equally applicable to all three solution methods described above. Such an approach cannot give any detail regarding the response of the generator voltage regulator or changes in machine characteristics with time. For a more detailed solution that considers time-dependent effects of machine impedance and voltage regulator action, the appropriate impedance and voltage terms in each expression must be continuously altered to accu-

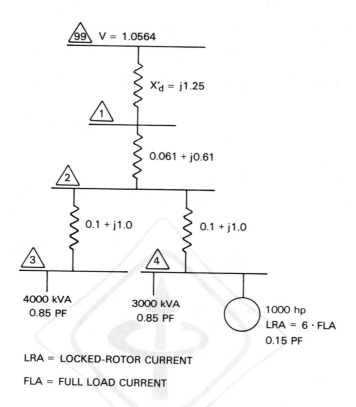

Fig 90
Revised Impedance Diagram Showing
Transient Reactance of Generator

rately reflect changes that occur in the circuit. This procedure is also applicable to any solution methods considered. Figure 91 shows a simplified representation of the machine parameters that must be repeatedly modified to obtain the correct solution.

Some type of reduced voltage starting is often used to minimize motor inrush current and thus reduce total voltage drop, when the associated reduction in torque accompanying this starting method is permissible. Representation used for the motor in any solution method for calculating voltage drop must be modified to reflect the lower inrush current. If auto-transformer reduced voltage starting is used, motor inrush will be reduced by the appropriate factor from Table 17. If, for example, normal inrush is 6 times full load current and an 80% tap auto-transformer starter is applied, the actual inrush multiplier used for determining the appropriate motor representation in the calculations is $(6)\ (0.64) = 4.2 \cdot$ full load current.

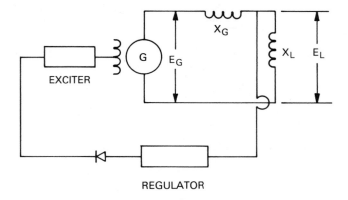

$$E_L = E_G \frac{(X_L)}{X_L + X_G}$$

where

X_G VARIES WITH TIME AS $X''_d \rightarrow X'_d \rightarrow X_d$,
E_G VARIES WITH TIME AS $T''_{do} \rightarrow T'_{do} \rightarrow T_{do}$
DEPENDING ON EXCITER/REGULATOR OUTPUT

**Fig 91
Simplified Representation of Generator
Exciter/Regulator System**

**Table 17
Auto-Transformer Line Starting Current**

Auto-Transformer Tap (% Line Voltage)	Line Starting Current (% Normal at Full Voltage)
50	25
65	42
80	64

Resistor or reactor starting limits the line starting current by the same current as motor terminal voltage is reduced (that is, 65% of applied bus voltage gives 65% of normal line starting current).

Y–start, Δ–run starting delivers 33% of normal starting line current with full voltage at the motor terminals. The starting current at any other voltage is, correspondingly, reduced by the same amount. Part winding starting allows 60% of normal starting line current at full voltage and reduces inrush accordingly at other voltages.

When a detailed motor speed-torque and accelerating time analysis is required, the following equations found in many texts apply (see Reference [18]). The equations in general apply to both induction and synchronous motors since the latter behave almost exactly as do induction machines during the starting period.

$$T \propto V^2$$

$$T = I_0 \alpha \tag{Eq 135}$$

$$I_0 = \frac{Wk^2}{2g} \text{ lb-ft-s}^2 \tag{Eq 136}$$

$$\omega^2 = \omega_0^2 + 2\alpha(\theta - \theta_0) \text{ rev/s} \tag{Eq 137}$$

$$\Delta\theta = \omega_0 t + \tfrac{1}{2} \alpha t^2 \text{ rev} \tag{Eq 138}$$

$$\alpha = \frac{T_n 2g}{Wk^2} \text{ rev/s}^2 \tag{Eq 139}$$

A simplified approximation for starting time is also available:

$$t \, (s) = \frac{Wk^2 \, (\text{rev/min}_1 - \text{rev/min}_2) \, (2\pi)}{60 g T_n} \tag{Eq 140}$$

where:

T = Average motor shaft output torque
V = Motor terminal voltage
I_0 = Moment of inertia
g = Acceleration due to gravity
ω = Angular velocity
α = Angular acceleration
t = Time in seconds to accelerate
T_n = Net average accelerating torque between rev/min_1 and rev/min_2
θ = Electrical angle in degrees
Wk^2 = Inertia

The basic equation for use with the equivalent circuit of Fig 86 is as follows (see References [9], [12], and [16]):

$$T = \frac{q_1 V^2 (r_2/s)}{\omega_s (r_1 + r_2/s)^2 + (X_1 + X_2)^2}$$

where:

T = Instantaneous torque
ω_s = Angular velocity at synchronous speed
$(r_1 + jX_1)$ = Stator equivalent impedance
$(r_2/s + jX_2)$ = Rotor equivalent impedance
q_1 = Number of stator phases (3 for a 3 ϕ machine)
V = Motor terminal voltage

9.6.3 The Simple Voltage Drop Determination. To illustrate this type of computer analysis, the system of Fig 88 will again be considered. It is assumed that bus 1 is connected to the terminals of a 12 MVA generator having 15% transient reactance (1.25 per unit on a 100 MVA base). Prior to starting, when steady-state load conditions exist, the impedance diagram of Fig 89 applies with the motor disconnected. The impedance diagram of Fig 90 applies when the 1000 hp motor on bus 4 is started.

Bus 99 in Fig 90 has been assigned a voltage of 1.056 per unit. This value can be confirmed using the expression for the internal machine transient driving voltage V given in the previous section with appropriate substitutions as follows:

$$V = (1.0 + j0.0) + (j1.25) (0.060114 - j0.042985)$$
$$= 1.0564 \text{ per unit voltage } \angle 4.08°$$

where values for the current through the X'_d element are expressed on a 100 MVA base and correspond to those that exist at steady state prior to motor starting with bus 1 operating at 1.0 per unit. The computer output report shown in Fig 92 shows steady-state load flow results for this case. All system loads are connected except the 1000 hp motor. Power flows are expressed in MW and Mvar. Quantities indicated for real and reactive power flow (and, accordingly, the circuit flow at 1.0 per unit voltage) through the line between bus 1 and bus 2 agree exactly with those used above to calculate V. Likewise, the bus 1 voltage is shown to be 1.0 per unit while the bus 99 voltage is 1.0564 per unit.

For convenience, the angle associated with the bus 99 reference voltage is assumed to be zero, which simply results in a corresponding shift in all other bus voltages. It is seen from the motor starting bus voltage computer report in Fig 93 that when this representation is used and subsequent motor starting calculations are made, the voltage at bus 4 is $0.7940 \angle -9.55°$ per unit. This voltage is well below the 0.85 criterion established earlier for proper operation of ac control devices.

9.6.4 Time-Dependent Bus Voltages. The load flow solution method for examining effects of motor starting allows a look at the voltage on the various system buses at a single point in time. A more exact approach is to model generator transient impedance characteristics and voltage sources closer to give results for a number of points in time following the motor starting event. Although the solution methods are applicable to multiple generator/motor systems as well, equations can be developed for a system of the form shown in Fig 94 to solve for generator, motor, and exciter field voltages as a function of time. The digital computer is used to solve several simultaneous equations that describe the voltage of each bus in a system at time zero and at the end of successive time intervals.

Figures 95 through 98 show in detail the type of input information required and the output obtained from a digital computer voltage drop study. The system shown in Fig 94 contains certain assumptions, which include the following:

(1) Circuit losses are negligible—reactances only used in calculations
(2) Initial load is constant kVA type
(3) Motor starting load is constant impedance type
(4) Motor starting power factor is in the range 0 to 0.25
(5) Mechanical effects, such as governor response, prime mover speed changes, and inertia constants, are negligible.

BUS NO. NAME	LINE FROM TO	VOLTAGE (PU)	ANGLE (DEGREES)	NET MEGAWATTS	NET MEGAVARS	NET MVA	TAP RATIO (PU)	CKT NO.	INTERMEDIATE MISMATCH
1 SWING		1.0000	0.0	6.0114	4.2985				
	1 - 2			6.0114	4.2985	7.3901	0.0	1	0.0 0.0
2 MAIN XFMR SEC		0.9707	-2.01	0.0	0.0				
	2 - 1			-5.9781	-3.9654	7.1736	0.0	1	
	2 - 4			2.5599	1.6795	3.0617	0.0	1	
3 5MVA XFMR SEC		0.9442	-4.00	-3.4000	-2.1070				
	3 - 2			-3.4000	-2.1070	4.0000	0.0	1	-0.000 -0.000
4 MTR STRT BUS		0.9511	-3.49	-2.5500	-1.5800				
	4 - 2			-2.5500	-1.5800	2.9998	0.0	1	-0.000 -0.000

GENERATION   6.0114   4.2985

LOAD   5.9500   3.6870

LINES	BUSES	LTCS
3	4	0

TOTAL-MISMATCH   -0.000   0.001

SYSTEM LOSSES   0.0612   0.6121

ITERATIONS   9

Fig 92

Load Flow Computer Output (Steady State)

BUS NO. NAME	LINE FROM TO	VOLTAGE (PU)	ANGLE (DEGREES)	NET MEGAWATTS	NET MEGAVARS	NET MVA	TAP RATIO (PU)	CKT NO.	INTERMEDIATE MISMATCH
1 MAIN XFMR PRI		0.9292	-4.89	0.0	0.0				
	1 - 99			-6.6904	-9.1669	11.3487	0.0	1	
	1 - 2			6.6911	9.1681	11.3501	0.0	1	0.001 0.001
2 MAIN XFMR SEC		0.8655	-7.40	0.0	0.0				
	2 - 1			-6.6001	-8.2581	10.5715	0.0	1	
	2 - 3			3.4230	2.3364	4.1444	0.0	1	
	2 - 4			3.1778	5.9227	6.7214	0.0	1	0.001 0.001
3 5MVA XFMR SEC		0.8354	-9.93	-3.4000	-2.1070				
	3 - 2			-3.4001	-2.1071	4.0001	0.0	1	-0.000 -0.000
4 MTR STRT BUS		0.7940	-9.55	-3.1173	-5.3195				
	4 - 2			-3.1175	-5.3196	6.1658	0.0	1	-0.000 -0.000
99 SWING		1.0564	0.0	6.6904	11.0313				
	99 - 1			6.6904	11.0313	12.9015	0.0	1	0.0 0.0

TOTAL-MISMATCH 0.001 0.002
GENERATION 6.6904 11.0313
LOAD 6.5173 7.4265
SYSTEM LOSSES 0.1742 3.6068
ITERATIONS 30

LINES 4   BUSES 5   LTCS 0

Fig 93

Load Flow Computer Output (Voltage Dip on Motor Starting)

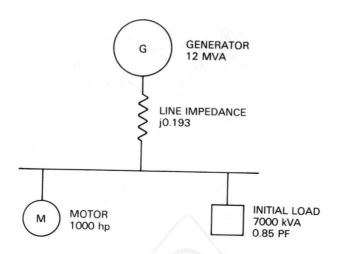

**Fig 94
Simplified System Model for
Generator Representation During Motor Starting**

Plotted results obtained from the computer compare favorably to those expected from an examination of Fig 83. In the particular computer program used to obtain this report, the excitation system models available are similar to those described in Reference [11]. Excitation system models are shown in simplified form in Fig 99. Continuously acting regulators of modern design permit full field forcing for minor voltage variations (as little as 0.5%), and these voltage changes have been modeled linearly for simplicity.

Variations in exciter field voltage (EFV) over each time interval considered are used to calculate system bus voltages at the end of these same intervals. A single main machine field circuit time constant is used in the generator representation, and Fromlich's approximation (see Reference [13]) for saturation effects is used when the voltage behind the generator leakage reactance indicates that saturation has been reached.

The tabulated output stops just short of full recovery since a more complex model is necessary to represent overshoot, oscillation, etc., beyond this point. Of primary concern in this type of study is the maximum voltage dip and the length of time to voltage recovery as a function of generator behavior and voltage regulator performance.

9.6.5 The Speed-Torque and Motor Accelerating Time Analysis. A simplified sample problem is presented for solution by hand. In this way, it is possible to appreciate how the digital computer aids in solving the more complex problems. The following information applies to the system shown in Fig 100.

(1) Motor hp = 1000 (induction)
(2) Motor rev/min = 1800
(3) Motor Wk^2 = 270 lb-ft^2
(4) load Wk^2 = 810 lb-ft^2

PROGRAM ASSUMPTIONS ----   1. ONLY REACTANCES ARE USED IN CALCULATIONS.
    2. EXCITER RESPONSE IS MODELED LINEARLY.
    3. GENERATOR SATURATION IS MODELED WITH FROMLICH'S EQUATION.
    4. THE INITIAL LOAD IS A CONSTANT KVA TYPE.
    5. MOTOR STARTING LOAD IS A CONSTANT IMPEDANCE TYPE.
    6. MOTOR STARTING PF IS IN THE RANGE 0.00 TO 0.25.
    7. MECHANICAL EFFECTS (LIKE GOVERNOR RESPONSE, PRIME MOVER SPEED
       CHANGES, ETC.) ARE NEGLECTED.

PROGRAM DATA BASE ----   GENERATOR kVA= 12000.00
    XD=1.300
    X'D=0.150
    XL=0.120
    T'DO=   5.000
    IFAG= 0.395
    IFNL= 0.450
    IF130= 0.692
    IFFL= 1.000
    C=0.750

    STATIC TYPE EXCITER
    CEILING VOLTAGE=3.60 PU
    SET POINT=1.00 PU
    EXCITER TIME CONSTANT=0.32
    REGULATOR TIME CONSTANT=0.0

    REACTANCE BETWEEN GENERATOR AND BUS=0.193

    HORSEPOWER OF STARTING MOTOR= 1000.0
    INRUSH CURRENT IN PU OF RATED=6.00
    RATED VOLTAGE=   1.
    OPERATING VOLTAGE=   1.

    INITIAL LOAD kVA= 7000.00
    P.F.=0.85

NOTE ----   ANSWERS GIVEN BELOW REFLECT THE TIME REQUIRED FOR THE GENERATOR
    VOLTAGE TO RECOVER FROM THE IMPACT OF THE MOTOR STARTING LOAD.
    THEY DO NOT CORRESPOND TO THE TIME REQUIRED FOR THE MOTOR TO START,
    NOR DO THEY IMPLY THAT SUFFICIENT TORQUE IS AVAILABLE TO ACCELERATE THE MOTOR.

Fig 95
Typical Output — Generator Motor Starting Program

TIME (SECONDS)	GENERATOR VOLTS (PU)	EXCITER FIELD VOLTS (PU)	MOTOR VOLTS (PU)
0 -	1.00	1.61	0.94
0.0	0.93	1.61	0.79
0.1	0.92	2.12	0.78
0.2	0.92	2.62	0.78
0.3	0.92	3.13	0.78
0.4	0.94	3.60	0.80
0.5	0.97	3.60	0.83
0.6	0.99	3.60	0.85
0.7	1.00	3.60	0.86

Fig 96
Typical Output — Generator Motor Starting Program

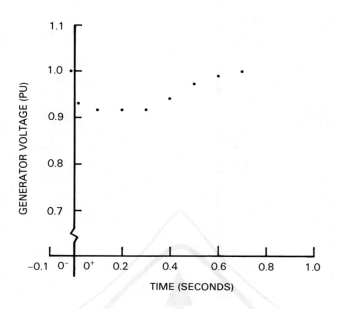

Fig 97
Typical Output — Plot of Generator Voltage Dip

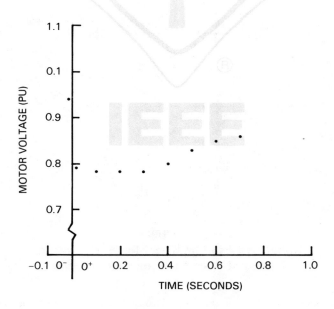

Fig 98
Typical Output — Plot of Motor Voltage Dip

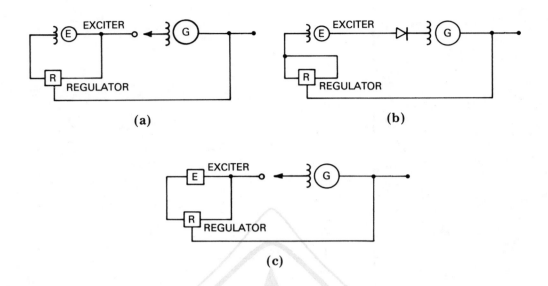

(a)   (b)

(c)

Fig 99
Simplified Representation of Typical
Regulator/Exciter Models for Use in Computer Programs

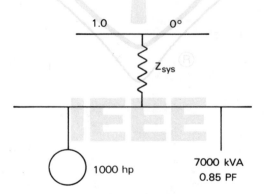

Fig 100
Simplified System Model for Accelerating
Time and Speed-Torque Calculations

Assuming Fig 85 describes the speed-torque characteristic of the motor and the load, it is possible to find an average value for accelerating torque over the time interval defined by each speed change. This can be done graphically for hand calculations, and the results are tabulated in Table 18.

**Table 18
Average Values for Accelerating
Torque Over Time Interval
Defined by a Speed Change**

speed	T_{motor}	T_{load}	T_{net}	T_{net}
0%	100%	30%	—	—
—	—	—	77.5%	2260.4 lb-ft^2
25%	120%	35%	—	—
—	—	—	100%	2916.7 lb-ft^2
50%	160%	45%	—	—
—	—	—	120%	2500.0 lb-ft^2
75%	190%	65%	—	—
—	—	—	62.5%	1822.9 lb-ft^2
95%	80%	80%	—	—

Applying the simplified formula for starting time provided earlier:

$$t_{0-25} = \frac{(270+810)(450-0)}{(308)(2260.4)} = 0.6981 \text{ second}$$

$$t_{25-50} = \frac{(1080)(900-450)}{(308)(2916.7)} = 0.5410 \text{ second}$$

$$t_{50-75} = \frac{(1080)(1350-900)}{(308)(3500.0)} = 0.4580 \text{ second}$$

$$t_{75-95} = \frac{(1080)(1710-1350)}{(308)(1822.9)} = 0.6925 \text{ second}$$

and, therefore, the total time to 95% of synchronous speed (or total starting time) is the sum of the times for each interval, or approximately 2.38 seconds. It can be seen how a similar technique can be applied to the speed-torque starting characteristic of a wound rotor motor (see Fig 84) to determine the required time interval for each step of rotor starting resistance. The results of such an investigation can then be used to specify and set timers that operate resistor switching contactors or program the control of a liquid rheostat.

The current drawn during various starting intervals can be obtained from a speed-current curve, such as the typical one shown in Fig 101. This example has assumed full voltage available to the motor terminals, which is an inaccurate assumption in most cases. Actual voltage available can be calculated at each time interval. The accelerating torque will then change by the square of the calculated voltage. This process can be performed by graphically plotting a reduced voltage speed-torque curve proportional to the voltage calculated at each time interval, but this becomes tedious in a hand calculation. Sometimes, in the interest of simplicity, a torque corresponding to the motor terminal voltage at the instant of the maximum voltage dip is used throughout the starting interval. More accurate results are possible with digital computer program analysis. A sample output report for the analysis is shown in Fig 102.

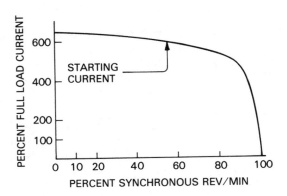

Fig 101
Typical Motor Speed-Current Characteristic

9.7 Summary. Several methods for analyzing motor starting problems have been presented. Types of motor starting studies available range from simple voltage drop calculation to the more sophisticated motor speed-torque and acceleration time study that approaches a transient stability analysis in complexity. Each study has an appropriate use and the selection of the correct study is as important a step in the solution process as the actual performance of the study itself. Examples presented here should serve as a guide for when to use each type of motor starting study, what to expect in the way of results, and how these results can be beneficially applied. The examples should also prove useful in gathering the required information for the specific type of study chosen. Experienced consulting engineers and equipment manufacturers can give valuable advice, information, and direction regarding the application of motor starting studies as well.

9.8 References

[1] ANSI/NFPA 70-1990, National Electrical Code Handbook.[12]

[2] IEEE C37.010-1979 (Reaff. 1988), IEEE Application Guide for AC High-Voltage Circuit Breakers Rated on a Symmetrical Current Basis (ANSI).[13]

[3] IEEE Std 141-1986, IEEE Recommended Practice for Electric Power Distribution for Industrial Plants (ANSI).

[4] IEEE Std 242-1986, IEEE Recommended Practice for Protection and Coordination of Industrial and Commercial Power Systems (ANSI).

---

[12] ANSI/NFPA publications can be obtained from the Sales Department, American National Standards Institute, 1430 Broadway, New York, NY 10018, or from Publication Sales, National Fire Protection Association, Batterymarch Park, Quincy, MA 02269.

[13] IEEE publications are available from the Institute of Electrical and Electronics Engineers, IEEE Service Center, 445 Hoes Lane, Piscataway, NJ 08855-1331.

```
MD  1  MOTOR TYPE- 1  TITLE-           BENCH MARK PROBLEM

MD  2  HP-  1000.      SPEED-1775 NEMA TYPE-A SYSTEM VOLTS- 4160.   RATED V- 4000.

MD  3  R1-         X1-         R2-         X2-         XO-

MD  4  NO. PHASES- 3 LOCKED ROTOR CURRENT-          FULL LOAD CURRENT- 145.
   * WARNING --- LOCKED ROTOR CURRENT NOT INPUT OR IS INVALID. EXECUTION CONTINUING.

MD  5  INRUSH MULTIPLIER- 6.0  STARTING PF - .15  MOTOR INERTIA WKSQ-  500.

MD  6  MOTOR STARTER DATA--- TYPE- 2 RST-         XST-         TAP

ST  7  IND. POINTS-
   * WARNING --- PROGRAM ASSUMING TYPICAL NEMA TYPE A TORQUE-SPEED CURVE.

SC  8  IND. POINTS-
   * WARNING --- PROGRAM ASSUMING A TYPICAL SPEED-CURRENT CURVE.

PF  9  IND. POINTS-
   * WARNING --- PROGRAM ASSUMING A TYPICAL SPEED-POWER FACTOR CURVE.

LD10  NO. POINTS-       LOAD TYPE-   NO LOAD
   * WARNING --- LOAD'S TORQUE-SPEED CURVE MUST BE INPUT, OR IS ASSUMED AS NO LOAD CURVE.  EXECUTION CONTINUING.

LD11  LOAD INERTIA WKSC-     GEARING RATIO
   * WARNING --- MOTOR TO LOAD GEAR RATIO NOT INPUT OR IS INVALID. PROGRAM ASSUMES -- 1.   EXECUTION CONTINUING.

SD12  SYSTEM IMPEDANCE RS-  .00101  XS-  .0101

SD13  SYSTEM GENERATION-              IMPEDANCE RG-         XG-

SD14  INITIAL SYSTEM LOAD     TYPE- 3 RL-         XL- 7000.                  PF- .85
```

Fig 102

Typical Output—Motor Speed-Torque and Accelerating Time Program

[5] NEMA Standards Publication ICS 1-1988, General Standards for Industrial Control and Systems; ICS 2-1988, Industrial Control Devices, Controllers and Assemblies; ICS 3-1988, Industrial Systems; ICS 4-1983 (Reaff. 1988), Terminal Blocks for Industrial Use; ICS 6-1988, Enclosures for Industrial Control and Systems.[14]

[6] NEMA Standards Publication MG 1-1987, Motors and Generators.

[7] CROFT, T., CARE, C., and WATT, J. *American Electrician's Handbook*, New York, McGraw-Hill, 1970.

[8] *Electrical Transmission and Distribution Reference Book*, Westinghouse Electric Corporation, East Pittsburgh, Pennsylvania, 1964.

[9] FITZGERALD, A., KINGSLEY, C., and KUSKO, A. *Electric Machinery*, New York, McGraw-Hill, 1971.

[10] HARBAUGH and PONSTINGL. How to Design a Capacitor Starting System for Large Induction and Synchronous Motors, IEEE IAS 1975 Annual Meeting— Conference Record.

[11] IEEE POWER GENERATION COMMITTEE. *Computer Representation of Excitation Systems*, Paper 31 TP 67-424, May 1, 1967.

[12] *Industrial Power Systems Handbook*, BEEMAN, D., ed., New York, McGraw-Hill, 1955.

[13] KIMBARK, E. *Power System Stability: Synchronous Machines*, Dover Publications, New York, 1956.

[14] MANNING, L. *Electrical Circuits*, New York, McGraw-Hill, 1966.

[15] NEUENSWANDER, J. *Modern Power Systems*, International, 1971.

[16] PETERSON, H. *Transients in Power Systems*, Dover Publications, New York, 1951.

[17] STAGG and EL-ABIAD. *Computer Methods in Power System Analysis*, New York, McGraw-Hill, 1971.

[18] WEIDMER, R. and SELLS, R. *Elementary Classical Physics*, 2 vols., Boston, Allyn and Bacon, Inc., 1965.

---

[14] NEMA publications are available from the National Electrical Manufacturers Association, 2101 L Street, NW, Washington, DC 20037.

Chapter 10
Harmonic Analysis Studies

10.1 Introduction. This chapter discusses the basic concepts involved in studies of harmonic analysis of industrial and commercial power systems. The need for such analysis, recognition of potential problems, corrective measures, required data, and benefits of using a computer as a tool in a harmonic analysis study will also be addressed in this chapter.

The main source of harmonics in power systems are static power converters, which include rectifiers for electrochemical and electrometallurgical processes, and static power converters that are used as power supplies for adjustable speed drives (both ac and dc). In addition, there are other nonlinear devices, such as arc furnaces and discharge lighting (fluorescent, high pressure sodium, etc.). Power supplies for electronic equipment (such as uninterruptible power supplies (UPS), numerical controlled machines and computers), and any load that is applied to a power system that requires other than a sinusoidal current will be a source of harmonic currents.

By modeling the power system impedances as a function of frequency, a study can be made to see the effect of the harmonic contributions from nonlinear loads on the voltage and currents in a power system.

10.2 History. From approximately 1910 to the 60's, the main nonlinear loads came from those few large users in the electrochemical and electrometallurgical industries. They developed a means of limiting the harmonic currents that their processes developed and thus minimized the effect on power systems and other users.

Small and medium-sized adjustable speed drives used motor-generator sets (MG sets) to feed dc motors and a few adjustable speed ac drives used wound-rotor motors. Still other variable speed drives were steam driven. For the MG sets, the mechanical linkage between the two systems transmitted power between them and at the same time electrically isolated each system from the other. However, these MG sets were bulky and tended to be high maintenance cost pieces of equipment.

The first attempt at electrical rectification was accomplished through mechanical means. A motor driven cam physically opened and closed switches at precisely the right instant on the voltage waveform to supply dc voltage and current to the load. At best, this approach was cumbersome since timing the switches and keeping them timed was extremely difficult. In addition, contact arcing plus mechanical wear also made this equipment a high maintenance item. Mechanical rectifiers were soon replaced by static equipment including mercury, selenium, and silicon diodes, and finally thyristors (SCR's).

With the invention and development of the thyristor, new low cost-effective equipment became available to allow standard dc or squirrel cage induction motors to drive pumps, fans, and machines with the ability to control the speed of these drives. The technology grew rapidly and the applications of these drives covered all aspects of process drives in all industries. These nonlinear type loads increased dramatically in just the 70's. This growth has continued and will continue.

Although solid-state power conversion appeared to be the panacea to the problems of the older methods, other system problems soon became noticeable, especially as the total converter load became a substantial portion of the total system power requirements.

The most noticeable effect in the basic rectification process with switching devices was the appearance of harmonic currents that flow between ac and dc systems. The currents equalize conflicting demand of both systems but cause prime problems on the ac or dc side.

The second most noticeable initial problem was the inherently poor power factor associated with static power converters, especially if operated with phase retard control of the output. Economics (utility demand billing) as well as system voltage regulation requirements made it desirable to improve the overall system power factor, which normally was accomplished using shunt power factor correction capacitors. However, when these capacitor banks were applied, other problems involving harmonic voltages and currents affecting these capacitors and other related equipment became prevalent.

Another problem was the excessive amount of interference induced into telephone circuits, due to mutual coupling between the electrical system and the communication system at these harmonic frequencies.

More recent problems involve the growing distortion of the ac supply voltages that affects the performance of computers, numerical controlled machines, and other sophisticated electronic equipment that are very sensitive to power line polution. These devices may respond incorrectly to normal inputs, give false signals, or possibly not respond at all. More recently, neutrals of four-wire systems (480/277V; 120/208V) have been the latest power system element affected by harmonics (see Reference [12][15]).

The widespread use of static power converters means that the control of the harmonic current generated by such loads is becoming increasingly important and must be addressed by applying the very specific engineering knowledge required.

The remainder of this chapter reviews these specific requirements.

10.3 General Theory

10.3.1 What Are Harmonics? Harmonics are voltages and/or currents present in an electrical system at some multiple of the fundamental (normally 60 Hz)

---

[15] The numbers in brackets correspond to those in the References at the end of this chapter.

frequency. Typical harmonic values produced by power converters are the fifth (300 Hz), seventh (420 Hz), eleventh (660 Hz), and so on.

To better understand harmonic-related problems, it is necessary to understand how and where harmonics are generated. Harmonics may be classified as:

- Characteristic harmonics
- Noncharacteristic harmonics

Characteristic harmonics are generally produced by power converters. Noncharacteristic harmonics are typically produced by arc furnaces and discharge lighting.

In converting ac power to dc power, a power converter effectively breaks or chops the ac current waveform by only allowing the current to flow during a portion of the cycle. The example in Figs 103 and 104 for a 6-phase, 6-pulse rectifier indicates the waveforms for the dc current and the corresponding ac line current for a (typical) highly inductive dc load.

The square (stepped) ac current, I_{ac}, represents a distorted sinusoidal waveform rich in harmonic content that can be expressed in terms of the fundamental frequency using Fourier analysis techniques (see Reference [19]). The Fourier series for this waveform is:

$$I_{ac} = \frac{2\sqrt{3}}{\pi} I_d \left(\cos\theta - \frac{1}{5}\cos 5\theta + \frac{1}{7}\cos 7\theta - \frac{1}{11}\cos 11\theta + \frac{1}{13}\cos 13\theta \ldots \right)$$

(Eq 142)

where:

$\theta = 2\pi f_1 t$
f_1 = Fundamental frequency (60 Hz)

The higher frequency terms are the harmonic components.

Fig 103
6-Phase, 6-Pulse Rectifier Schematic

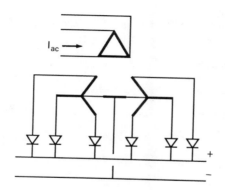

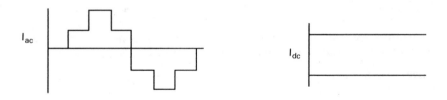

Fig 104
6-Phase, 6-Pulse Rectifier Input (I_{ac})
and Output (I_{dc}) Current Waveforms

A similar Fourier analysis of the distorted sinusoidal waveforms of other harmonic-generating equipment as mentioned previously will yield similar harmonic current components.

Arc furnaces differ from drives and rectifiers because harmonic voltages are generated instead of harmonic currents. The arc resistance, and therefore the voltage/current characteristics, are continually varying due to movement of the scrap, bubbling of the molten metal, magnetic repulsion of the arc in one phase from the other two phases, and so forth. In addition, magnetic repulsion forces between furnace leads cause swinging of these leads, which result in variation of the reactance of these leads. The overall result of the nonlinearity of the arc and arc furnace parameters results in the generation of some harmonic voltages along with harmonic currents in the secondary circuit. Because of the unpredictable nature of the arc, harmonic magnitudes are not as readily determined unless measurements are taken. For new installations, typical data is available based on worst-case conditions. Of particular concern is that the significant harmonics generated are the orders 2 through 9, inclusive. It should be noted that, due to the single phase relationships of the three electrodes, there will be a difference in the triplet harmonics (third, sixth, ninth) between the three phases, which will not be trapped by the delta winding of the furnance transformer and which will flow in the primary circuit.

With harmonic sources connected to a power system, harmonic currents will flow in the ac system, provided a path exists. If this occurs, harmonic-related problems can have significant detrimental effects in the ac system, such as:

- Capacitor heating/failures
- Telephone interference
- Rotating equipment heating
- Relay misoperation
- Transformer heating
- Switchgear failure
- Fuse blowing

Various parameters particular to each system determine the magnitude of these harmonic problems.

10.3.2 Resonance. The application of capacitors for power factor improvement in power systems where harmonic-generating equipment is connected necessitates the consideration of the potential problem of an excited harmonic resonance condition.

For ideal circuit elements, inductive reactance increases directly with increasing frequency and capacitive reactance decreases directly with increasing frequency. At the resonance frequency of any inductive-capacitive (LC) circuit, the inductive reactance equals the capacitive reactance.

There are two forms of resonance that must be considered: series resonance and parallel resonance. For the series circuit in Fig 105, the total impedance at the resonance frequency reduces to the resistance component only. In the case where this component is small, current magnitudes at the exciting frequency will be high.

Figure 106 is a plot of impedance versus frequency of the series circuit.

From a practical viewpoint, series resonance conditions are source limited. As long as the capacity of the harmonic source (MVA) is not excessively large compared to the rating of the capacitor bank (MVAc), the harmonic current is typically within limits. Figure 107 illustrates a practical installation where series resonance may be a problem. Although there are no harmonic sources present in the local system of Fig 108, all of the much larger utility system harmonic currents at the offending harmonic frequency could also result in serious problems — both with the power factor capacitor bank and the step-down power transformer forming a tuned series filter circuit.

Parallel resonance is similar to series resonance in that, at its exciting frequency, the capacitance reactance equals the inductive reactance. However, its parallel impedance is significantly different. Figure 110 is a plot of impedance versus frequency for the parallel circuit of Fig 109. At the resonance frequency f_r, the impedance is very high and, when excited from a source at this frequency, a high circulating current will flow in the capacitance-inductance loop — although the source current is small in comparison.

To illustrate parallel resonance further, select 60 Hz reactance of 0.60 ohms and -30.23 ohms for X_L and X_C, respectively. The parallel impedance of the capacitor and inductor at 60 Hz can be calculated to be $+0.61$ ohms. For illustration purposes, inject one ampere (60 Hz) current I_f into the circuit. Using Ohm's law, the

Fig 105
Series Circuit

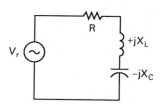

Fig 106
Impedance versus Frequency

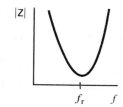

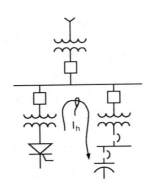

Fig 107
Series Circuit (Utility Source
Contains No Harmonics)

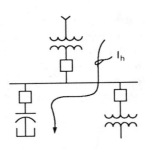

Fig 108
Series Circuit (Utility Source
Contains Harmonics)

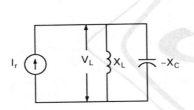

Fig 109
Parallel Circuit

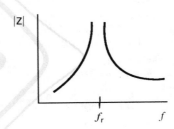

Fig 110
Impedance versus Frequency

voltage across the load V_L due to the fundamental frequency is then 0.61 volts. The current in the capacitor is –0.02 amperes; the current in the inductor is +1.02 amperes.

For this same circuit, assume a current source equal in magnitude to 1/7 ampere ($I_f/7$) at the seventh harmonic (420 Hz). At this frequency, the inductive reactance becomes 4.22 ohms and the capacitive reactance becomes –4.4 ohms. The parallel combination can be calculated to be 92.4 ohms. Again, using Ohm's law, the voltage across the load (V_L) is now 13.20 volts with the current through the capacitor equal to 3 amperes, while the current through the inductor is +3.13 amperes.

The most common condition leading to parallel resonance in industrial power systems is when the source inductance resonates with a power factor capacitor bank at a frequency excited by the facility's harmonic sources. For this condition, the harmonics are amplified by the resonant condition and are only limited by the damping of the parallel circuit.

In actual electrical systems utilizing power factor correction capacitors, either type of resonance or a combination of both may occur if the resonant point happens to be close to one of the frequencies generated by harmonic sources in the system. The result may be the flow of excessive amounts of harmonic current and/or the appearance of excessive harmonic overvoltages. Possible consequences of such an occurrence are excessive capacitor fuse operation, capacitor failures, unexplained protective relay tripping, telephone interference, or overheating of other electrical equipment.

10.4 Modeling. To analyze a system for resonance effects requires the calculation of the various harmonic currents throughout the system and the calculation of the harmonic voltages these currents cause. At each frequency, the system impedances are different. For all conditions, the circuit configuration remains the same. Rectifiers and other similar harmonic-generating equipment are represented as current sources at each harmonic frequency. With reference to the previous Fourier expansion, the maximum theoretical harmonic current magnitude from each converter is equal to the fundamental frequency full load current magnitude divided by the order of the harmonic.

Arc furnaces and other noncharacteristic harmonic sources are represented as harmonic voltage sources or, using Norton's theorem, are converted to an equivalent harmonic current source.

Harmonic current magnitudes are also functions of the number of converter pulses. The magnitudes of system harmonic voltages are produced by the harmonic currents flowing in the harmonic impedances of the ac system. The order of the characteristic harmonic currents is NP $\pm$ 1, where "N" is any integer and "P" is the number of converter pulses. Thus, for a 6-pulse converter, the order of harmonics are 5th, 7th, 11th, 13th, 17th, 19th, etc. For a 12-pulse rectifier, the order of harmonics are 11th, 13th, 23rd, 25th, 35th, 37th, etc. For all cases,

N equals any integer.
NP + 1 harmonics are positive sequence quantities.
NP – 1 harmonics are negative sequence quantities.
NP harmonics are zero sequence quantities for power converters having P number of pulses.

This procedure of using a higher number of phases or pulses for lower order harmonic cancellation is referred to as *phase multiplication* (discussed in 10.5). Although phase multiplication theoretically will cancel normal characteristic harmonics below the NP $\pm$ 1 order; in practice, both current magnitude and phase angles will deviate enough to allow only incomplete cancellation. Most literature on the subject indicates that 10% to 25% of the maximum harmonic magnitude will remain (see Reference [22]). To be as realistic as possible, this factor, sometimes referred to as a *harmonic cancellation factor* (HCF), should be taken into account.

Additional reduction of the harmonic current magnitude is due to the series inductive reactance between the harmonic source and the utility supply. The larger this inductive reactance is (commutating reactance), the more it impedes that particular harmonic generation. For example, the maximum fifth harmonic current magnitude available is $1/h = 1/5 = 0.2$ per unit or 20% (commutating reactance = 0). However, with a significant commutating reactance, the actual magnitude may only be 17%. This reduction is referred to as *commutating reactance factor* (CRF). Graphs of actual per unit I_h versus commutating reactance also as a function of phase control angle for thyristor converters are available in Fig 46 in IEEE Std 444-1973 [4].

The final factor reducing a particular harmonic current is the per unit loading factor (LDF). If a converter is only 50% loaded (fundamental component of current), then the harmonic current will only be 50% of its maximum value on that system. For drives utilizing phase retard control, this loading factor also is proportional to the ac fundamental current component but is not necessarily proportional to the dc, kW, or hp output. As drives are phased back for speed control, the harmonic content generated will increase due to phase retard and alter the CRF, whereas the LDF will be a function of the load on the shaft.

The graph in Fig 111 is typical for thyristor drives with the kVA value approximately constant over the entire range of speeds. However, the loading factor is defined as the fundamental component only. For low speeds, the kvar component of the kVA term is very rich in harmonic content. The assumption of constant kVA and maximum loading at maximum phase retard (minimum speed) will yield conservative results in evaluating the maximum harmonic current magnitudes generated.

The total harmonic current value injected into the system by a particular device is:

$$I_h = (I_{FL}/h)(LDF)(HCF)(CRF)$$

where:

I_{FL} = Fundamental full load current of the device
h = Order of harmonic

Where values for the individual harmonic reduction factor are not available, typical data has been established in Reference [11] (pending IEEE Std 519-1981 revision).

10.4.1 Analysis Techniques. With the basic system connections (single-line diagram) and impedances established, a harmonic analysis study should analyze

**Fig 111
Typical Thyristor Driver Characteristics**

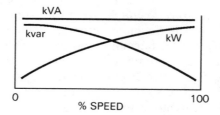

the system under steady-state conditions for normal power flow and harmonic current flow (sometimes referred to as harmonic load flow) for all the harmonics being modeled and for as many system switching conditions as required. A typical range of modeled harmonic frequencies may be from the fifth (300 Hz) to the 37th (2220 Hz) harmonic. The harmonic resonant point at a particular location will probably differ under each separate switching condition so all normal modes of operation should be modeled as a separate case in the analysis.

For larger complex electrical systems, or when trying to closely match the model with measured data, modeling both the resistance and reactance elements of electrical system components as a function of frequency may be warranted. For example, the ac component of resistance of transformers and reactors due to eddy current losses varies as the square of the frequency (IEEE C57.110-1986 [2][16]).

Since each switching condition requires many solved cases of the system for each of the harmonics being modeled, the use of digital computer programs provides an excellent tool for solving the multitude of calculations required for each switching condition solution. The number of calculations required for all but the simplest systems normally prohibits the use of hand calculations.

For complex systems that can exhibit a large number of switching configurations, it is sometimes advantageous to initially reduce the number of switching configurations to be modeled by the use of resonance scans (sometimes referred to as impedance scans). A resonance scan calculates the impedance of the system as a function of frequency as seen from the harmonic source and plots its magnitude accordingly. By noting the frequency of the maximum (parallel resonance) and minimum (series resonance) impedance values (peaks and valleys), switching conditions not resonant near the harmonic frequencies generated may be eliminated from further consideration. Only those switching conditions resonant near the generated harmonic frequencies need to be evaluated. Figure 112(a) is a sample resonance scan for a specific switching condition illustrating a parallel resonant point near 5.2 and a series resonant point near 6.0. Figures 112(b) and 112(c) show the respective single-line diagram and its respective impedance diagram.

Power factor correction capacitors of certain ratings are designed to continuously carry 135% of their nameplate rated (fundamental) kVA or kVAc, 110% of their rated voltage, and 180% of their rated current (see IEEE Std 18-1980 [3] and NEMA-CP-1-1989 [7][17]). This overload capability provides a margin for system overvoltages and/or harmonic voltages that may occur. The total loading of a bank may be calculated as the sum of the kVA loadings of the fundamental and of each harmonic. This may be expressed as:

$$kVAc = \sum_h (V_h I_h) = \sum_h (V_h^2 / X_h) \qquad \text{(Eq 143)}$$

---

[16] IEEE publications are available from the Institute of Electrical and Electronics Engineers, IEEE Service Center, 445 Hoes Lane, Piscataway, NJ 08855-1331.

[17] NEMA publications are available from the National Electrical Manufacturers Association, 2101 L Street, NW, Washington, DC 20037.

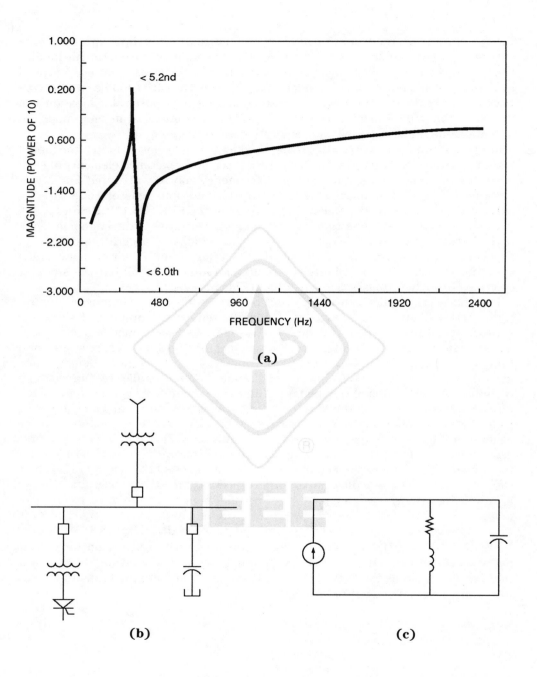

Fig 112
Sample Resonance Scan
(a) Scan, (b) Related Single-Line Diagram,
(c) Related Impedance Diagram

where:

h = Fundamental or order of harmonic (including the first or fundamental)
V = Fundamental or harmonic voltage
I = Fundamental or harmonic current
X = Fundamental or harmonic reactance

Capacitors must also have sufficient dielectric strength to withstand the antici-pated peak voltages resulting from the fundamental and the harmonics. This peak voltage is pessimistically calculated as the arithmetic sum of all the component voltages (not an RMS value).

$$V_{peak} = \sum_{h=1} V_h \qquad \text{(Eq 144)}$$

The peak values are used because of the random phase relationships existing between the various harmonic components.

The third factor to consider is the total RMS current that connections, bushings, and other components of the capacitor bank must carry. The RMS current is calculated as:

$$I_{rms} = \left(\sum_h I_h^2 \right)^{1/2} \qquad \text{(Eq 145)}$$

Additional problems can arise due to harmonics in motors, lighting ballast trans-formers, and other similar equipment. These problems are essentially excessive heating due to circulating harmonic currents throughout the system. To evaluate this effect, RMS voltages rather than peak values are required. Therefore, RMS voltages should also be calculated and printed throughout the system. The total RMS voltage may be calculated using the following equation:

$$V_{rms} = \left(\sum_h V_h^2 \right)^{1/2} \qquad \text{(Eq 146)}$$

In order to evaluate the impact of harmonics on a power system, detailed results of a computerized harmonic analysis study must include the four main points mentioned above, e.g., total capacitor bank kVAc loading, peak voltages, RMS cur-rents, and RMS voltages. The voltage and current values (peak and RMS) should be calculated at all critical system locations susceptible to harmonic problems where the appropriate values may be compared to the rating of the device being evaluated.

When evaluating harmonic limits in systems, especially at the point of common coupling (PCC) with the utility supply, two distortion factors (DF) are defined as follows (see IEEE Std 519-1981 [5] and Reference [11]):

VDF = Voltage distortion factor
CDF = Current distortion factor

Sometimes these terms are also referred to as total harmonic distortion (THD) factors and expressed as V_{THD} and I_{THD}.

$$\text{VDF} = V_{\text{THD}} = \frac{1}{V_1} \left(\sum_{h=2}^{\infty} V_h^2 \right)^{1/2} \qquad \text{(Eq 147)}$$

$$\text{CDF} = I_{\text{THD}} = \frac{1}{I_1} \left(\sum_{h=2}^{\infty} I_h^2 \right)^{1/2} \qquad \text{(Eq 148)}$$

where:

V_1, I_1 = Fundamental RMS values of voltage and current
V_h, I_h = Harmonic RMS values of voltage and current

Current guidelines for THD at the PCC are listed in IEEE Std 519-1981 [5]. This guide and its pending update (see Reference [11]) list specific limits for individual harmonics as well as THD. These are practical limits aimed at ensuring that other users supplied by the same utility system will not be adversely effected. Subsection 10.7 lists these limits.

10.5 Solutions to Harmonic Problems. The primary solution to any harmonic-related problem is accomplished by shifting the system resonant point to some other frequency not generated by the electrical equipment in the system or injected by the utility system. The simplest and least expensive method is to alter or bypass system operating conditions and procedures that will lead to harmonic resonance. If this approach is impractical or undesirable, then quite often additional apparatus is required.

The remedial measures involving additional equipment generally used to minimize harmonic effects include shunt L-C filters located at the harmonic source and tuned to series resonance at the troublesome harmonics. This approach provides a low impedance path for the harmonic currents to flow with very little flowing back into the rest of the ac system. However, a separate filter may be required for every major harmonic source and, therefore, may be prohibitively expensive.

In other cases where power factor correction capacitors in the ac system cause resonance at the generated harmonics, their location or size may be changed to eliminate the resonance, or series reactors may be added to detune them at the troublesome resonance frequency.

There are essentially three separate filtering schemes that will accomplish adequate filtering. Economic considerations as well as the particular filtering requirements for each case will determine which scheme is most desirable. Quite often, utility requirements for harmonic content injected into their system will dictate which scheme is to be used. Figures 113(a), 114(a), and 115(a) illustrate these schemes with Figs 113(b), 114(b), and 115(b) showing their respective impedance versus frequency plots.

The least expensive and, therefore, most desirable of these three schemes is the low pass (single frequency) filter of Fig 113. Generally a carefully selected tuning reactor will be sufficient to alleviate harmonic resonance. The careful selection of the tuning reactor is stressed. If it is incorrectly selected, the harmonic frequency for which it is tuned will probably be lowered to acceptable levels but another

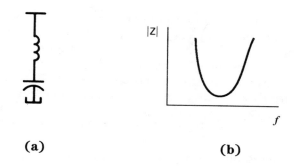

(a)   (b)

Fig 113
(a) Low Pass (Single Frequency) Filter
(b) Impedance versus Frequency Plot

(a)   (b)

Fig 114
(a) Broad Band Filter
(b) Impedance versus Frequency Plot

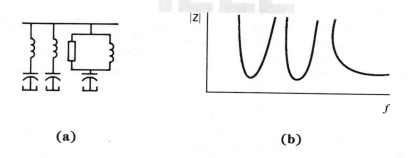

(a)   (b)

Fig 115
(a) Combination Single Frequency (Two Filters)
and Broad Band Filter
(b) Impedance versus Frequency Plot

harmonic frequency may then become dominant with the resonant point only being shifted to another harmonic frequency present on the system.

The series resonance frequency of the harmonic filter is typically selected slightly below the troublesome harmonic. For example, fifth harmonic filters are typically tuned at 4.7 to 4.8. This permits a small margin for error within the system modeling and manufacturing tolerances of the filter components. It also allows for a few individual capacitor fuses within the bank to blow, thereby detuning the filter without shifting the system resonance point so close to resonance that serious problems occur.

Figure 114 (broadband filter) will not filter out an individual harmonic as much as the low pass filter but will filter more than one troublesome harmonic. This may eliminate multiple filter requirements as in Fig 115, which results in an economic savings.

Figures 114 and 115 are generally used when more stringent harmonic content requirements are in effect. In Fig 115, the two lowest and most troublesome harmonics are filtered out individually with one high pass filter used to filter all higher order of harmonics above these two harmonics.

For new installations or major expansions of existing facilities that have a significant amount of harmonic producing equipment connected, there is an additional method of reducing lower order harmonics if a harmonic analysis is incorporated within the design stages. The concept of phase multiplication mentioned previously and the inherent cancellation of lower order harmonics can be employed to greatly reduce harmonic magnitudes. From an economic and manufacturing viewpoint, it is impractical to construct converter transformers with more than two secondary windings on each leg. However, transformer connections and additional phase shifters can be selected to achieve higher pulse operation of the total converter installation with a number of rectifiers.

In Fig 116, both rectifier transformers are individually three-phase units but are phase shifted 30° with respect to each other. When viewed from bus A and when they are both equally loaded, they collectively appear to be a 6-phase, 12-pulse system. Similarly, in Fig 117, buses C and D appear to have 6-phase, 12-pulse rectification but, due to the differing connections of the power transformers, the system becomes a 12-phase, 24-pulse system viewed from bus B. This configuration will greatly reduce all generated harmonics below the 23rd (due to phase multiplication cancellation below the NP ± 1 order) compared to 6-pulse systems.

Standard delta or wye connections of transformers permit only 30° phase shifts, but zig-zag or polygon connections can be used to obtain others (as in Fig 117). Another method using these special transformer connections to accomplish phase multiplication may entail different phase shifting of the rectifier transformers themselves. For example, one rectifier rated 12-phase, 12-pulse but with 6 phases shifted – 30°, and 6 phases at +0° with respect to the primary system will appear to be a 24-phase, 24-pulse system when operated with another identically rated and loaded 12-phase, 12-pulse rectifier, which has – 15° and +15° phase shifts.

It must be stressed that lower order harmonic cancellation using phase multiplication techniques is only applicable when each component is equally rated and loaded. For the case where not all units are equally loaded or some are offline, then

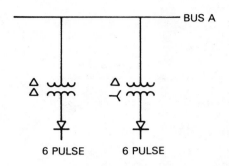

Fig 116
12-Pulse System

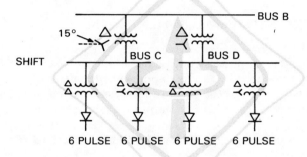

Fig 117
24-Pulse System

only partial lower order harmonic cancellation occurs. A harmonic analysis study should also address probable unbalanced loading conditions with only partial lower order harmonic cancellation considerations. The specific information and data required for the analysis should be obtained from the equipment vendor.

Often an installation that utilizes phase multiplication for harmonic reduction may not need any filtering at all or may require only small tuning reactors as opposed to a complex RLC filtering scheme. The net economic effect can be significant.

10.5.1 Examples. The partial single-line diagram in Fig 118 was taken from an actual harmonic analysis study performed in the design stages for an electrochemical plant.

This particular system required the addition of tuning reactors in series with each capacitor bank to minimize harmonic resonance problems.

Tables 19 and 20 are abbreviated copies of the computer solutions for this system. The first solution modeled the system without any special harmonic filters to

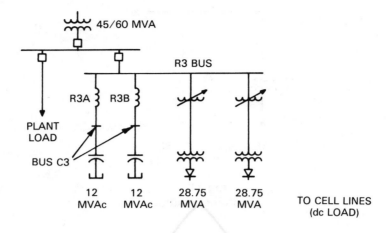

Fig 118
Partial Single-Line Diagram

determine if a resonant condition would exist after application of the proposed capacitor banks. As indicated by the harmonic profile at bus C3 (two banks), the system is resonant very close to the seventh harmonic. A comparison of capacitor bank ratings to the values listed in the computer printout indicated that filtering was indeed required.

The application of tuning reactors R3A and R3B tuned to the 4.7th harmonic was sufficient to suppress this resonant condition to acceptable levels. The second computer solution simulated the power system with the tuning reactors installed and indicates the seventh harmonic levels were almost eliminated with the totalized quantities greatly reduced. The RMS voltage at the terminals of the capacitor bank (bus C3) was lowered from 1.766 per unit to 1.102 per unit of the system base value. The current and kVAc ratings were well within limits. The voltage ratings of the proposed banks were increased to 109% of the nominal system voltage to reflect the increased voltage at the terminals of the capacitor bank as a result of the voltage rise across the reactor.

The values listed under bus R3 are those of the main bus. Satisfactory system harmonic filtering is also achieved.

The previous example illustrates how effective filtering can be when properly designed. However, filtering may not always be the best solution. In another electrochemical plant with a very large complex electrical system, special harmonic filtering was not implemented. The existing system utilized power factor correction capacitors at each major rectifier location. A plant expansion required an additional rectifier and capacitor bank. Initial filter design simulation corrected harmonic problems for the new bank but shifted the resonant point to another location, which caused harmonic problems with existing capacitor banks. The practical and economical solution in this case was to oversize the new capacitor bank's voltage rating to compensate for harmonic content without special filtering.

Table 19
First Computer Solution — Without Filters

HARMONIC ANALYSIS STUDY

SOLUTION FOR SWITCHING CONDITION 195

BUS NAMES	C3	C2	C1	R3	R2	R1
5 TH HARMONIC BUS VOLTAGES AND SHUNT LOAD CURRENTS						
BUS VOLTAGES	0.135	0.009	0.009	0.135	0.009	0.009
LOAD CURRENT	0.673	0.044	0.043	0.673	0.044	0.043
7 TH HARMONIC BUS VOLTAGES AND SHUNT LOAD CURRENTS						
BUS VOLTAGES	1.414	0.115	0.117	1.414	0.115	0.117
LOAD CURRENT	9.896	0.805	0.819	9.896	0.805	0.819
11 TH HARMONIC BUS VOLTAGES AND SHUNT LOAD CURRENTS						
BUS VOLTAGES	0.009	0.004	0.004	0.009	0.004	0.004
LOAD CURRENT	0.101	0.048	0.046	0.101	0.048	0.046
13 TH HARMONIC BUS VOLTAGES AND SHUNT LOAD CURRENTS						
BUS VOLTAGES	0.004	0.002	0.002	0.004	0.002	0.002
LOAD CURRENT	0.055	0.030	0.028	0.055	0.030	0.028
17 TH HARMONIC BUS VOLTAGES AND SHUNT LOAD CURRENTS						
BUS VOLTAGES	0.002	0.000	0.000	0.002	0.000	0.000
LOAD CURRENT	0.035	0.008	0.008	0.035	0.008	0.008
19 TH HARMONIC BUS VOLTAGES AND SHUNT LOAD CURRENTS						
BUS VOLTAGES	0.001	0.000	0.000	0.001	0.000	0.000
LOAD CURRENT	0.023	0.005	0.005	0.023	0.005	0.005
TOTALIZED QUANTITIES INCLUDING ALL ABOVE HARMONICS PLUS FUNDAMENTAL						
BUS V(ARITH)	2.617	1.183	1.184	2.617	1.183	1.184
BUS V(RMS)	1.766	1.056	1.057	1.766	1.056	1.057
LOAD I(RMS)	9.975	1.325	1.333	9.975	1.325	1.333
LOAD kVA	15.184	1.196	1.199	0.000	0.000	0.000

Sometimes the judicious sizing and placement of a power factor capacitor bank within a given power system will effectively provide harmonic filtering without special filters. For example, the 1.5 MVA transformer in Fig 107 or the 1.725 MVA transformer in Fig 119 could serve as an inductive-resistive harmonic filter. However, if this is the case, the loading of these transformers may have to be derated to allow for harmonic content (see IEEE C57.110-1986 [2] and Reference [14]).

As a result of the load flow analysis in Chapter 6, a 300 kVAc power factor capacitor (PFC) bank was added at bus 39. The only nonlinear load in the system is connected to bus 5 and, as indicated previously, may lead to a potential series

Table 20
Second Computer Solution — With Filters

HARMONIC ANALYSIS STUDY

SOLUTION FOR SWITCHING CONDITION 195

BUS NAMES	C3	C2	C1	R3	R2	R1
5 TH HARMONIC BUS VOLTAGES AND SHUNT LOAD CURRENTS						
BUS VOLTAGES	0.062	0.020	0.020	0.008	0.003	0.003
LOAD CURRENT	0.294	0.093	0.096	0.294	0.093	0.096
7 TH HARMONIC BUS VOLTAGES AND SHUNT LOAD CURRENTS						
BUS VOLTAGES	0.014	0.006	0.006	0.017	0.008	0.008
LOAD CURRENT	0.094	0.043	0.043	0.094	0.043	0.043
11 TH HARMONIC BUS VOLTAGES AND SHUNT LOAD CURRENTS						
BUS VOLTAGES	0.002	0.005	0.005	0.010	0.012	0.012
LOAD CURRENT	0.023	0.049	0.049	0.023	0.049	0.049
13 TH HARMONIC BUS VOLTAGES AND SHUNT LOAD CURRENTS						
BUS VOLTAGES	0.001	0.003	0.003	0.008	0.009	0.010
LOAD CURRENT	0.014	0.036	0.036	0.014	0.036	0.036
17 TH HARMONIC BUS VOLTAGES AND SHUNT LOAD CURRENTS						
BUS VOLTAGES	0.001	0.000	0.000	0.008	0.004	0.004
LOAD CURRENT	0.010	0.005	0.005	0.010	0.005	0.005
19 TH HARMONIC BUS VOLTAGES AND SHUNT LOAD CURRENTS						
BUS VOLTAGES	0.000	0.000	0.000	0.006	0.003	0.003
LOAD CURRENT	0.007	0.004	0.004	0.007	0.004	0.004
TOTALIZED QUANTITIES INCLUDING ALL ABOVE HARMONICS PLUS FUNDAMENTAL						
BUS V(ARITH)	1.180	1.137	1.137	1.131	1.120	1.121
BUS V(RMS)	1.102	1.100	1.100	1.050	1.050	1.050
LOAD I(RMS)	1.095	1.057	1.058	1.095	1.057	1.058
LOAD kVA	1.174	1.157	1.157	0.076	0.057	0.057

resonance condition that is seldom evaluated, yet could lead to possible problems (see Fig 119).

To check the series resonance frequency of this system and to determine if harmonic considerations need to be included in the 300 kVAc capacitor design, the following equation applies (see Reference [9]):

$$H_S = \sqrt{\frac{MVA_{TX}}{(MVAc)(Z_{TX})} - \frac{(MW_L)^2}{(MVAc)^2}}$$

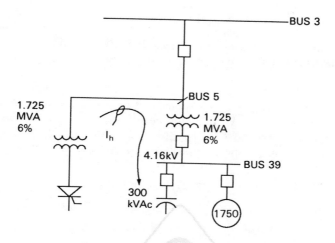

Fig 119
Partial Single-Line Diagram

where:

H_S = Series resonance point
MVA_{TX} = Transformer MVA
Z_{TX} = Transformer nameplate impedance in per unit
MW_L = MW of the load at the power factor capacitor bank location (MW_L is a damping factor).

Utilizing the system data, the following H_S points are calculated:

With damping: $H_S = \sqrt{\dfrac{1.725/2}{(0.3)(0.06)} - \dfrac{(1.5)^2}{(0.3)^2}}$

$$= 4.79$$

Without damping: $H_S = \sqrt{\dfrac{1.725/2}{(0.3)(0.06)}}$

$$= 6.922$$

If this capacitor is energized without the load on bus 39, then the calculated H_S of 6.922 would apply. Under this condition, it is very probable that most of the seventh harmonic current will flow directly to the 300 kVAc capacitor. The majority of the other harmonics will flow primarily to the remainder of the system. Therefore, the total harmonic loading of this 300 kVAc PFC can be approximated as:

$$I_{rms} = \sqrt{I_1^2 + I_7^2} = \sqrt{(41.64)^2 + (34.14)^2} = 53.85 \text{ A}$$

$$\%I_{RATED} = 100 \times \frac{I_{rms}}{I_1} = 100 \times \frac{53.85}{41.64} = 129\%$$

where:

I_1 = 60 Hz capacitor's full load amperes reflected to 4.16 kV

I_7 = Maximum seventh harmonic current of the nonlinear load reflected to 4.16 kV

Similarly, with the capacitor bank energized in conjunction with its load, the calculated H_S of 4.79 applies. Under this condition, most of the fifth harmonic current would be expected to flow to the 300 kVAc capacitor bank. The total harmonic loading for this condition can be approximated as:

$$I_{rms} = \sqrt{I_1^2 + I_5^2} = \sqrt{(41.64)^2 + (47.88)^2} = 63.45 \text{ A}$$

$$\%I_{RATED} = 100 \times \frac{I_{rms}}{I_1} = 100 \times \frac{63.45}{44.64} = 152\%$$

where:

I_5 = Maximum fifth harmonic current of the nonlinear load reflected to 4.16 kV

While either of these conditions yields less than the 180% capacitor current limit, the RMS current will be above typical individual capacitor fusing factors and usual bank protective relay settings. The design of this bank must take these harmonic currents into account.

10.6 When Is a Harmonic Study Required? For power systems where the following conditions are met, a harmonic study is usually required:

(1) Application of capacitor banks to systems where 20% or more of the total load is comprised of converters or other harmonic-generating equipment

(2) History of harmonic-related problems including excessive capacitor fuse operation

(3) During the design stage of a facility composed of capacitor banks and harmonic-generating equipment

(4) Restrictive power company requirements that limit harmonic injection back into their system to very small magnitudes

(5) Plant expansions that add significant harmonic-generating equipment operating in conjunction with capacitor banks

Occasionally, when harmonics appear to be the cause of system problems, it is desirable to determine the system harmonic resonance point. To determine this resonance point, the short-circuit capacity at each capacitor bank location is required. A close approximation of this parallel resonance point is given by the equation:

$$H_r = \sqrt{\frac{MVA_{sc}}{MVAc}}$$

where:

H_r = Resonance point in per unit of fundamental frequency
MVA_{sc} = Short-circuit capacity
$MVAc$ = Mvar rating of the unfiltered capacitor bank at that location (see Reference [22])

This equation is very useful for an initial evaluation. If the resonance point is close to one of the harmonic frequencies present in the system, then possible harmonic-related problems could occur.

It is not uncommon to encounter a system where it is more practical to take harmonic measurements on a system as a diagnostic measure rather than to perform a detailed, time-consuming harmonic analysis study. In other cases, measurements are used to obtain a harmonic spectrum as input data by which the system model is confirmed prior to performance of a detailed harmonic analysis study (this is especially desirable for arc furnace installations). In order to ensure that harmonic measurements will produce reliable results, careful consideration must be given to both test equipment and procedures being applied (see References [23] and [24]). The test results may often produce a solution or at least identify the cause of a harmonic problem, so that the need for a detailed harmonic study is either eliminated or the study is greatly simplified.

10.6.1 Data Required. The following data are required for a typical study:

(1) Single-line diagram of power system to be studied
(2) The short-circuit capacity and X/R ratio of the utility power supply system. The existing harmonic voltage spectrum of the utility system at the PCC (external to the system being modeled).
(3) Subtransient reactance and kVA of all rotating machines. Where possible, all machines on a given bus should be lumped together into one composite equivalent machine.
(4) Percent reactance and resistance of all lines, cables, bus work, current limiting reactors, and saturable reactors on a given kVA base and the rated voltage of the circuit in which the circuit element is located
(5) The connections, percent impedance, and kVA of all power transformers
(6) The three-phase connections, kvar, and unit kV ratings of all shunt capacitors and shunt reactors. Nearby utility capacitor banks may also need to be identified and modeled.
(7) Nameplate ratings, number of phases, pulses, and converter connections, whether they are diodes or thyristors and, if thyristors, the maximum phase delay angle, per unit loading, and loading cycle of each converter unit connected to the system. Actual manufacturer's test sheets on each transformer are also helpful but not absolutely mandatory. If this information is not readily available, the kVA rating of the converter transformer may be used for establishing the harmonic current being injected into the system.
(8) Specific system configurations and operating procedures for the converter circuits being studied
(9) Maximum expected voltage for the system supplying the converter loads
(10) For arc furnace installations, secondary lead impedance from the transformer to the electrodes plus a loading cycle to include arc MW, secondary

voltages, secondary current furnace transformer taps, and transformer connections.

(11) Power company imposed harmonic limits at the PCC. Frequently, those limits specified in IEEE Std 519-1981 [5] are used.

10.7 Distortion Limits

This subsection lists the suggested distortion limits as listed in IEEE Std 519-1981 [5]. In addition, the distortion limits in the pending update of IEEE Std 519-1981 [11] are also listed.

10.7.1 Pending IEEE Std 519-1981 Revision. [18] There are now two criteria that are used to evaluate harmonic distortion. The first is a limitation in the harmonic current that a user can inject into the utility system. The second criteria is the quality of the voltage that the utility must furnish the user. The interrelationship of these criteria shows that the harmonic problem is a system problem and not tied just to the individual load that requires the harmonic current. Tables 22 and 23 list these criteria:

10.8 References

[1] IEEE C37.99-1980 (Reaff. 1985), IEEE Guide for Protection of Shunt Capacitor Banks (ANSI).

[2] IEEE C57.110-1986, IEEE Recommended Practice for Establishing Transformer Capability When Supplying Nonsinusoidal Load Currents (ANSI).

[3] IEEE Std 18-1980, IEEE Standard for Shunt Power Capacitors (ANSI).

[4] IEEE Std 444-1973, IEEE Standard Practices and Requirements for Thyristor Converters for Motor Drives.

[5] IEEE Std 519-1981, IEEE Guide for Harmonic Control and Reactive Compensation of Static Power Converters.

Table 21
Voltage Distortion Limits for Medium Voltage and High Voltage Power Systems

Power System Voltage Level	Dedicated* System Converter	General Power System
Medium voltage 2.4–69 kV	8%	5%
High voltage 115 kV and above	1.5%	1.5%

*A dedicated system is one servicing only converters or loads not affected by voltage distortion.

---

[18]When IEEE Std 519-1981 is revised and published, the revision will supersede IEEE Std 519-1981, and it will become a Recommended Practice.

Table 22
Harmonic Current Limits for Nonlinear Loads at the PCC with Other Loads at Voltages of 2.4 to 69 kV (IEEE Std 519-1981 Revision (Pending))

	MAXIMUM HARMONIC CURRENT DISTORTION IN % OF FUNDAMENTAL					
	HARMONIC ORDER (ODD HARMONICS)					
I_{sc}/I_L	<11	$11<h<17$	$17<h<23$	$23<h<35$	$35<h$	THD
<20*	4.0	2.0	1.5	0.6	0.3	5.0
20–50	7.0	3.5	2.5	1.0	0.5	8.0
50–100	10.0	4.5	4.0	1.5	0.7	12.0
100–1000	12.0	5.5	5.0	2.0	1.0	15.0
>1000	15.0	7.0	6.0	2.5	1.4	20.0

NOTES: Even harmonics are limited to 25% of the odd harmonic limits above.

*All power generation equipment is limited to these values of current distortion, regardless of actual I_{sc}/I_L.

where:

I_{sc} = Maximum short-circuit current at PCC.
I_L = Maximum load current (fundamental frequency) at PCC.

For PCC's from 69 to 138 kV, the limits are 50% of the limits above. A case-by-case evaluation is required for PCC's of 138 kV and above.

Table 23
Harmonic Voltage Limits for Power Producers (Public Utilities or Co-Generators) (IEEE Std 519-1981 Revision (Pending))

	HARMONIC VOLTAGE DISTORTION IN % AT PCC		
	2.3–69 kV	69–138 kV	>138 kV
Maximum for individual harmonic	3.0	1.5	1.0
Total harmonic distortion (THD)	5.0	2.5	1.5

NOTE: High voltage systems can have up to 2% THD where the cause is an HVDC terminal that will attenuate by the time it is tapped for a user.

[6] IEEE Std 597-1983, IEEE Practices and Requirements for General Purpose Thyristor Drives (ANSI).

[7] NEMA-CP-1-1989, NEMA Standards Publication, CP-1-1988 Shunt Capacitors.

[8] ADAMSON, C. and HINGORANI, N. G. *High Voltage Direct Current Power Transmission*, 1960, Garraway LTD, London.

[9] ARRILLAGA, J., BRADLEY, D. A., and BODGER, P. S. *Power Systems Harmonics*, John Wiley & Sons, New York, 1985.

[10] BORREBACH, E. J. "The Effect of Arc Furnace Loads on Power Systems," *IEEE/IAS Conference Record*, paper no. MI-ThuAM3 1085, Oct. 7–10, 1974.

[11] DUFFEY, C. K. and STRATFORD, R. P. "Update of Harmonic Standard IEEE Std 519-1981, IEEE Recommended Practices and Requirements for Harmonic Control in Electric Power Systems," *IEEE/IAS/PCIC Conference Record*, paper no. PCIC-88-7, 1988.

[12] FREUND, A. "Double the Neutral and Derate the Transformer — Or Else!," *EC&M Magazine*, Dec. 1988.

[13] GONZALEZ, D. A. and McCALL, J. C. "Design of Filters to Reduce Harmonic Distortion in Industrial Power Systems," *IEEE/IAS Transactions*, vol. IA, no. 3, May/June 1987, pp. 504–511.

[14] IEEE Publication no. 84TH0115-6 DWR, "Sine-Wave Distortion in Power Systems and the Impact on Protective Relaying," IEEE Power Engineering Society.

[15] KIMBARK, E. *Direct Current Transmission*, John Wiley & Sons, Inc., 1971.

[16] LUDBROOK, A. "Harmonic Filters for Notch Reduction," *IEEE/IAS Transactions*, vol. 24, no. 5, Sept./Oct. 1988.

[17] RAMEY, D. G., SHANKLE, D. F., and McFADDEN, R. H. "Electrical Transients in Arc Furnace Installation," *IEEE/IAS/I&CPS Conference Record*, May 2, 1972.

[18] RICE, DAVID E. "Adjustable Speed Drive and Power Rectifier Harmonics — Their Effect on Power System Components," *IEEE/IAS Transactions*, Jan./Feb. 1986.

[19] SCHIEMAN, R. G. "Electrical Circuit Problems Caused by SCR Drives," IEEE Pulp and Paper Conference, June 17–19, 1970.

[20] SCHIEMAN, R. G. and SCHMIDT, W. C. "Power Line Pollution by 3-Phase Thyristor Motor Drives," IEEE/IAS Conference, 1976 Annual Meeting.

[21] SHIPP, D. D. "Harmonic Analysis and Suppression for Electrical Systems Supplying Static Power Converters and Other Non-Linear Loads," *IEEE/IAS Transactions*, Sept./Oct. 1979.

[22] STEEPER, D. E. and STRATFORD, R. P. "Reactive Compensation and Harmonic Suppression for Industrial Power Systems Using Thyristor Converters," *IEEE/IAS Conference Record*, paper no. SPC-THU-PMI 1087, Oct. 7–10, 1974.

[23] SUBJAK, J., Jr. and McQUILKIN, J. "Harmonics — Causes, Effects, Measurements, and Analysis — Update," IEEE/IAS Cement Industry Technical Conference, May 1989.

[24] TOTH, J., III and VELAZQUEZ, D. J. "Benefits of an Automated Data Acquisition System Applied to On-Line Harmonic Measurements," *IEEE/IAS Transactions*, Sept./Oct. 1986.

Chapter 11
Switching Transient Studies

11.1 Power System Switching Transients

11.1.1 Introduction. An electrical transient occurs on a power system each time a circuit change is made. This circuit change may be the result of a normal switching operation, such as breaker opening or closing, a light switch is turned on or off, etc. Bus transfer switching operations along with abnormal conditions such as inception and clearing of system faults also cause transients.

The phenomena involved in power system transients can be classified into two major categories:

(1) Interaction between magnetic and electrostatic energy stored in inductors and capacitors, respectively.

(2) Interaction between the mechanical energy stored in rotating machines and electrical energy stored in circuits.

Unlike the first category, which is comprised solely of electromagnetic transients, the second one deals with electromechanical transients and will not be discussed in this book. These kinds of transients are considered in the chapters on short-circuit and stability studies.

Most power system transients are oscillatory in nature and are characterized with transient periods of oscillation. The transient period is usually very short when compared with the power frequency. Yet, these transient periods are extremely important because at such times the circuit components and electrical equipment are subjected to the greatest stresses resulting from abnormal voltages and currents. While over-voltages may result in flashovers or insulation breakdown, over-currents may damage power equipment due to excessive heat dissipation and/or excessive electromagnetic forces. Flashovers usually cause temporary power outages due to tripping of the protective devices while insulation breakdown leads to permanent equipment damage.

For this reason, a clear understanding of the circuit during transient periods is essential in the formulation of steps required to minimize and sometimes prevent the damaging effects of switching transients.

11.1.2 Circuit Elements. Among the components found in power system circuits are the resistor, inductor, and capacitor. All circuit devices whether in a utility system, industrial plant, or commercial building, possess each of these components. The elements used to approximate these components are: resistance (R), inductance (L), and capacitance (C).

Ohm's law

$$v(\mathrm{t}) = Ri(\mathrm{t})$$ (Eq 151)

indicates that the current $i(\mathrm{t})$ and the voltage $v(\mathrm{t})$ are directly proportional to each other.

The remaining two elements, L and C, are characterized by their ability to store energy. The term "inductance" means the property of an element to store electromagnetic energy in a magnetic field. This energy storage is accomplished by establishing a magnetic flux within the ferromagnetic material. For a linear time-invariant inductor, the magnetic flux is specified as the product of the inductance and the terminal current. Thus

$$\phi(\mathrm{t}) = Li(\mathrm{t})$$ (Eq 152)

where $\phi(\mathrm{t})$ is the magnetic flux in webers, L is the inductance in henries, and $i(\mathrm{t})$ is the current in amperes. By Faraday's law, the voltage at the terminals of the inductor is the time derivative of the flux, namely

$$v(\mathrm{t}) = \frac{d\phi}{dt}$$ (Eq 153)

Combining this relationship with Eq 152 gives the voltage-current relation of a linear inductor, i.e.,

$$v(\mathrm{t}) = L\,\frac{di}{dt}$$ (Eq 154)

Finally, the term "capacitance" means the property of an element that stores electrostatic energy. In a typical capacitance element, this energy storage is accomplished by accumulating charges between two surfaces that are separated by an insulating material. The stored charge in a linear capacitor is related to the terminal voltage by

$$q(\mathrm{t}) = Cv(\mathrm{t})$$ (Eq 155)

where C is the capacitance in farads when the units of $q(\mathrm{t})$ and $v(\mathrm{t})$ are in coulombs and volts, respectively. Since the electrical current flowing through a particular point in a circuit is the time derivative of the electrical charge, Eq 155 can be differentiated with respect to time to yield a relationship between the terminal current and the terminal voltage. Thus

$$i(\mathrm{t}) = \frac{dq}{dt} = C\,\frac{dv}{dt}$$ (Eq 156)

Under steady-state conditions, the energy stored in the elements L and C oscillates between the inductances and capacitances in the circuit at the power frequency. When there is a sudden change in the circuit, such as a switching event, a fault, etc., a redistribution of energy takes place to accommodate the new system condition. This redistribution of currents and voltages cannot occur instantaneously for the following reasons:

(1) The electromagnetic energy stored in an inductor is $E=(1/2)LI^2$. For a constant inductance, the change in the magnetic energy requires a change in current. But the change in current in an inductor is opposed by an emf of magnitude $v(t)=Ldi/dt$. For the current to change instantaneously ($dt=0$), an infinite voltage is required. Since this is unrealizable in practice, the change in energy in an inductor requires a finite time period.

(2) The electrostatic energy stored in a capacitor is given by $E=(1/2)CV^2$ and the voltage-current relationship is given by $i(t)=Cdv/dt$. For a linear time-invariant capacitor, an instantaneous change in voltage ($dt=0$) requires an infinite current that cannot be achieved in practice. Therefore, the change in voltage in a capacitor also requires a finite time period.

These two concepts plus the recognition that the rate of energy generated must be equal to the sum of the rate of energy dissipated and the rate of energy stored (principle of energy conservation) is basic to the understanding and analysis of transients in power systems.

11.1.3 Analytical Techniques. The classical method of treating transients consists in setting up and solving the differential equation or equations, which must satisfy the system response at every instant. The equations describing the response of such systems can be formulated as linear time-invariant differential equations with constant coefficients. The solution of these equations consists of two parts: the homogeneous solution, which describes the transient response of the system and the particular solution, which describes the steady-state response of the system to the forcing function or stimulus.

As discussed and described in Chapter 3, the analytical solution of these linear differential equations is often obtained by the Laplace transform. This method does not require the evaluation of constants of integration and is, therefore, useful in the solution of complex circuits, where the classical method is quite cumbersome.

11.1.4 Transient Analysis Based on Laplace Transform. Although they do not represent the types of problems regularly encountered in power systems, the transient analysis of the simple RL and RC circuits described in Chapter 3 of this book are useful illustrative examples of how the Laplace method can be used for solving transient problems. Practical circuits, however, are far more complicated, so that, even after simplification, they often retain many circuit elements in series-parallel combination that will require several differential or integrodifferential equations to describe the behavior of the circuit. These equations must be solved simultaneously to evaluate the variables of interest. To do this efficiently, the Laplace transform method is used.

11.1.4.1 LC Transients. In this chapter, we will look at more general types of circuits that are described by higher order differential equations.

The first class of circuits that we will consider is the double-energy transient or LC circuit. In double-energy electric circuits, energy storage and release take place in the magnetic field of inductors and in the electric field of the capacitors. The interchange of these two forms of energy may, under certain conditions, produce oscillations. The theory of these oscillations is of great importance in electric power systems.

We will start with the circuit shown in Fig 120 in which the circuit elements are represented as transform impedances as described in Chapter 3.

Let us examine the response of the circuit to a step input of voltage due to the closing of the switch at $t=0$. We will assume that the capacitor is initially charged to $Vc(0^-)$ with the polarity as indicated. Kirchoff's voltage law around the circuit gives the following equation:

$$\frac{V}{s} = I(s)sL - LI(0^-) + \frac{I(s)}{sC} - \frac{Vc(0^-)}{s} \qquad \text{(Eq.157)}$$

Since there could be no current flowing in the circuit before the switch closing, the term $LI(0^-)$ can be ignored. Solving for the current in Eq 157 yields

$$I(s) = \frac{V + Vc(0^-)}{L} \left[\frac{1}{s^2 + \omega_o^2} \right] \qquad \text{(Eq 158)}$$

where ω_o^2 is the natural frequency of the circuit, namely $1/(LC)^{1/2}$. From the table of inverse Laplace transforms (see Chapter 3), the transient response in the time domain is

$$i(t) = \frac{V + Vc(0^-)}{Z_0} \left[\text{Sin}(\omega_o t) \right] \qquad \text{(Eq 159)}$$

where Z_0 is the surge impedance of the circuit defined by

$$Z_0 = \left[\frac{L}{C} \right]^{1/2} \qquad \text{(Eq 160)}$$

Clearly, the transient response indicates that the current oscillates sinusoidally with a frequency governed by the circuit parameters L and C only.

Fig 120
Double-Energy Network

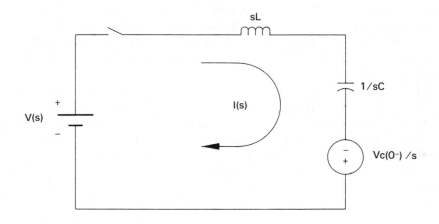

Another interesting feature about the time response for the current in the circuit is that the magnitude of the current is inversely proportional to the surge impedance of the circuit Z_0, which is also a function of L and C only. The importance of this parameter to the analysis of transient problems will be demonstrated later in the chapter.

Quite often, however, in power system analysis, we are interested in the voltage across the capacitor. Referring to Fig 120, the capacitor voltage is

$$Vc(s) = \frac{I(s)}{sC} - \frac{Vc(0^-)}{s} \qquad \text{(Eq 161)}$$

where $Vc(0^-)$ is the initial charge in the capacitor. Solving for $I(s)$ in the above equation, the result is

$$I(s) = sCVc(s) + CVc(0^-) \qquad \text{(Eq 162)}$$

Since the current $I(s)$ is common to both elements L and C, we can replace Eq 162 into Eq 158 to obtain the voltage across the capacitor. Thus, after rearranging the terms, we have:

$$Vc(s) = \frac{V\omega_o^2}{s(s^2 + \omega_o^2)} + \frac{sVc(0^-)}{(s^2 + \omega_o^2)} \qquad \text{(Eq 163)}$$

From the table of inverse Laplace transforms, the transiet response is found to be

$$Vc(t) = V\left\{1 - Cos(\omega_o t)\right\} + Vc(0^-)Cos(\omega_o t) \qquad \text{(Eq 164)}$$

The above equation is plotted in Fig 121 for various values of initial capacitor voltage. Examination of these curves indicates that, without damping, the capacitor voltage swings as far above the source voltage V as it starts below. In a real

Fig 121
Capacitor Voltage

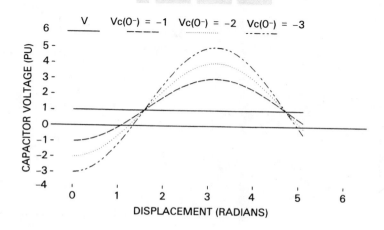

circuit, however, this will not be the case since circuit resistance will introduce losses and will damp the oscillations. Treatment of the effect of resistance that was not included in this example is presented next.

11.1.4.2 Damping. The examined LC circuit did not include losses, but nearly every practical electrical component has some losses arising primarily from circuit resistance.

As a first attempt to a switching transient problem, losses are usually neglected. In addition to greatly reducing the complication of the calculations, this procedure leads to solutions that give more conservative answers. Once the behavior of the circuit is understood, system losses can then be considered.

Consider the parallel RLC circuit flow in Fig 122, in which the circuit elements are represented by their transform admittances.

Many practical transient problems found in power systems can be simplified to this form for the purpose of analysis and still yield acceptable results. Assuming zero initial conditions and a constant current source, the equation describing the current in the parallel branches in the Laplace notation is

$$\frac{I(s)}{s} = V(s)G + sCV(s) + \frac{V(s)}{sL} \tag{Eq 165}$$

Solving for the voltage when $I(s)$ is a constant, we have

$$V(s) = \frac{I}{C}\left[\frac{1}{s^2 + s/RC + 1/LC}\right] \tag{Eq 166}$$

which can be written as

$$V(s) = \frac{I}{C}\left[\frac{1}{(s + r_1)(s + r_2)}\right] \tag{Eq 167}$$

where r_1 and r_2 are the roots of the characteristic equation defined as

$$r_1, r_2 = -\frac{1}{2RC} \pm \frac{1}{2}\left[(1/RC)^2 - (4/LC)\right]^{1/2} \tag{Eq 168}$$

Fig 122
Parallel RLC Circuit

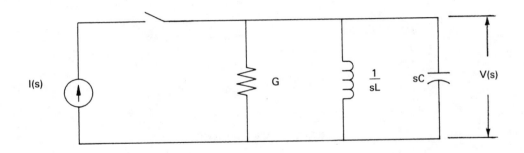

At the resonance frequency ω_o, the reactive power to the inductance and capacitance are equal, and the source has to supply only the true power required by the resistance in the circuit.

The ratio between the magnitude of the reactive power, P_R, of either the inductance or the capacitance at resonance, and the magnitude of true power, P_T of the circuit, is known as the quality factor or Q of the circuit. Therefore, for the parallel circuit, we have

$$Q_P = \frac{P_R}{P_T} = \frac{V^2 B}{V^2 G} = \frac{R}{X} \tag{Eq 169}$$

where B is the parallel susceptance of either the inductor or capacitor, G is the parallel conductance, and V is the voltage across the element.

Since in a parallel circuit the voltage is common to all components, substituting $\omega_o L$ for $1/B$ and R for $1/G$ in Eq 169 yields

$$Q_P = \frac{R}{\omega_o L} \tag{Eq 170}$$

Since the natural frequency of the circuit is

$$\omega_o = \frac{1}{\sqrt{LC}} \tag{Eq 171}$$

then, substituting Eq 171 into Eq 170 yields

$$Q_P = R(C/L)^{1/2} \tag{Eq 172}$$

Rearranging Eq 168 yields

$$r_1, r_2 = -\frac{1}{2RC} \pm \frac{1}{2RC} \left[1 - 4R^2 C/L\right]^{1/2} \tag{Eq 173}$$

and substituting Eq 172 into above expression, the result is

$$r_1, r_2 = -\frac{1}{2RC} \pm \frac{1}{2RC} \left[1 - 4Q_P^2\right]^{1/2} \tag{Eq 174}$$

Depending on the values of the circuit parameters, the quantity inside the brackets in Eq 174 may be positive, zero, or negative. For positive values, that is $4Q_P^2 < 1$, the roots are real, negative, and unequal. For this case, the inverse Laplace transform of Eq 167 is

$$v(t) = \frac{IR e^{-\alpha t}}{\beta_1} \left[e^{+\omega_D t} - e^{-\omega_D t}\right] \tag{Eq 175}$$

where α is the damping coefficient defined as $\alpha = (2RC)^{-1}$, $\beta_1 = (1 - 4Q_P^2)^{1/2}$, and ω_D is the damped natural angular frequency defined as $\omega_D = \alpha \beta_1$.

Substituting $2\text{Sinh}(\omega_D t)$ for the exponential function inside the brackets yields

$$v(t) = \frac{2IR e^{-\alpha t}}{\beta_1} \left[\text{Sinh}(\omega_D t)\right] \tag{Eq 176}$$

For the case where the quantity $4Q_P^2 = 1$, then the roots are equal, negative, and real. The solution of Eq 175 is

$$v(t) = \frac{1}{C} \, te^{-\alpha t} \qquad \text{(Eq 177)}$$

Finally, when the quantity inside the brackets in Eq 174 is less than zero, that is, $4Q_P^2 > 1$, the roots are complex and unequal. Therefore, the solution is

$$v(t) = \frac{2IRe^{-\alpha t}}{\beta_2} \, [\,\text{Sin}(\omega_D t)\,] \qquad \text{(Eq 178)}$$

where $\beta_2 = (4Q_P^2 - 1)^{1/2}$.

Let us now consider the series RLC circuit shown in Fig 123.

With $V(s)$ as a constant, the equation describing the current in the circuit in Laplace notation is

$$I(s) = \frac{V}{s} \left[\frac{1}{R + sL + 1/sC} \right] \qquad \text{(Eq 179)}$$

Rearranging the terms, we obtain for the current

$$I(s) = \frac{V}{L} \left[\frac{1}{s^2 + sR/L + 1/LC} \right] \qquad \text{(Eq 180)}$$

Fig 123
Series RLC Circuit

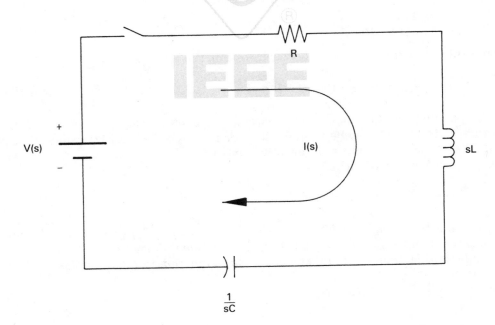

The expression inside the brackets is similar to the expression for the parallel RLC circuit shown in Eq 166. The only difference is the coefficient of s.

Rewriting, we have

$$I(s) = \frac{V}{L} \left[\frac{1}{(s + r_1)(s + r_2)} \right] \qquad \text{(Eq 181)}$$

where r_1 and r_2 are the roots of the characteristic equation defined as

$$r_1, r_2 = -\frac{R}{2L} \pm \frac{R}{2L} \left[1 - 4L/R^2C \right]^{1/2} \qquad \text{(Eq 182)}$$

Again, we define the quality factor of the series RLC circuit, Q_S, as the ratio of the magnitude of the reactive power of either the inductance or the capacitance at the resonant frequency to the magnitude of the true power in the circuit, we have

$$Q_S = \frac{P_R}{P_T} \qquad \text{(Eq 183)}$$

With $P_R = I^2X$ and $P_T = I^2R$, then

$$Q_S = \frac{I^2X}{I^2R} \qquad \text{(Eq 184)}$$

But since in a series circuit the current is common to all components, then

$$Q_S = \frac{\omega_o L}{R} \qquad \text{(Eq 185)}$$

where $\omega_o L = X$. Also, since $\omega_o = 1/\sqrt{LC}$, then Eq 185 can be written as

$$Q_S = \frac{(L/C)^{1/2}}{R} = \frac{Z_0}{R} \qquad \text{(Eq 186)}$$

Here we note that the above expression is a ratio between surge impedance and resistance, which is the reciprocal of the expression developed for the parallel RLC circuit described by Eq 172, that is,

$$Q_S = \frac{1}{Q_P} \qquad \text{(Eq 187)}$$

Substituting Eq 186 into Eq 182 results in

$$r_1, r_2 = -\frac{R}{2L} \pm \frac{R}{2L} \left[1 - 4Q_S^2 \right]^{1/2} \qquad \text{(Eq 188)}$$

The above equation has already been solved for the parallel RLC circuit. To obtain the expression as a function of time, we simply substitute Q_S for Q_P and (V/R) for (IR) in Eqs 176 and 178 and (V/L) for (I/C) in Eq 177. Thus, for the case where the quantity inside the brackets is less than 1, that is, $4Q_S^2 < 1$, then

$$i(t) = \frac{2Ve^{-\alpha t}}{R\beta_1} \left[\text{Sinh}(\omega_D t) \right] \qquad \text{(Eq 189)}$$

For the case where the quantity $4Q_S{}^2 = 1$, then the roots are equal, negative, and real. The solution of Eq 188 is

$$i(t) = \frac{V}{L} t e^{-\alpha t} \tag{Eq 190}$$

Finally, when the quantity $4Q_S{}^2 > 1$, the roots are complex and unequal. Therefore, the solution is

$$i(t) = \frac{2V e^{-\alpha t}}{R\beta_2} [\mathrm{Sin}(\omega_D t)] \tag{Eq 191}$$

where α is the damping coefficient defined as $\alpha = (2L/R)^{-1}$, $\beta_1 = (1 - 4Q_S{}^2)^{1/2}$, $\beta_2 = (4Q_S{}^2 - 1)^{1/2}$, and ω_D is the damped natural frequency defined as $\omega_D = \alpha\beta$.

11.1.5 Normalized Damping Curves. The response of the parallel and series RLC circuits to a step input of current or voltage, respectively, can be expressed as a family of normalized transient curves, which can then be used to estimate the response of simple switching transient circuits to a step input of either voltage or current. To do this, we proceed as follows:

(1) To per unitize the solutions, we use the undamped response of a parallel LC circuit as the base. Thus, for the voltage

$$v(t) = \frac{I}{\omega_o C} \mathrm{Sin}(\omega_o t) \tag{Eq 192}$$

and for the current in a series LC circuit

$$i(t) = \frac{V}{\omega_o L} \mathrm{Sin}(\omega_o t) \tag{Eq 193}$$

The maximum voltage or current occurs when $\omega_o t = \pi/2$,

$$v(t) = \frac{I}{\omega_o C} \text{ and } i(t) = \frac{V}{\omega_o L} \tag{Eq 194}$$

(2) Letting the quantity $\omega_o t$ be the displacement θ in radians or $t = (\theta/\omega_o)$, we can substitute the quantity (αt) in Eqs 176, 177, and 178 with the expression $(\theta/2Q_P)$. Using the same procedure, we can also replace the quantity $(\omega_D t)$ in the same equations with the expression $(\theta\beta/2Q_P)$.

(3) Finally, dividing Eqs 176, 177, and 178 by the right side of the expression for $v(t)$ in Eq 194, we obtain a set of normalized curves for the voltage on the parallel RLC circuit as a function of the dimensionless quantities Q_P and the displacement θ.

Thus, for $4Q_P^2 < 1$:

$$f(Q_P,\theta) = \frac{2Q_P e^{-(\theta/2Q_P)}}{\beta_1} [\mathrm{Sinh}(\theta\beta_1/2Q_P)] \tag{Eq 195}$$

$$f(Q_P,\theta) = \theta e^{-(\theta/2Q_P)} \tag{Eq 196}$$

$$f(Q_P, \theta) = \frac{2Q_P e^{-(\theta/2Q_P)}}{\beta_2} \ [\operatorname{Sin}(\theta \beta_2/2Q_P)]$$

(Eq 197)

where the β's are defined by Eqs 175 and 178.

Equations 195, 196, and 197 are plotted in Figs 124 and 125 for various values of Q_P. For series RLC circuits, we divide the same equations of step 3 above by the right side of the expression for $i(t)$ and substitute Q_S for Q_P.

11.1.6 Switching Transient Examples. In the previous paragraphs, we examined some simple circuits that can be used to model many switching problems in electrical power systems. Very often, practical switching transient problems can be reduced to either parallel or series RLC circuits for the purpose of evaluating the response of the network to a particular stimuli on a first trial basis. To gain familiarity with the normalized curves developed in the previous paragraph, we will now examine some typical switching problems in power systems.

Consider, for example, a 1000 kVA unloaded transformer which, when excited from the 13.8 kV side with its rated voltage of 13.8 kV, it takes a no-load current of 650 mA at a power factor of 10.4%. The test circuit shown in Fig 126 was then set up.

The battery voltage V and the resistance R are chosen such that, with the switch closed, the battery delivers 10 mA. For this example, we will assume a shunt capacitance of 2.8 nF per phase. We would like to know the voltage across the capacitance to ground, when the switch is suddenly opened and interrupts the flow of current.

Fig 124
Normalized Damping Curves
$$1 \le Q_P \le 10$$

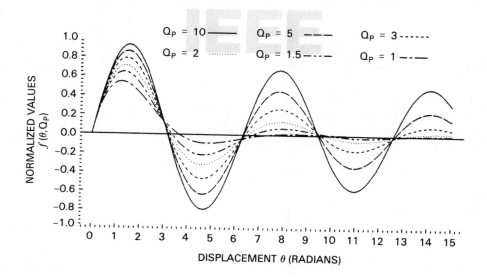

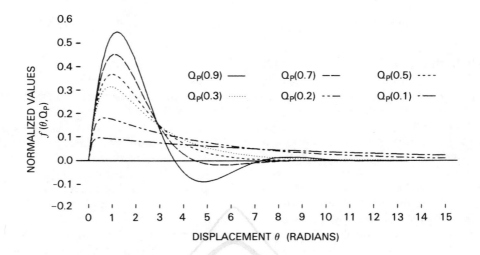

Fig 125
Normalized Damping Curves
$$0.1 \leq Q_P \leq 0.9$$

From the information provided, the no-load current of the transformer is

$I_{NL} = 0.067794 - j0.64644$ amperes

The magnetizing reactance X_M and inductance L_M per phase are, respectively,

$X_M = 13.8/(\sqrt{3} \times 0.64644) = 12.325$ kΩ

$L_M = X_M/2\pi f = 32.693$ H

With a shunt capacitance of 2.8 nF, the shunt capacitive reactance is

$X_{SH} = (2\pi f C_{SH})^{-1} = 947.351$ kΩ

The effects of the shunt capacitive reactance at the power frequency are negligible since $X_{SH} > X_M$. The resistance R_C is

$R_C = 13.8 \left(\sqrt{3} \times 0.06794\right)^{-1} = 117.272$ kΩ

Using delta-wye transformation, the circuit in Fig 126 can be redrawn as shown in Fig 127.

In Laplace transform notation, the equation describing the circuit at $t \geq 0^+$ is

$$0 = \frac{IL(0^-)}{Ls} + \frac{Vc(s)}{s} + sCVc(s) - CVc(0^-) + \frac{Vc(s)}{R}$$

where:

$L = 3L_M/2 = 49.035$ H
$C = 2C_{SH}/3 = 1.867$ nF
$R = 3R_C/2 = 175.908$ kΩ

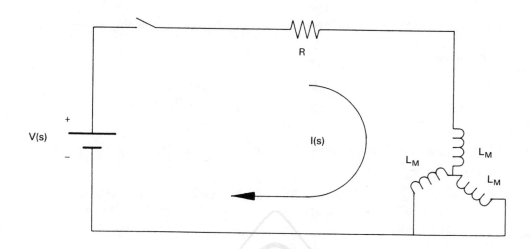

Fig 126
Test Setup of Unloaded Transformer

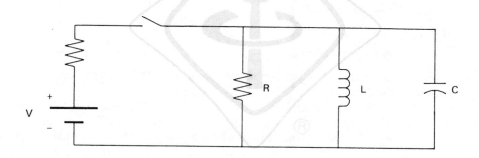

Fig 127
Equivalent RLC Circuit for Unloaded
Transformer

Since dc steady-state was obtained before the switch opened, the term $CVc\,(0^-)$ is zero and the initial current in the inductor at $t = 0^+$ is $IL\,(0^-) = 10$ mA. Solving for $Vc\,(s)$ and after rearranging the terms, we obtain

$$Vc\,(s) = \frac{-IL\,(0^-)}{C}\left[\frac{1}{s^2 + s/RC + 1/LC}\right]$$

The time-response solution for this expression has already been obtained and is shown in Eqs 176, 177, and 178, depending on the values of the circuit parameters. We could insert the values for R, L, and C into one of those equations and obtain the capacitor voltage for this problem. But this has also been done through the normalized curves shown in Figs 124 and 125.

Therefore, to obtain the answers from this particular problem, we proceed as follows:

(1) The surge impedance of the circuit is

$$Z_0 = (L/C)^{1/2} = 162.076 \text{ k}\Omega$$

(2) Without damping, the peak-transient voltage would be

$$V_{peak} = -I(0^-)Z_0 = -10 \text{ mA}(162076) = -1621\text{V}$$

(3) But since there is damping, the quality factor of the circuit Q_P is

$$Q_P = R/Z_0 = 1.09$$

(4) From the curves shown in Fig 124 and with $Q_P \approx 1.09$, the maximum per unit voltage is 0.57. Therefore, from step 2 above, the maximum voltage (peak) with damping is

$$V_{max} = -1621(0.57) = -923.97 \text{ V}$$

(5) The maximum peak occurs at approximately $\theta = 1.2$ radians. Since $\omega_o t = \theta$ and $\omega_o = (LC)^{-1/2}$, or

$$\omega_o = 3305 \text{ radians/second}$$

and

$$t = 1.2/3305 = 363.083 \ \mu\text{sec}$$

Let us now examine another practical case, dealing with capacitor bank switching, depicted in Fig 128.

Capacitor C_1 is rated 30 Mvar, three-phase at 13.8 kV. C_2 is initially uncharged and is rated 10 Mvar, three-phase at 13.8 kV. The inductance L of 35 μH is the inductance of the cable connecting capacitor C_2 to the bus. We would like to know the size of the resistor required to limit the inrush current to 5800 amperes during energization.

**Fig 128
Capacitor Bank Switching**

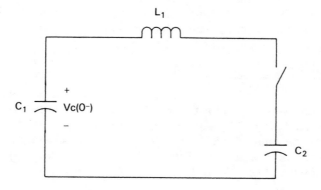

From the problem statement, we have

$$X_{C1} = 13.8^2/30 = 6.34 \ \Omega$$
$$C_1 = (2\pi f X_{C1})^{-1} = 417.86 \ \mu F$$
$$X_{C2} = 13.8^2/10 = 19.044 \ \Omega$$
$$C_2 = 139.287 \ \mu F$$

Assuming the worst case, that is, C_1 charged to peak system voltage or

$$V_{C1}(0^-) = 13.8\sqrt{2}/\sqrt{3} = 11.268 \text{ kV}$$

and with a surge impedance of

$$Z_0 = \left[L(C_1 + C_2)/C_1 C_2\right]^{1/2} = 0.579 \ \Omega$$

then, with no damping, the inrush current would be

$$I = V_{C1}(0^-)/Z_0 = 19.467 \text{ kA}$$

Redrawing the circuit of Fig 128 to show the necessary addition of a resistor to limit the inrush current yields the circuit of Fig 129. The problem requires that the inrush current should not exceed 5800 amperes. This represents a per unit value of

$$I_{PU} = 5800/19467 = 0.30$$

Referring to Fig 125, and remembering that we are dealing with a series circuit and must therefore replace Q_P with Q_S, we see that $Q_S = 0.30$ will reduce the current to 5800 amperes or less. Now, since

$$Q_S = Z_0/R = 0.30$$

then

$$R = Z_0/Q_S = 1.93 \ \Omega$$

Fig 129
Equivalent Circuit for Capacitor Switching

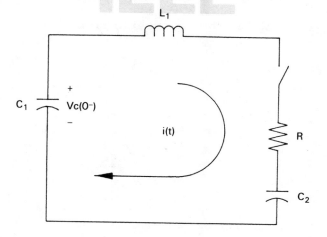

Therefore, to limit the inrush current to 5800 amperes a 1.93 ohm resistor must be placed in series with capacitor C_2.

11.1.7 Transient Recovery Voltage. Circuit breakers provide the mechanism to interrupt the short-circuit currents during a system fault. When the breaker contacts open, the fault current is not interrupted instantaneously. Instead, an electric arc forms between the breaker contacts, which is maintained as long as there is enough current flowing. Since the fault current varies sinusoidally at the power frequency, the arc will extinguish at the first current zero.

However, at the location of the arc, there are still hot, ionized gases and, if voltages develop across the contacts of the breaker, it is possible that the arc will reignite. Unfortunately, most power system circuits are predominantly inductive and, therefore, circuit interruption is a race between the increase of dielectric strength of the breaker or switch and the recovery voltage. The latter is essentially a characteristic of the circuit itself.

For inductive circuits, we know that the current lags the voltage by approximately 90°. Thus, when the current is zero, the voltage is at its maximum. This means that, immediately after interruption of the arc, a rapid buildup of voltage across the breaker contacts may cause the arc to reignite and reestablish the circuit. The rate by which the voltage across the breaker rises depends on the inductance and capacitance of the circuit.

The simplest form of single-phase circuit that can be useful to illustrate this phenomenon is that shown in Fig 130.

In the circuit, L is the inductance limiting the short-circuit current and C is the natural capacitance of the circuit in the vicinity of the circuit breaker. It may include capacitance to ground through bushings, current transformers, etc. The voltage source is assumed to vary sinusoidally and, since it is at its peak at the time the short-circuit current is interrupted, it can be expressed as

$$v(t) = V\sqrt{2}\,\mathrm{Cos}(\omega t) \qquad\qquad\text{(Eq 198)}$$

where ω is the power frequency in radians per second.

Fig 130
Simplified Diagram to Illustrate TRV

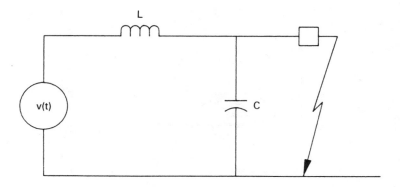

If the switch is opened, the flow of current is interrupted at first current zero and a voltage known as the transient recovery voltage (TRV) will appear across the breaker contacts, which is essentially the voltage across the capacitor. This voltage is zero during the fault but, when the circuit breaker clears the fault, the voltage across the contacts builds up to approximately twice the peak of the voltage at the power frequency.

The equivalent circuit of Fig 130 may be analyzed by means of the Laplace transform. The network equation in the s-domain for $t \geq 0^+$ is

$$V(s) = I(s)sL - LI(0^-) + \frac{I(s)}{sC} \qquad \text{(Eq 199)}$$

Solving the above equation for $I(s)$ with $I(0^-) = 0$ yields

$$I(s) = \frac{V(s)}{L} \left[\frac{s}{s^2 + \omega_o^2} \right] \qquad \text{(Eq 200)}$$

Substituting $sCVc(s)$ for $I(s)$, the result is

$$Vc(s) = V(s)\omega_o^2 \left[\frac{1}{s^2 + \omega_o^2} \right] \qquad \text{(Eq 201)}$$

The Laplace of the driving function described by Eq 198 is

$$V(s) = V\sqrt{2} \left[\frac{s}{s^2 + \omega^2} \right] \qquad \text{(Eq 202)}$$

Combining Eqs 201 and 202, the recovery voltage or the voltage across the capacitor is

$$Vc(s) = V\sqrt{2}\omega_o^2 \left[\frac{s}{(s^2 + \omega^2)(s^2 + \omega_o^2)} \right] \qquad \text{(Eq 203)}$$

From the table of Laplace transforms, the transient response is

$$Vc(t) = \frac{V\sqrt{2}}{[1 - (\omega/\omega_o)^2]} \left[\cos(\omega t) - \cos(\omega_o t) \right] \qquad \text{(Eq 204)}$$

The events before and after the fault are depicted in Fig 131.

Without damping as described by Eq 204, the recovery voltage reaches a maximum of $2\sqrt{2}V$ at half cycle of the natural frequency, after the switch is opened. This is true when the natural frequency is high as compared with the fundamental frequency and when losses are insignificant. Losses will reduce the maximum value of Vc, as shown in Fig 131. Upon interruption of the fault current by the circuit breaker, the source attempts to charge the capacitor voltage to the potential of the supply. As a matter of fact, without damping, the capacitor voltage will overshoot the supply voltage by the same amount as it started below. If the natural frequency of the circuit is high (L and C very small), the voltage across the breaker contacts will rise very rapidly. If this rate-of-rise exceeds the dielectric strength of the medium between the contacts, the breaker will not be able to sustain the voltage and reignition will occur.

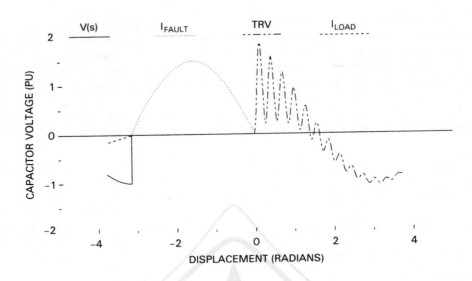

Fig 131
Transient Recovery Voltage

11.1.8 Summary. The material covered thus far is by no means an exhaustive discussion of electrical transients in power systems. The objective of the foregoing material was to provide the reader with the basic techniques required to perform simple switching transient analysis of very elementary circuits. We have seen that, even for simple series or parallel RLC circuits, the mathematical expressions are quite cumbersome and very difficult to solve analytically. It is evident that any slight increase in circuit complexity will result in expressions very difficult to handle and solve by conventional methods.

Typical industrial power distribution systems will involve many series and parallel circuit combinations with very complex relationships. To set down and solve analytically the equations representing such a system would be a formidable task. This is when solutions by computer methods are most appropriate. Two of the most common computer methods are analog and digital. The analog computer makes use of scaled-down components, i.e., resistors, inductors, and capacitors to model a particular system. The digital computer, on the other hand, utilizes computer programs (software packages) developed especially for the purpose of transient analysis.

11.2 Switching Transient Studies

11.2.1 Introduction. Unlike classical power system studies, i.e., short circuit, load flow, etc., switching transient studies are conducted quite infrequently in industrial power distribution systems. Capacitance switching in industrial and utility systems account for most of such investigations, to assist in the resolution of certain transient behavioral questions in conjunction with the application or failure of a particular piece of equipment.

Two basic approaches present themselves in the determination and prediction of switching transient duties in electrical equipment: direct transient measurements (to be discussed later in this chapter) and computer modeling. The latter can be divided into transient network analyzer (TNA) and digital computer modeling.

In 11.1, some useful insights regarding the physical aspects prevailing in a circuit during a transient period were obtained with a minimum of mathematical complications. In fact, experienced transient analysts use known circuit response patterns based on a few basic fundamentals, to assess the general transient behavior of a particular circuit and to judge the validity of more complex switching transient results. Indeed, simple configurations composed of linear circuit elements can be processed by hand as a first approximation. Beyond these relatively simple arrangements, the economics and effective determination of electrical power system transients, requires the utilization of TNA's or digital computer programs. These two approaches to the solution of complex switching transients in power systems are the subject on this section of the chapter. They will include excerpts from actual switching transient studies.

11.2.2 Switching Transient Study Objectives. The basic objectives of switching transient investigations are to identify and quantify transient duties that may arise in a system as a result of intentional or unintentional switching events, and to prescribe economical corrective measures, if necessary. The results of a switching transient study can affect the operating procedures as well as the equipment in the system. The following include some specific broad objectives, one or more of which are included in a given study:

(1) Identify the nature of transient duties that can occur for any realistic switching operation. This includes determining the magnitude, duration, and frequency of the oscillations.

(2) Determine if abnormal transient duties are likely to be imposed on equipment by the inception of faults and their subsequent removal.

(3) Recommend corrective measures to mitigate excessive transients, such as resistor insertion, tuning reactors, appropriate system grounding, and application of surge arresters and surge protective capacitors.

(4) Recommend alternative operating procedures, if necessary, to minimize transient duties.

(5) Document the study results on a case-by-case basis in readily understandable form for those responsible for system design and operation. Such documentation usually includes reproduction of waveshape displays and interpretation of at least the limiting cases.

11.2.3 Control of Switching Transients. The philosophy of mitigation and control of switching transients revolves around:

(1) Minimizing the number and severity of the switching events

(2) Limitation of the rate of exchange of energy that prevails among system elements during the transient period

(3) Extraction of energy

(4) Shifting the resonant points to avoid amplification of particular offensive frequencies

(5) Provision of energy reservoirs to contain released or trapped energy within safe limits of current and voltage

(6) Provision of discharge paths for high frequency currents resulting from switching

In practice, this is usually accomplished through one or more of the following methods:

(1) Temporary insertion of resistance between circuit elements, for example, the insertion of resistors in circuit breakers

(2) Inrush control reactors

(3) Damping resistors in filter and surge protective circuits

(4) Tuning reactors

(5) Surge capacitors

(6) Filters

(7) Surge arresters

(8) Necessary switching only, with properly maintained switching devices

(9) Proper switching sequences

11.2.4 Transient Network Analyzer (TNA)

11.2.4.1 Introduction. Through the years, a small number of TNA's have been assembled for the purpose of conducting transient analysis in power systems. A typical TNA is made of scaled-down power system component models, which are interconnected in such a way as to represent the actual system under study. The inductive, capacitive, and resistive characteristics of the various power system components are simulated by inductors, capacitors, and resistors in the analyzer. Usually, these have the same ohmic value as the actual components of the system frequency. The analyzer generally operates in the range of 20 to 100 V_{RMS} line-to-neutral, which represents 1.0 per unit voltage on the actual system.

The model approach of the TNA finds its virtue in the relative ease with which individual components can duplicate their actual power system counterparts as compared with the difficulty of mathematically representing combinations of nonlinear interconnected elements in an analytical solution. Furthermore, the switching operation that produces the transients is under the direct control of the operator and the circuit can easily be changed to show the effect of any parameter variation.

11.2.4.2 Modeling Techniques. Typical hardware used in a TNA to model the actual system components will be described now. It should be fully recognized that any specific set of components can be modeled in more than one way and considerable judgment on the part of the TNA staff is necessary to select the optimum model for a given situation. Also, it should be recognized that, while there is a great similarity among the components of the various TNA's in existence today, there are also unique hardware approaches to any given system. The following is a general description of some of the hardware models.

(1) Transmission lines — are modeled basically as a four-wire system, with three wires associated with the phase conductors and the fourth wire encompassing the effects of shield wire and earth return.

(2) Circuit breakers — consist of a number of independent mercury wetted sealed relays. The instant of both closing and opening of each individual

relay can be controlled by the operator or by the computer system. The model has the capability of simulating breaker actions like pre-striking, re-striking, and reignition.

(3) Shunt reactors — can be totally electronic or analog with variable saturation characteristics and losses.

(4) Transformers — are a critical part of the TNA. This is because many temporary over-voltages include ferroresonance of the line-transformer switch circuit, which makes the nonlinear magnetic representation of the transformer very critical. The model consists of an array of inductors configured and adjusted to represent the actual transformer.

(5) Arresters — both silicon carbide and metal oxide, can be modeled. The models for both types of arresters can be totally electronic.

(6) Secondary arc — available in some TNA facilities, is a model that can simulate a fault arc and its action after the system circuit breakers are cleared.

(7) Power sources — can be three-phase motor-generator sets or three-phase electronic frequency converters. The short-circuit impedance of these sources is such that they appear as an infinite bus on the impedance base of the analyzer.

(8) Synchronous machines — can be totally electronic or analog models and are used to study the effects of load rejection or other events that could be strongly affected by the action of the synchronous machine.

(9) Static var systems — include an electronic control circuit, a thyristor controlled reactor, and a fixed capacitor with harmonic filters. The control logic circuit monitors the three-phase voltages and currents and can be set to respond to either the voltage level, the power factor, or some combination of the two.

(10) Series capacitor protective devices — are used in conjunction with series compensated ac transmission lines. When a fault occurs, the voltage on the series capacitor rises to a high value unless it is bypassed by protective devices, such as power gap or metal oxide varistors. The TNA can represent both of these devices.

11.2.5 Capacitor Bank Switching — TNA Case Study

11.2.5.1 Introduction. The following describes a case study in which the customer planned to install a total of 75 Mvar of switched capacitor banks at a 115 kV substation. The plan included two separately switched 37.5 Mvar banks to compensate for var loading and voltage drop that would occur in the system when power was being imported from other sources. Since this was the customer's first experience with capacitor installation above 34.5 kV, the customer requested a TNA study to determine the transient over-voltages that could result during energization of the capacitor banks.

11.2.5.2 Study Objectives. The primary objective of this investigation was to determine if any switching surge over-voltage problems could be experienced when the proposed capacitor banks are added to the 115 kV substation. The system was modeled in the TNA to determine the switching surge voltages that can be generated during normal and abnormal switching conditions for the specific purpose of determining:

(1) The influence of the capacitor banks on the existing surge arresters and the application of protective devices at the buses where the capacitors will be located (see Fig 132)
(2) If preinsertion resistors are required for the capacitor bank breakers
(3) Current limiting reactor requirements for both capacitor banks
(4) If any magnification of the capacitor switching transient voltages at remote system locations is a possibility
(5) If the system is susceptible to resonance due to added capacitor banks

11.2.5.3 Study Results. The system investigated is depicted in Fig 132. The proposed capacitor banks are connected to the 115 kV bus through the circuit breakers A and B. The entire investigation consisted of 32 different system configurations and switching operations. Due to space limitations, however, only the results of two of these cases will be presented here, namely the energization of both capacitor banks.

The results of three-phase restrike, fault initiation, line energization, etc., which were part of the study, will also not be presented because of space limitations.

There are two or more output pages for each case investigated. The first page tabulates the system voltages recorded for the various system conditions as identified in the table's headings. They include both the temporary pre-switching, energizing, and post-switching voltages, as shown in Fig 133 (a), (b), and (c).

The succeeding pages display the statistical distribution curves of the transient voltages and/or the oscillograms of the voltage, current, and/or waveforms taken during the investigation, as shown in Figs 134, 135, and 136 for case 1.

The results of case 2, that is, energization of capacitor bank 2, are shown in Fig 137 (a), (b), and (c), through Fig 141.

11.2.5.4 Discussion. A maximum transient voltage of 1.38 pu and 1.64 pu was obtained during energization of each of two 37.5 Mvar banks at the 115 kV bus (66.5 kV line-to-ground), locations 4 and 5. The 1.64 pu (154 kV peak line-to-neutral) was recorded in case 1, where the first of the two banks was energized. In case 2, the 1.38 pu (130 kV peak line-to-neutral) transient voltage was recorded as a result of energizing the second bank. In each of the two cases, the system was operating under normal conditions and the capacitor switches did not include any closing resistors or current-limiting reactors. Transient voltages of these magnitudes are generally not considered to be of sufficient magnitude to cause a 96 kV rated conventional gapped-type arrester to operate or to cause any undue stress to either a 90 kV or 96 kV rated metal oxide type arrester, connected at the line-to-ground system voltage of 66.5 kV.

The switching operations of both capacitor banks did not cause any serious transient over-voltages at remote locations in the system and no resonant conditions were detected.

11.2.6 Electromagnetic Transients Program (EMTP)

11.2.6.1 Introduction. The electromagnetic transients program (EMTP) is a software package that can be used for single-phase and multiphase networks to calculate either steady-state phasor values or electromagnetic switching transients. The results can be either printed or plotted.

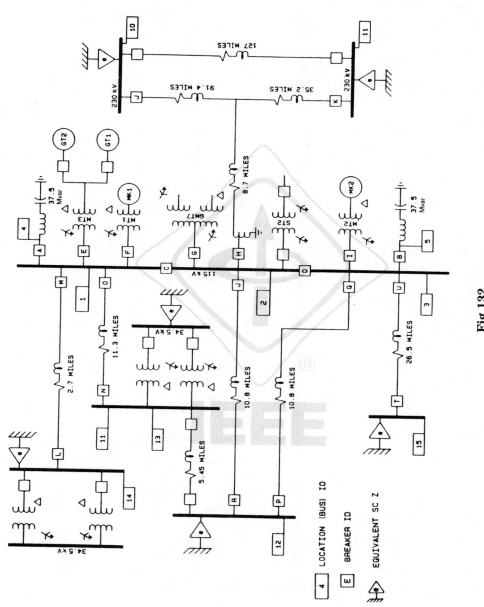

**Fig 132
System Single-Line Diagram**

TEMPORARY LINE-TO-NEUTRAL VOLTAGE
CREST PER UNIT QUANTITIES
PRE-SWITCH VOLTAGES: BREAKER A OPEN.
CAPACITOR BANK #2 OUT OF SERVICE

LOCATION	A	B	C
1	1.05	1.05	1.05
2	1.05	1.05	1.05
3	1.05	1.05	1.05
4	0.01	0.01	0.01
5	0.01	0.02	0.01
10	1.05	1.05	1.05
11	1.05	1.05	1.05
12	1.05	1.05	1.05
13	1.05	1.05	1.05
14	1.05	1.05	1.05
15	1.05	1.05	1.05

(a)
Pre-Switching Voltages – Case 1

TEMPORARY LINE-TO-NEUTRAL VOLTAGE
CREST PER UNIT QUANTITIES
(* - DENOTES NONSINUSOIDAL)
POST-SWITCH VOLTAGES: BREAKER A CLOSED.
CAPACITOR BANK #2 OUT OF SERVICE

LOCATION	A	B	C
1	1.05	1.06	1.05
2	1.05	1.05	1.05
3	1.05	1.05	1.05
4	1.05	1.05	1.05
5	0.01	0.01	0.01
10	1.06	1.06	1.06
11	1.06	1.06	1.06
12	1.05	1.05	1.05
13	1.06	1.05	1.05*
14	1.05	1.05	1.05
15	1.06	1.06	1.06

(b)
Energizing Voltages – Case 1

SWITCHING LINE-TO-NEUTRAL VOLTAGE
CREST PER UNIT QUANTITIES
ENERGIZING CAPACITOR BANK #1
CAPACITOR BANK #2 OUT OF SERVICE

LOCATION	A	B	C
1	1.40	1.10	1.64
2	1.40	1.10	1.64
3	1.40	1.10	1.64
4	1.40	1.10	1.64
5	0.01	0.01	0.01
10	1.16	1.12	1.34
11	1.14	1.12	1.33
12	1.36	1.08	1.69
13	1.52	1.10	1.80
14	1.36	1.07	1.60
15	1.66	1.23	1.86

(c)
Post-Switching Voltages – Case 1

Fig 133
System Voltages – Case 1

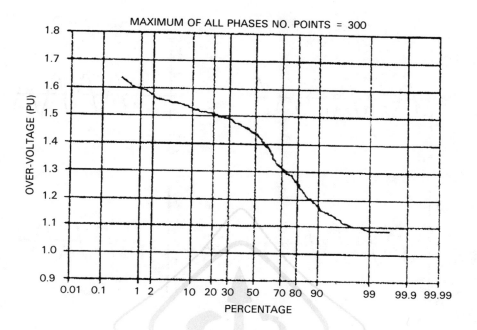

Fig 134
Probability Distribution – Case 1

11.2.6.2 Network and Device Representation. The program allows for arbitrary connection of the following elements:

(1) Lumped resistance, inductance, and capacitance
(2) Multiphase π circuits, when the elements R, L, and C become symmetric matrixes
(3) Transposed and untransposed distributed parameter transmission lines with wave propagation represented either as distortionless, or as lossy through lumped resistance approximation
(4) Nonlinear resistance with a single-valued, monotonically increasing characteristics
(5) Nonlinear inductance with single-valued, monotonically increasing characteristics
(6) Time varying resistance
(7) Switches with various switching criteria to simulate circuit breakers, spark gaps, diodes, and other network connection options
(8) Voltage and current sources representing standard mathematical functions, such as sinusoidals, surge functions, steps, ramps, etc. In addition, point-by-point sources as a function of time can be specified by the user.
(9) Single- and three-phase, two- or three-winding transformers

11.2.7 Capacitor Bank Switching—EMTP Case Study

11.2.7.1 Introduction. As part of a modernization program that included the addition of two paper machine drives to the existing system, it was determined that

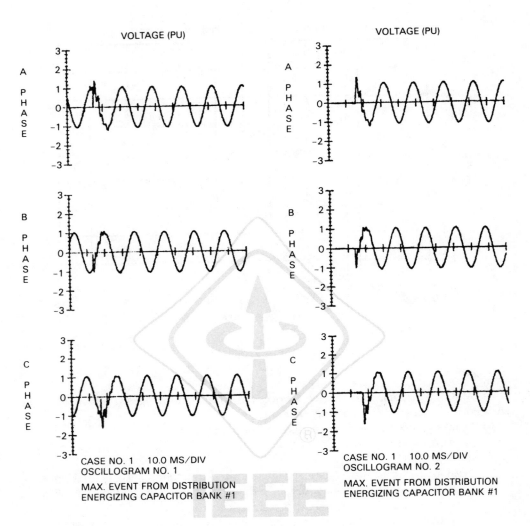

Fig 135
Voltage Oscillograms, Locations 1 and 4 – Case 1

a 10 Mvar capacitor bank was required to improve the plant power factor and the system voltage stabilization. Further analysis also indicated the need for a tuning reactor in series with the capacitor bank in order to minimize the effects of harmonic resonance problems.

Because of recent plant outages caused by what appeared to be normal switching operations, and because the proposed capacitor bank would require frequent switching to meet system voltage and power factor requirements, the customer requested that a switching transient investigation be conducted to determine the voltage and current waveforms associated with the switching of the proposed capacitor bank.

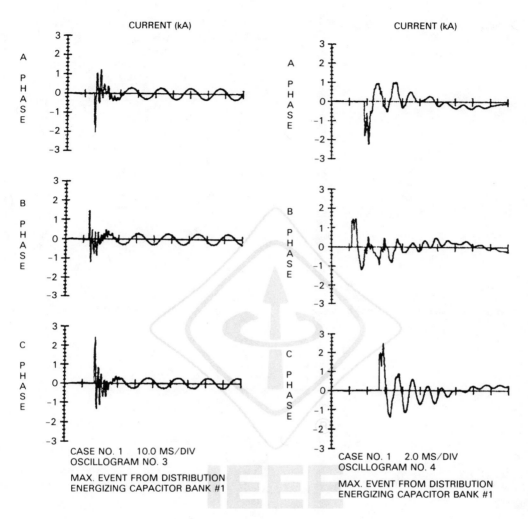

Fig 136
Current Oscillograms, Location 4 – Case 1

11.2.7.2 Study Objectives. The objectives of the study were to assist the customer in evaluating the effect of filter bank switching transients and in determining the solution to minimize these effects on the electrical system and equipment. Specifically, the study addressed the transient voltages and current waveforms during energization of the filter bank and the effects that these transients might have on the slip energy recovery drive and on the proposed dc drives for the new paper machines.

11.2.7.3 Circuit Model and Cases Studied. The study circuit and pertinent system parameters used in the study are depicted in Fig 142. Table 24 describes the cases studied.

TEMPORARY LINE-TO-NEUTRAL VOLTAGE
CREST PER UNIT QUANTITIES
(* - DENOTES NONSINUSOIDAL)
PRE-SWITCH VOLTAGES: BREAKER B OPEN.
CAPACITOR BANK #1 IN SERVICE

LOCATION	A	B	C
1	1.06	1.04	1.05
2	1.05	1.05	1.05
3	1.05	1.05	1.05
4	1.05	1.05	1.05
5	0.03	0.04	0.04
10	1.05	1.05	1.05
11	1.05	1.05	1.05
12	1.06	1.05	1.04
13	1.05	1.05	1.05*
14	1.05	1.05	1.05
15	1.05	1.06	1.04

(a)
Pre-Switching Voltages – Case 2

TEMPORARY LINE-TO-NEUTRAL VOLTAGE
CREST PER UNIT QUANTITIES
(* - DENOTES NONSINUSOIDAL)
PRE-SWITCH VOLTAGES: BREAKER B CLOSED.
CAPACITOR BANK #1 IN SERVICE

LOCATION	A	B	C
1	1.06	1.06	1.05
2	1.06	1.05	1.05
3	1.06	1.05	1.05
4	1.06	1.06	1.04
5	1.06	1.05	1.05
10	1.08	1.07	1.07
11	1.06	1.07	1.06
12	1.06	1.06	1.05
13	1.06	1.07	1.05*
14	1.06	1.06	1.05
15	1.07	1.07	1.06

(b)
Energizing Voltages – Case 2

SWITCHING LINE-TO-NEUTRAL VOLTAGE
CREST PER UNIT QUANTITIES
ENERGIZING CAPACITOR BANK #2
CAPACITOR BANK #1 IN SERVICE

LOCATION	A	B	C
1	1.10	1.12	1.38
2	1.10	1.12	1.38
3	1.10	1.12	1.38
4	1.10	1.12	1.38
5	1.10	1.12	1.38
10	1.13	1.12	1.18
11	1.11	1.10	1.16
12	1.09	1.09	1.40
13	1.14	1.11	1.42
14	1.10	1.11	1.38
15	1.14	1.12	1.39

(c)
Post-Switching Voltages – Case 2

Fig 137
System Voltages – Case 2

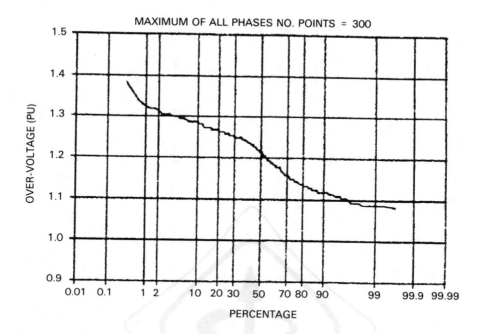

Fig 138
Probability Distribution – Case 2

Various system configurations were investigated to determine, by computer simulation, the transient voltage waveforms associated with the energization of the filter bank. The switching operation for the cases investigated (as listed in Table 24) was initiated when the phase-to-phase voltage (Va–b) at the STPT bus was at its peak ($t = 8.4$ msec). When resistor pre-insertion is used, it remains in the circuit for a period of three cycles and then is shorted out by a second switching operation, as depicted in the single-line diagram shown in Fig 142.

11.2.7.4 Study Results and Discussion. Selected transient voltage waveforms that were calculated and plotted by the program for cases 01, 08, and 09 are shown in Figs 143 through 148. Tables 25 and 26 summarize the results of all cases, for the worst peak over-voltages calculated, in kV and in pu, respectively.

The following are some observations:

(1) Removal of the 325 kvar capacitor bank on bus L135 (case 02) eliminates the high frequency oscillations (1000 Hz) experienced in case 01.

(2) The transients are substantially reduced when the 10 Mvar filter bank is divided into two 5 Mvar banks (cases 04 and 05).

(3) The transient decay is faster when pre-insertion resistors are used (cases 05, 08, and 09).

(4) The magnitude of the transient over-voltages is greatly reduced when the 10 Mvar bank is divided into two 5 Mvar banks and when resistor pre-insertion (5.2 Ω) is used during energization (cases 08 and 09).

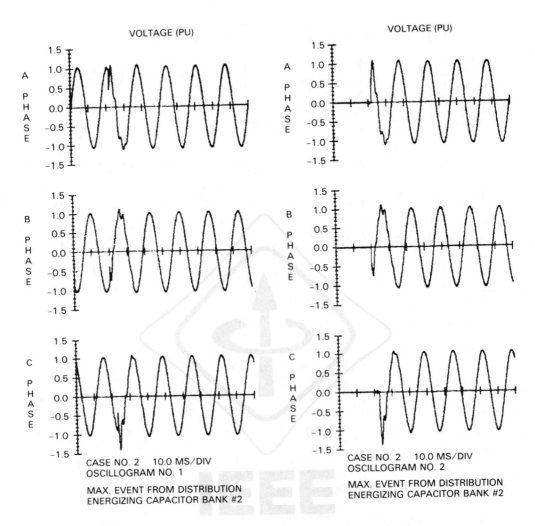

Fig 139
Voltage Oscillograms, Locations 3 and 5 – Case 2

11.2.8 Summary. Complete switching transient study documentation includes not only detailed individual case study results for transient responses associated with various arrangements and conditions surveyed, but also analysis, recommendations, and conclusions of the study. The study report also includes a complete listing of parameters (R, L, and C) of various system components, characteristics of protective devices, and a description of any unusual or special representations used in the study.

11.2.9 Switching Transient Problem Areas. Switching of predominantly reactive equipment represents the greatest potential for creating excessive transient

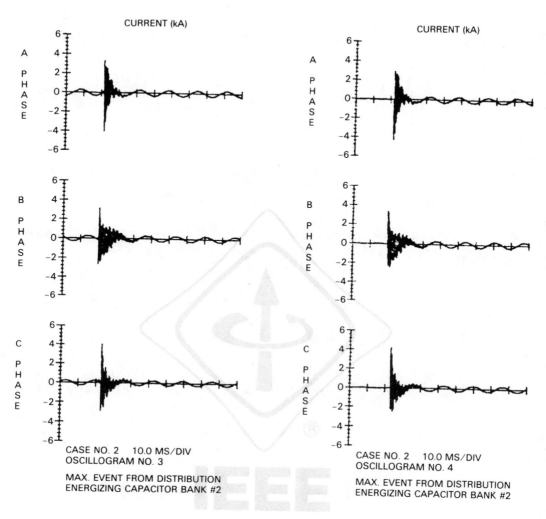

**Fig 140
Current Oscillograms, Locations 4 and 5 – Case 2**

duties. Principal offending situations are switching capacitor banks with inadequate or malfunctioning switching devices and energizing and deenergizing transformers with the same switching deficiencies. Capacitors can store, trap, and suddenly release relatively large quantities of energy. Similarly, highly inductive equipment possesses an energy storage capability that can also release large quantities of electromagnetic energy during a rapid current decay. Since transient voltages and currents arise in conjunction with energy redistribution following the switching event, the greater the energy storage in associated system elements, the greater the transient magnitudes become.

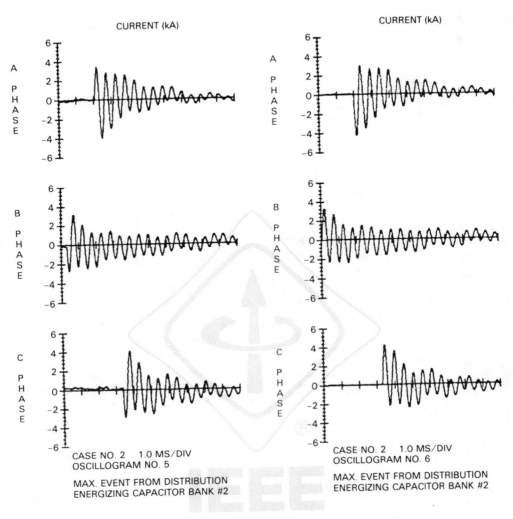

**Fig 141
Current Oscillograms, Locations 4 and 5 – Case 2
Expanded Time Scale**

   Generalized switching transient studies have provided many important criteria to enable system designers to avoid excessive transients in most common circumstances. The criteria for proper system grounding to avoid transient over-voltages during a ground fault are a prime example. There are also several not-too-common potential transient problem areas that are analyzed on an individual basis. The following is a partial list of transient-related problems, which can and have been analyzed through computer modeling:

   (1) Energizing and deenergizing transients in arc furnace installations
   (2) Ferroresonance transients

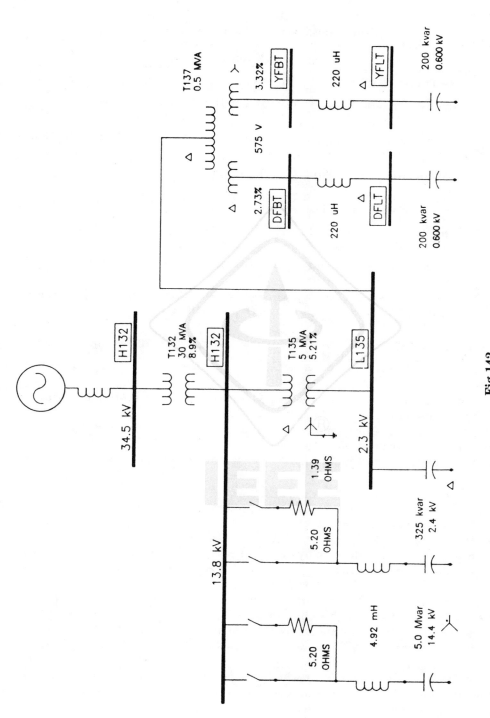

Fig 142

System Single-Line Diagram

Table 24
Filter Energization — Cases Studied

Case	Description
01	ENERGIZATION OF 10 Mvar FILTER BANK.
02	ENERGIZATION OF 10 Mvar FILTER BANK, WITH 325 kvar CAPACITOR BANK DISCONNECTED.
03	10 Mvar FILTER BANK DIVIDED INTO TWO 5 Mvar BANKS, ENERGIZATION OF FIRST 5 Mvar FILTER BANK.
04	ENERGIZATION OF SECOND 5 Mvar FILTER BANK.
05	ENERGIZATION OF 10 Mvar FILTER BANK WITH RESISTOR PRE-INSERTION; (R = 2.6 OHMS FOR 3 CYCLES)
06	ENERGIZATION OF 10 Mvar FILTER BANK WITH RESISTOR PRE-INSERTION; (R = 26 OHMS FOR 3 CYCLES)
07	ENERGIZATION OF 10 Mvar FILTER BANK WITH RESISTOR PRE-INSERTION; (R = 13 OHMS FOR 3 CYCLES)
08	10 Mvar FILTER BANK DIVIDED INTO TWO 5 Mvar BANKS; ENERGIZATION OF FIRST 5 Mvar FILTER BANK, WITH RESISTOR PRE-INSERTION; (R = 5.2 OHMS FOR 3 CYCLES)
09	ENERGIZATION OF SECOND 5 Mvar FILTER BANK, WITH RESISTOR PRE-INSERTION; (R = 5.2 OHMS FOR 3 CYCLES)

(3) Lightning and switching surge response of motors, generators, transformers, transmission towers, cables, etc.
(4) Lightning surges in complex station arrangements to determine optimum surge arrester location
(5) Propagation of switching surge through transformer and rotating machine windings
(6) Switching of capacitors
(7) Restrike phenomena during line dropping and capacitor deenergization
(8) Neutral instability and reversed phase rotation
(9) Energizing and reclosing transients on lines and cables
(10) Switching surge reduction by means of controlled closing of circuit breaker, resistor pre-insertion, etc.
(11) Statistical distribution of switching surges
(12) Transient recovery voltage on distribution and transmission systems

The studies presented in this chapter have been primarily based on closing or opening of electrical circuits, and, therefore, are not generally applicable to transfer switching in emergency and standby power systems. Here, significant transients often occur when inductive loads are rapidly transferred between two out-of-phase sources. Transients can also occur when four-pole transfer switches are both used for line and neutral switching as may be necessary for separately derived systems. Typical solutions for such problem areas often require transfer switch designs that

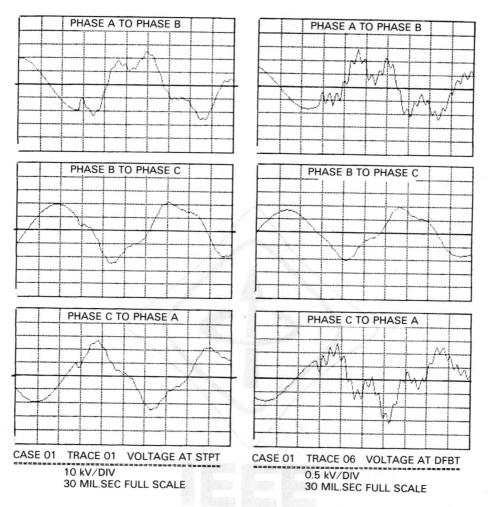

Fig 143
Voltage Oscillograms at STPT and DFBT Buses – Case 01

include in-phase monitors and overlapping neutral conductor switching. For further reading on this subject, see IEEE Std 446-1987 (ANSI) [2].[19]

The behavior of transformer and machine windings under transient conditions is also an area of great concern. Due to the complexities involved, it would be almost impossible to cover the subject in this chapter. For those interested, Chapter 11 of [B4] covers the subject in greater detail. References [4] and [5] also cover transients in transformers and rotating machines.

---

[19]The numbers in brackets correspond to those in the References at the end of this chapter. IEEE publications are available from the Institute of Electrical and Electronics Engineers, IEEE Service Center, 445 Hoes Lane, Piscataway, NJ 08855-1331.

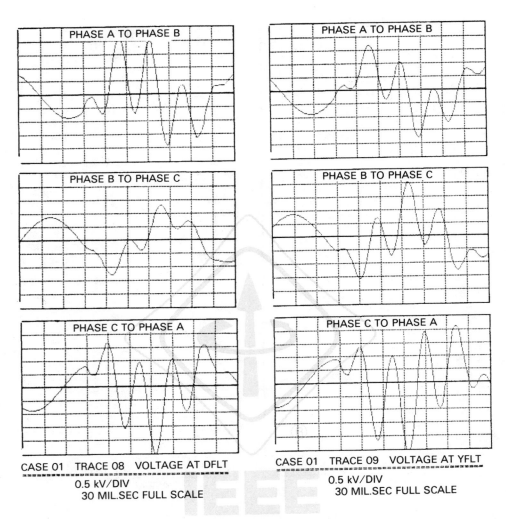

Fig 144
Voltage Oscillograms at DFLT and YFLT Buses – Case 01

11.3 Switching Transients — Field Measurements

11.3.1 Introduction. The choice of measurement equipment, auxiliary equipment selection, and techniques of setup and operation are in the domain of practiced measurement specialists. No attempt will be made here to delve into such matters in detail, except from the standpoint of conveying the depth of involvement entailed by switching transient measurements and from the standpoint of planning a measurement program to secure reliable transient information of sufficient scope for the intended purpose.

Field measurements seldom, if ever, include fault switching, and often recommended corrective measures are not in place to be used in the test program except

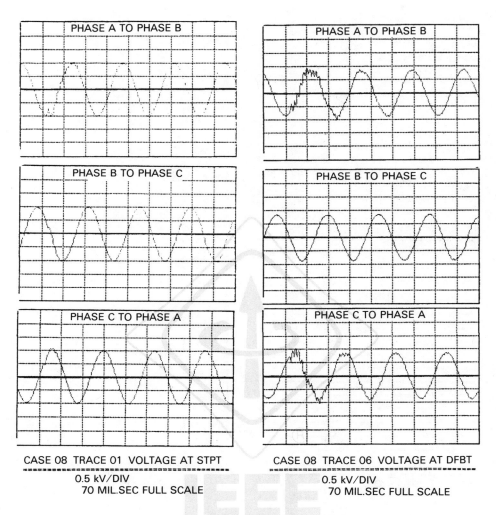

CASE 08 TRACE 01 VOLTAGE AT STPT
0.5 kV/DIV
70 MIL.SEC FULL SCALE

CASE 08 TRACE 06 VOLTAGE AT DFBT
0.5 kV/DIV
70 MIL.SEC FULL SCALE

Fig 145
Voltage Oscillograms at STPT and DFBT Buses – Case 08

on a followup basis. For systems still in the design stage or when fault switching is required, the transient response is usually obtained with the aid of a TNA or a digital computer program.

There are basically three types of transients to consider in field measurements:
(1) Transients under our control
(2) Recurrent transients
(3) Random transients

The first category includes those switching transients where we open or close a breaker or switch at our discretion and are usually able to anticipate the consequences. The second group covers the transients occurring regularly, for example,

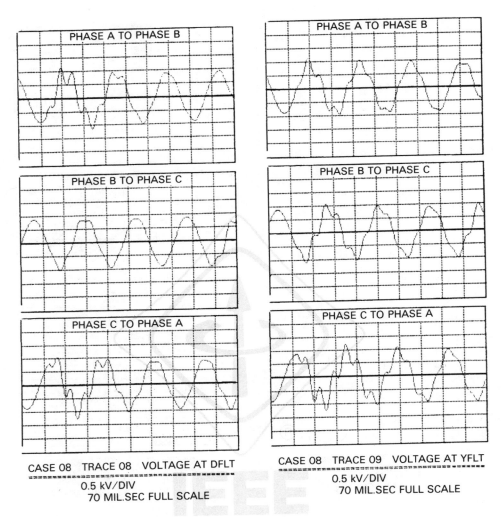

Fig 146
Voltage Oscillograms at DFLT and YFLT Buses – Case 08

commutation transients. Finally, random transients are those of usually unknown origin, generated by extraneous operations beyond our control. These may include inception and interruption of faults, lightning strikes, etc. To detect and/or record random transients, it is necessary to monitor the system continuously.

11.3.2 Signal Derivation. The ideal result of a transient measurement, or for that matter, any measurement at all, is to obtain a perfect replica of the transient voltage, current, or whatever as a function of time. The majority of the measurements aim at this ideal result. Quite often, the transient quantity to be measured is not obtained directly. However, we cannot take measurements in a system without disturbing it to some extent, since the transient quantity to be measured cannot be

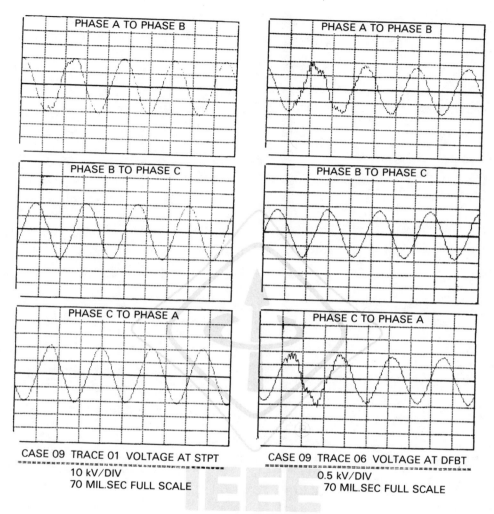

Fig 147
Voltage Oscillograms at STPT and DFLT Buses – Case 09

obtained directly. For example, if a shunt is used to measure current, we are, in reality, measuring the voltage across the shunt to which the current gives rise. This voltage is frequently assumed to be proportional to the current when, in fact, this is not always true with transient currents. Or, if the voltage to be measured is too great to be handled safely, appropriate attenuation must be used. In steady-state measurements, such errors are usually insignificant. But in transient measurements, this is more difficult to do.

Therefore, since switching transients involve natural frequencies of very wide range (of several orders of magnitude), signal sourcing must be by special current transformers, noninductive resistance dividers, noninductive shunts, or compen-

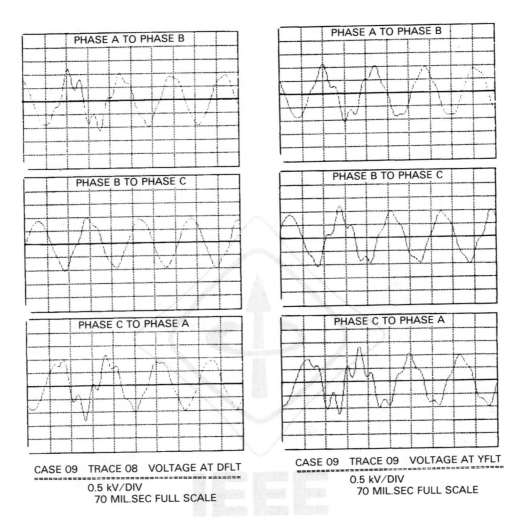

CASE 09   TRACE 08   VOLTAGE AT DFLT

0.5 kV/DIV
70 MIL.SEC FULL SCALE

CASE 09   TRACE 09   VOLTAGE AT YFLT

0.5 kV/DIV
70 MIL.SEC FULL SCALE

**Fig 148
Voltage Oscillograms at DFLT and YFLT Buses – Case 09**

sated capacitor dividers to minimize errors. While conventional CT's and PT's can be suitable for harmonic measurements, their frequency response is usually inadequate for switching transient measurements.

11.3.3 Signal Circuits, Terminations, and Grounding. Due to the very high currents with associated high magnetic flux concentrations, which may accompany transient phenomena in power systems, it is essential that signal circuitry be extremely well shielded and constructed to be as interference-free as possible.

Double-shielded low loss coaxial cable is satisfactory for this purpose. Additionally, it is essential that signal circuit terminations be made carefully with high quality

Table 25
Summary of Maximum Calculated Voltages in kV

Bus	Study Case								
	01	02	03	04	05	06	07	08	09
STPT	26.57	25.71	25.22	22.92	22.30	26.73	25.30	21.45	22.11
L135	5.70	4.79	4.83	4.29	5.02	5.43	4.71	4.40	3.95
L135G	3.70	2.54	2.78	2.07	2.62	2.96	2.74	2.25	2.20
L136	0.95	0.93	0.89	0.81	0.86	NA	0.87	0.80	0.98
L136G	0.53	0.49	0.47	0.47	0.45	NA	0.52	0.43	0.44
DFBT	1.62	1.43	1.33	1.14	1.15	NA	1.25	1.04	1.06
YFBT	1.53	1.40	1.39	1.17	0.96	NA	1.28	0.98	1.02
DFLT	2.57	2.52	2.37	1.75	1.52	NA	1.89	1.27	1.32
YFLT	2.70	2.64	2.74	1.73	1.51	NA	2.00	1.26	1.34

Table 26
Summary of Maximum Calculated Voltages in pu

Bus	Study Case								
	01	02	03	04	05	06	07	08	09
STPT	1.36	1.32	1.29	1.17	1.14	1.37	1.30	1.10	1.13
L135	1.75	1.47	1.49	1.32	1.54	1.67	1.45	1.35	1.21
L135G	1.63	1.35	1.48	1.10	1.39	1.58	1.46	1.20	1.17
L136	1.39	1.36	1.31	1.19	1.27	NA	1.28	1.18	1.44
L136G	1.36	1.25	1.20	1.20	1.14	NA	1.32	1.10	1.13
DFBT	1.99	1.76	1.63	1.40	1.41	NA	1.53	1.28	1.31
YFBT	1.88	1.72	1.71	1.43	1.18	NA	1.57	1.20	1.25
DFLT	3.16	3.10	2.92	2.15	1.87	NA	2.32	1.56	1.62
YFLT	3.33	3.25	3.37	2.13	1.86	NA	2.46	1.55	1.65

hardware and assure proper impedance match in order to avoid spurious reflections.

It is desirable that signal circuits and instruments be laboratory tested as an assembly before field measurements are undertaken. This testing should include the injection of a known steep wave into the input end of the signal circuit and comparison of this waveshape with that of the receiving instruments. Only after a close agreement between the two waveshapes is achieved should the assembly be approved for switching transient field measurements. These tests also aid overall calibration.

All the components of the measurement system should be grounded via a continuous conducting grounding system of lowest practical inductance to minimize internally induced voltages. The grounding system should be configured to avoid ground loops that can result in injection of noise. Where signal cables are unusually long, excessive voltages can become induced in their shields but industrial switching transient measurement systems have not, as yet, involved such cases.

11.3.4 Equipment for Measuring Transients. The complement of instruments used depends on the circumstances and purpose of the test program. Major items comprising the total complement of display and recording instrumentation for transient measurements are one or more of the following:

(1) One or more oscilloscopes, including a storage type scope with multichannel switching capability. When the presence of the highest speed transients, that is, those with front times of less than a microsecond is suspected, a high speed, single trace surge test oscilloscope with direct CRT (cathode ray tube) connections is sometimes used to record such transients with the least possible distortion.

(2) Multichannel magnetic light beam oscillograph with high input impedance amplifiers

(3) Peak-holding digital readout memory voltmeter (sometimes called "peak picker") that is manually reset

The storage scope should have at least a three-channel capability to permit simultaneous display of the three-phase voltage signals. An additional channel is desirable as a spare or to display another signal of interest. The single trace surge test oscilloscope with direct CRT input is capable of producing the highest resolution of specific signals of interest on faster sweep speeds, normally from 10 to 200 μs/divisions.

From the standpoint of conducting switching transient field measurements, one of the most difficult aspects is securing an acceptable and reliable triggering method for the storage scope when multichannel switching is used to record more than one signal. Considerable experimenting may be necessary in order to catch the transient activity, due to its short duration. One successful approach in some tests on systems with an open (nonshielded) bus has been to use a simple wire antenna connected to the external trigger of the scope. The antenna will sense airborne signals emanating from the power circuit bus in concert with the initiation of switching. Associated sweep speeds of 200 to 1000 μs/divisions have been found generally most useful for recording all but the very fastest switching transient voltages.

The magnetic oscillograph displays all voltages and signals being monitored. Current signals derived from special current transformers or shunts are fed directly to oscillograph galvanometers through the appropriate damping resistors. Voltage signals derived from capacitive dividers are isolated from low impedance galvanometers by high input impedance (megohm range) oscillograph amplifiers. Oscillograph records are virtually indispensable in the efficient interpretation of transient phenomena recorded through the orientation of the broad perspective that they provide. Such oscillograph traces also give records of slow speed switching transients, system oscillations, and harmonics.

Finally, the peak-holding voltmeter allows a valuable quantitative on-the-spot evaluation of the severity of transients produced by a particular switching operation. This permits a quick comparison between test runs and also can be left on for extended periods, sometimes unattended, to obtain a reading of the highest transmission occurring during that period.

The occurrence of most electrical transients is quite unpredictable. To detect and/or record random disturbances, it is necessary to monitor the circuit on a continuous basis. There are many instruments available in the market today for this purpose. Most of these instruments are computer based, that is, the information can be captured digitally and later retrieved for display or computer manipulation. These instruments vary in sophistication depending on the type and speed of transient measurements that are of interest.

11.4 Typical Circuit Parameters for Transient Studies

11.4.1 Introduction. Compared to conventional power system studies, switching transient analysis data requirements are often more detailed and specific. These requirements remain basically unchanged regardless of the basic analysis tools and aids that are employed, whether they are digital computer or transient network analyzers.

To determine the transient response of a circuit to a specific form of excitation, it is first necessary to reduce the network to its simplest form composed of R's, L's, and C's. After solving the circuit equations for the desired unknown, values must be assigned to the various circuit elements in order to determine the response of the circuit.

11.4.2 System and Equipment Data Requirements. The following generalized data listed encompass virtually all information areas required in an industrial power system switching transient study:

(1) Single-line diagram of the system showing all circuit elements and connection options

(2) Utility information, for each tie, at the connection point to the tie. This should include: (a) impedances R, X_L, X_C, both positive and zero sequence representing minimum and maximum short-circuit duty conditions; (b) maximum and minimum voltage limits; and (c) description of reclosing procedures and any contractual limitations, if any

(3) Individual power transformer data, such as, rating; connections; no-load tap voltages; LTC voltages, if any; no-load saturation data; magnetizing current; positive and zero sequence leakage impedances; and neutral grounding details

(4) Capacitor data for each bank, connections, neutral grounding details, description of switching device and tuning reactors, if any

(5) Impedances of feeder cables or lines, that is, R, X_L, and X_C (both positive and zero sequence)

(6) Information about other power system elements, such as, (a) surge arrester type, location and rating; (b) grounding resistors or reactors, rating and impedance of buffer reactors; (c) rating, subtransient and transient reactance of rotating machines, grounding details, etc.

(7) Operating modes and procedures

The material presented in the following pages is a compendium of parameter values, such as R_S, L_S, and C_S, for typical power system components that can be used in lieu of actual values. Most of the tabulated values were obtained from IEEE C37.011-1979 [1]. (This standard is in the process of being updated by the TRV Working Group of the IEEE Switchgear Committee.)

Table 27
Approximate Positive Sequence Reactance Values for Standard
25- to 60-Cycle, Self-Cooled, Two-Winding Power Transformers

Rated High Voltage	Rated Low Voltage	Percent Reactance					
		Fully Insulated		With Reduced Neutral Insulation		Reduced One Insulation Class with Reduced Neutral Insulation	
		Min.	Max.	Min.	Max.	Min.	Max.
2400 – 15 000	440 – 15 000	4.5	7.0				
15 001 – 25 000	440 – 15 000	5.5	8.0				
25 001 – 34 500	440 – 15 000	6.0	8.0				
	15 001 – 25 000	6.5	9.0				
34 501 – 46 000	440 – 15 000	6.5	9.0				
	25 001 – 34 500	7.0	10.0				
46 001 – 69 000	440 – 34 500	7.0	10.5				
	34 501 – 46 000	8.0	11.0				
69 001 – 92 000	440 – 34 500	7.5	10.5	7.0	10.0		
	34 501 – 69 000	8.5	12.5	8.0	11.5		
92 001 – 115 000	440 – 34 500	8.0	12.0	7.5	10.5	7.0	10.0
	34 501 – 69 000	9.0	14.0	8.5	12.5	8.0	11.5
	69 001 – 92 000	10.0	15.5	9.5	14.0	9.0	13.0
115 001 – 138 000	440 – 34 500	8.5	13.0	8.0	12.0	7.5	10.5
	34 501 – 69 000	9.5	15.0	9.0	14.0	8.5	12.0
	69 001 – 115 000	10.5	17.0	10.0	16.0	9.5	14.0
138 001 – 161 000	440 – 46 000	9.0	14.0	8.5	13.0	8.0	12.0
	46 001 – 92 000	10.5	16.0	9.5	15.0	9.0	14.0
	92 001 – 132 000	11.5	18.0	10.5	17.0	10.0	16.0
161 001 – 196 000	440 – 46 000	10.0	15.0	9.0	14.0	8.5	13.0
	46 001 – 92 000	11.5	17.0	10.5	16.0	9.5	15.0
	92 001 – 161 000	12.5	19.0	11.5	18.0	10.5	17.0
196 001 – 230 000	440 – 46 000	11.0	16.0	10.0	15.0	9.0	14.0
	46 001 – 92 000	12.5	18.0	11.5	17.0	10.5	16.0
	92 001 – 161 000	14.0	20.0	12.5	19.0	11.5	18.0

Table 28
Outdoor Bushing Capacitance to Ground

kV	A Rating	Range in pF	kV	A Rating	Range in pF
15.0	600	160–180	115.0	800	250–450
	1200	190–220		1200	250–430
				1600	250–430
23.0	400	200–450			
	600	280	138.0	800	250–450
	1200	190–450		1200	250–420
	2000	280–650		1600	250–460
	3000	370–560			
	4000	500–620	161.0	800	260–440
				1200	260–440
34.5	400	200–390		1600	260–440
	600	150–220			
	1200	170–390	196.0	800	350–550
	2000	240–360		1200	350–550
	3000	350–620		1600	350–550
			330.0	1600	530
46.0	400	180–330	345.0	820–2000	
	600	150–280		BIL: 1050	550
	1200	170–330		1175	500
	2000	200–330		1300	450
69.0	400	180–270			
	600	250			
	1200	160–290	500.0	800–2000	
	2000	210–320		BIL: 1425	500
				1550	500
				1675	520

11.5 References

[1] IEEE C37.011-1979 (Reaff. 1988), IEEE Application Guide for Transient Recovery Voltage for AC High-Voltage Circuit Breakers Rated on a Symmetrical Basis (ANSI).

[2] IEEE Std 446-1987, Recommended Practice for Emergency and Standby Power Systems for Industrial and Commercial Applications (ANSI).

[3] GILL, J. D. Transfer of Motor Loads Between Out-of-Phase Sources, *IEEE Transactions on Industry Applications*, vol. IA-15, no. 4, Jul./Aug. 1979, pp. 376–381.

[4] MAZUR, A., KERSZENBAUM, I., and FRANK, J. Maximum Insulation Stress Under Transient Voltages in the HV Barrel-Type Winding of Distribution and Power Transformers, *IEEE Transactions on Industry Applications*, vol. 24, no. 3, May/Jun. 1988.

Table 29
Synchronous Machine Constants

		Approximate Reactances in Percentage of Machine kVA Rating						Open-Circuit Time Constant T_{do} (sec)
		X_d	X'_d	X''_d	X_2	X_0	X_{eq}	
Turbine generators, two-pole	Average	115	15	9	11	3	75	4
	Range	95–145	12–21	7–14	9–16	1–8	60–100	3–7
Turbine generators, four-pole	Average	115	23	14	16	5	75	6
	Range	95–145	20–28	12–17	14–19	1.5–14	60–100	4–9
Waterwheel generator, without amortisseur windings	Average	100	35	30	50	7	65	5
	Range	60–145	20–45	17–40	30–65	4–25	40–100	2–10
Waterwheel generators, with amortisseur windings	Average	100	35	22	22	7	65	5
	Range	60–145	20–45	13–35	13–35	4–25	40–100	2–10
Synchronous condensers	Average	180	40	25	25	8	70	8
	Range	150–220	30–60	20–35	20–35	2–15	60–90	5–12
Salient-pole motors, high speed	Average	80	25	18	19	5	50	2.5
	Range	65–90	15–35	10–25	10–25•	2–15	40–60	1–4
Salient-pole motors, low speed	Average	110	50	35	35	7	70	2.5
	Range	80–150	40–70	25–45	25–45	4–27	50–100	1–4

NOTE: With the exception of X''_d for turbine generators and the column for X_{eq}, the above figures represent the approximate average and range of machine constants for both rated voltage and rated current conditions. The figures given for X''_d for turbine generators represent rated voltage values. The values given for X_{eq} are representative figures for machines of normal design operating at their full-load ratings.

Table 30
Instrument Transformer Capacitance
(Primary Winding to Ground and to Secondary
with Its Terminals Shorted and Grounded)

Insulation Class kV	Capacitance in pF		Current Transformers
	Potential Transformers		
	Line-to-Line	Line-to-Neutral	
15	260	— —	— —
25	250 – 440	270 – 800	180 – 260
34.5	310 – 440	270 – 900	160 – 250
46	350 – 430	300 – 970	170 – 220
69	360 – 440	340 –1300	170 – 260
115	470 – 520	480 – 610	210 – 320
138	490 – 550	530 – 660	— —
161	510 – 580	510 – 700	310 – 380
196	— —	580 – 820	330 – 390
230	600 – 680	600 – 810	350 – 420
345	— —	920	— —

[5] WHITE, E. L. Surge-Transference Characteristics of Generator/Transformer Installations, *Proceedings of the IEE*, vol. 116 (1969), pp. 575.

11.6 Bibliography

[B1] DOMMEL, H. W. *Electro-Magnetic Transients Program (EMTP) Manual*, Portland, Boneville Power Administration, 1984.

[B2] FICH, S. *Transient Analysis in Electrical Engineering*, New York, Prentice-Hall, Inc., 1971.

[B3] GOLDMAN, S. *Laplace Transform Theory and Electrical Transients*, New York, Dover Publications, 1966.

[B4] GREENWOOD, A. *Electrical Transients in Power Systems*, New York, John Wiley & Sons, Inc., 1971.

[B5] MILLER, R. *Algebraic Transient Analysis*, San Francisco, Reinhart Press, 1971.

[B6] PETERSON, H. A. *Transients in Power Systems*, New York, General Electric Company, 1951.

[B7] *Transient Network Analyzer Manual*, New York, General Electric Company, 1978.

[B8] WADHA, C. L. *Electrical Power Systems*, New Delhi: Wiley Eastner Limited, 1983.

Table 31
Generator Armature Capacitance to Ground

Generator Size in MVA	Total Three-Phase Winding Capacitance to Ground in μF
(1) Steam Turbine Driven:	
Conventionally Cooled	
(two-pole 3600 rev/min)	
15 up to 30	0.17 – 0.36
30 up to 50	0.22 – 0.44
50 up to 70	0.27 – 0.52
70 up to 225	0.34 – 0.87
225 up to 275	1.49
(four-pole 1800 rev/min)	
125 up to 225	0.04 – 1.41
Conductor Cooled: Gas	
(two-pole 3600 rev/min)	
100 up to 300	0.33 – 0.47
Conductor Cooled: Liquid	
(two-pole 3600 rev/min)	
190 up to 300	0.27 – 0.67
300 up to 850	0.49 – 0.68
(four-pole 1800 rev/min)	
250 up to 300	0.37 – 0.38
300 up to 850	0.71 – 0.94
Above 850	1.47
(2) Hydro Driven:	
720 to 360 rev/min	
10 to 30 MVA	0.26 – 0.53
225 to 85 rev/min	
25 to 100 MVA	0.90 – 1.64

NOTE: There is no direct correlation between generator MVA size, size limit, and capacitance limits. For instance, a 50 MVA generator may have an armature capacitance to ground anywhere from 0.27 to 0.52 μF, depending on machine design.

Table 32
Phase Bus Capacitance

A Rating	Isolated Phase Bus		Segregated Phas Bus
	15 kV Class 110 kV BIL pF/ft	23 kV Class 150 kV BIL pF/ft	15 kV Class 110 kV BIL pF/ft
1200	8.9 – 14.3	8.0 – 12.4	10.0
2000	10.2 – 14.3	9.0 – 12.4	10.0 – 10.2
2500	10.2 – 14.3	9.0 – 12.4	
3000	10.2 – 14.3	9.0 – 12.4	10.0 – 10.2
3500	10.2 – 14.3	9.0 – 12.4	
4000	14.0 – 14.3	12.4 – 13.5	10.0 – 12.6
4500	14.0 – 14.3	12.7 – 13.5	
5000	14.0 – 19.0	12.7 – 15.8	12.5 – 14.9
5500	14.0 – 19.0	12.7 – 15.8	
6000	14.0 – 19.0	13.5 – 15.8	15.0 – 17.1
6500	14.0 – 19.0	13.5 – 15.8	
7000	17.3 – 22.6	14.4 – 17.6	17.1
7500	17.3 – 22.6	14.4 – 17.6	
8000	21.7	17.6	—
9000	21.7	18.1	—
10 000	21.7	18.1	—
11 000	23.7	20.5	—
12 000	23.7	20.5	—

Table 33
Typical Values of Inductance
Between Capacitor Banks

Rated Maximum Voltage (kV)	Inductance per Phase of Bus (μH/ft)	Typical Inductance Between Banks (μH)
15.5 and below	0.214	10–20
38	0.238	15–39
48.3	0.256	20–40
72.5	0.256	25–50
121	0.261	35–70
145	0.261	40–80
169	0.268	60–120

Table 34
Typical Transmission Line Characteristics
of 69 to 230 kV

Nominal Voltage (kV) (Line-to-Line)	69	115	138	161	230
X_1 (Ω/mi)	0.783	0.759	0.771	0.771	0.785
R_1 (Ω/mi)	0.340	0.224	0.194	0.137	0.107
$|Z_1|$ (Ω/mi)	0.854	0.792	0.795	0.783	0.792
X_1/R_1	2.30	3.40	3.98	5.64	7.36
X_0	2.37	2.30	2.48	2.22	2.35
R_0	1.22	0.755	0.586	0.591	0.576
$|Z_0|$	2.67	2.42	2.55	2.30	2.42
X_0/R_0	1.95	3.05	4.23	3.76	4.08
X_0/X_1	3.03	3.03	03.22	2.88	2.99
X_1 ($1/B_1$) (MΩ/mi)	0.184	0.178	0.183	0.174	0.177
B_1 (μMho/mi)	5.43	5.63	5.46	5.73	5.63
B_0 (μMho/mi)	3.38	3.09	3.41	3.63	3.33
Surge Impedance Pos. Seq. (ohms)	397.	375.	381.	370.	375.
Surge Impedance Zero Seq. (ohms)	889.	885.	865.	795.	852.
Surge Impedance Loading (SIL) (MVA)	0.216	35.3	49.9	70.1	141.
Charging Current (A/mi)	0.216	0.373	0.435	0.533	0.748
Charging MVA (MVA/mi)	0.0258	0.0774	0.104	0.149	0.298
Line Current at SIL (A)	100.	177.	209.	252.	354.
I^2R Loss at SIL (kW/mi)	3.43	7.00	8.45	8.60	13.4
Equiv. Spacing (ft)	13.8	13.8	15.1	19.8	24.9

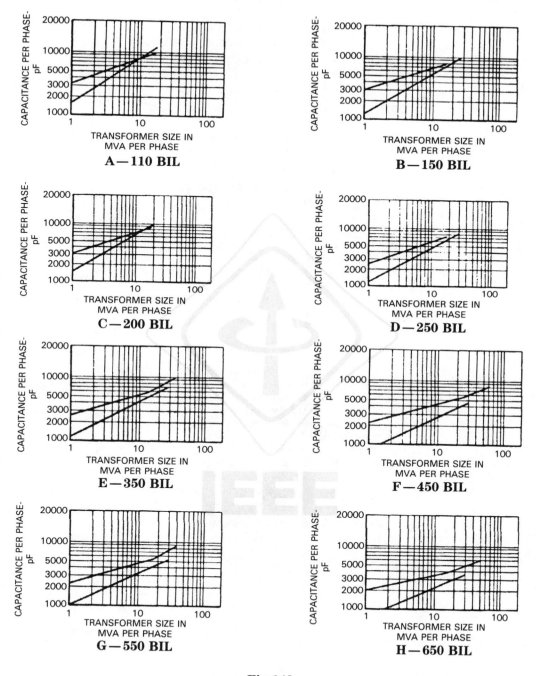

Fig 149
Typical Values of Transformer Winding Capacitance to Ground

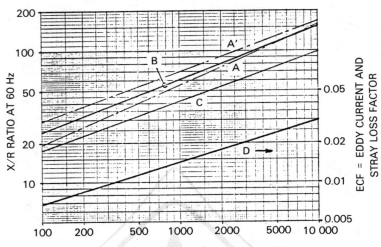

APPROXIMATE X/R RATIO AND RESISTANCE
OF CASE-IN-CONCRETE REACTORS

REACTOR THREE-PHASE SELF kVA (K_3)

$3 \times$ SELF kVA OF SINGLE-PHASE REACTOR $= 3 \times I^2X \times 10^{-3} = K3$

A — Nominal X/R ratio for aluminum conductor reactor $= 2.31K_3^{0.461}$
A' — "High Q" X/R ratio for aluminum conductor reactor $= 5.09K_3^{0.379}$
B — Maximum X/R ratio for copper conductor reactor $= 3.88K_3^{0.398}$
C — Nominal 60 Hz X/R ratio for tuning reactor $= 3.07K_3^{0.377}$
D — Eddy current and stray loss factor for tuning reactor
$= (1.16 \times 10^{-3}) (K_3^{0.353})$:
R_n = Resistance at fn (any frequency) $= R_{60}(1+n^2ECF)/(1+ECF)$,
where R_{60} = 60 Hz resistance and $n = fn/60$

Fig 150
Typical X/R Ratio and Resistance of Reactors

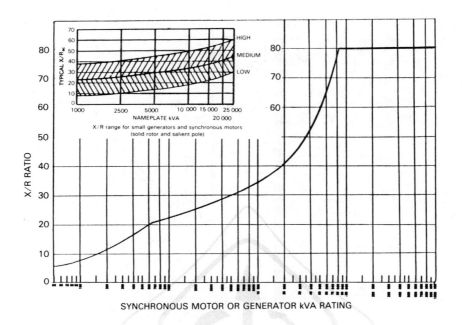

Fig 151
Typical X/R Ratio of Generators

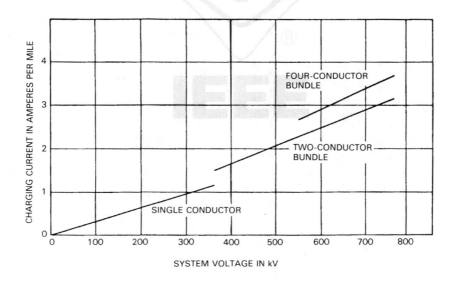

Fig 152
Typical Charging Current for Cable

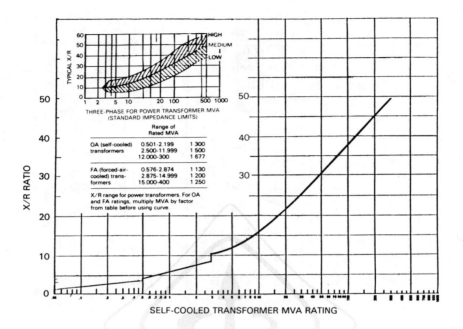

Fig 153
Typical X/R Ratio of Transformers

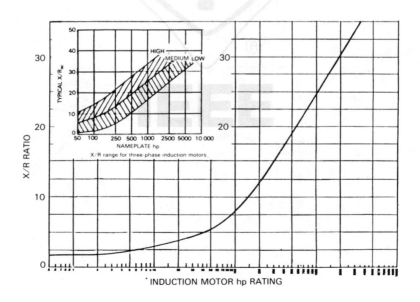

Fig 154
Typical X/R Ratio of Induction Motors

Chapter 12
Reliability Studies

12.1 Introduction. An important aspect of power system design involves consideration of service reliability requirements of loads to be supplied and service reliability provided by any proposed system. System reliability assessment and evaluation methods based on probability theory allow the reliability of a proposed system to be assessed quantitatively. Such methods permit consistent, defensible, and unbiased assessments of system reliability that are not otherwise defensible, and that are not otherwise possible.

Quantitative reliability evaluation methods permit reliability indexes for any electric power system computed from knowledge of the reliability performance of the constituent components of the system. Thus, alternative system designs can be studied to evaluate the impact on service reliability and cost of changes in component reliability, system configuration, protection and switching scheme, or system operating policy including maintenance practice. A detailed treatment of reliability evaluation methods is given in IEEE Std 493-1980 [1].[20]

12.2 Definitions. The definitions presented here provide much of the required nomenclature for discussions of power system reliability.

availability. A term that applies either to the performance of individual components or to a system. Availability is the long-term average fraction of time that a component or system is in service satisfactorily performing its intended function. An alternative and equivalent definition for availability is the steady-state probability that a component or system is in service.

component. A piece of equipment, a line or circuit, or a section of a line or circuit, or a group of items that is viewed as an entity for purposes of reliability evaluation.

expected interruption duration. The expected, or average, duration of a single load interruption event.

---

[20]The numbers in brackets correspond to those in the References at the end of this chapter. IEEE publications are available from the Institute of Electrical and Electronics Engineers, IEEE Service Center, 445 Hoes Lane, Piscataway, NJ 08855-1331.

exposure time. The time during which a component is performing its intended function and is subject to failure.

failure. Any trouble with a power system component that causes any of the following to occur:
  (1) Partial or complete plant shutdown, or below-standard plant operation
  (2) Unacceptable performance of user's equipment
  (3) Operation of the electrical protective relaying or emergency operation of the plant electrical system
  (4) Deenergization of any electric circuit or equipment
A failure on a public utility supply system can cause the user to have either of the following:
  (1) A power interruption or loss of service
  (2) A deviation from normal voltage or frequency of sufficient magnitude or duration
A failure on an in-plant component causes a forced outage of the component, that is, the component is unable to perform its intended function until repaired or replaced. The terms *failure* and *forced outage* are often used synonymously.

failure rate (forced outage rate). The mean number of failures per unit of exposure time for a component. Usually, *exposure time* is expressed in years and *failure rate* is given in terms of failures per year.

forced unavailability. The long-term average fraction of time that a component or system is out of service as a result of failures.

interruption. The loss of electric power supply to one or more loads.

interruption frequency. The expected average number of power interruptions to a load per unit time, usually expressed as interruptions per year.

outage. The state of a component or system when it is not available to properly perform its intended function.

repair time. The clock time from the time of component failure to the time when the component is restored to service, either by repair of the failed component or by substitution of a spare component for the failed component. It is not the time required to restore service to a load by putting alternate circuits into operation. It includes time for diagnosing the trouble, locating the failed component, waiting for parts, repairing or replacing, testing, and restoring the component to service. The terms *repair time* and *forced outage duration* can be used synonymously.

scheduled outage. An outage that results when a component is deliberately taken out of service at a selected time, usually for purposes of construction, maintenance, or repair.

scheduled outage duration. The time period from the initiation of a scheduled outage until construction, preventive maintenance, or repair work is completed and the affected component is made available to perform its intended function.

scheduled outage rate. The mean number of scheduled outages per unit of exposure time for a component.

switching time. The period from the time a switching operation is required because of a component failure until that switching operation is completed. Switching operations include: throwover to an alternate circuit, opening or closing a sectionalizing switch or circuit breaker, re-closing a circuit breaker following a trip-out from a temporary fault, etc.

system. A group of components connected or associated in a fixed configuration to perform a specified function of distributing power.

unavailability. The long-term average fraction of time that a component or system is out of service due to failures or scheduled outages. An alternative definition is the steady-state probability that a component or system is out of service. Mathematically, unavailability = (1 - availability).

12.3 System Reliability Indexes. The two basic system reliability indexes that have proven most useful and meaningful in power distribution sytstem design are load interruption frequency and expected duration of load interruption events. These indexes can be readily computed using the methods in IEEE Std 493-1980 [1]. The two basic indexes of interruption frequency and expected interruption duration can be used to compute other indexes that are also useful:

(1) Total expected average interruption time per year, or other time period
(2) System availability or unavailability as measured at the load supply point in question
(3) Expected energy demanded, but unsupplied, per year

Note that the disruptive effect of power interruptions is often nonlinearly related to the duration of the interruption. Thus, it is often desirable to compute not only an overall interruption frequency but also frequencies of interruptions categorized by the appropriate durations.

12.4 Data Needed for System Reliability Evaluations. Data needed for quantitative evaluation of system reliability depends to some extent on the nature of the system being studied and the detail of the study. In general, however, data on the performance of individual components together with the times required to perform various switching operations are required.

System component data generally required are summarized as follows:

(1) Failure rates (forced outage rates) associated with different modes of component failure
(2) Expected average time to repair or replace failed component
(3) Scheduled maintenance outage rate of component
(4) Expected average duration of a scheduled outage event

If possible, component data should be based on historical performance of components in the same environment as those in the proposed system being studied. The reliability surveys conducted by the Power Systems Reliability Subcommittee provide a source of component data when such specific data is not available.

Switching time data needed includes:

(1) Expected times to open and close a circuit breaker
(2) Expected times to open and close a disconnect or throwover switch
(3) Expected time to replace a fuse link
(4) Expected times to perform such emergency operations as cutting in clear, installing jumpers, etc.

Switching times should be estimated for the system being studied based on experience, engineering judgment, and anticipated operating practice.

12.5 Method for System Reliability Evaluation. The general method for system reliability evaluation, which is recommended, has evolved over a number of years. The method is well suited to the study and analysis of electric power distribution systems as found in industrial plants and commercial buildings. The method is systematic and straightforward and lends itself to either hand or computer computation. An important feature of the method is that system weak points can be readily identified, both numerically and nonnumerically, thereby focusing design attention on those sections of the system that contribute most to service unreliability.

The procedure for system reliability evaluation is outlined as follows:

(1) Assess the service reliability requirements of the loads and processes supplied and determine appropriate service interruption definition or definitions
(2) Perform a failure modes and effects analysis (FMEA) identifying and listing those component failures and combinations of component failures that result in service interruptions and constitute minimal cut-sets of the system
(3) Compute interruption frequency contribution, expected interruption duration, and the probability of each of the minimal cut-sets of (2)
(4) Combine results of (3) to produce system reliability indexes

The above steps are discussed in more detail in later sections.

12.5.1 Service Interruption Definition. The first step in any electric power system reliability study should be a careful assessment of the power supply quality and continuity required by the loads that are served. This assessment should be summarized and expressed in a service interruption definition used in the succeeding steps of the reliability evaluation procedure. The interruption definition specifies the reduced voltage level (voltage dip) together with the minimum duration of such reduced voltage period that results in substantial degradation or complete loss of function of the load or process being served. Frequency reliability studies are conducted on a *continuity* basis in which case interruption definitions reduce to a minimum duration specification with voltage assumed to be zero during the interruption.

12.5.2 Failure Modes and Effects Analysis. The FMEA for power distribution systems amount to the determination and listing of those component outage events or combinations of component outages that result in an interruption of service at the load point being studied according to the interruption definition adopted. This analysis must be made considering the different types and models of outages that components can exhibit and the reaction of the system's protection scheme to these events. Component outages are categorized as:

(1) Forced outages or failures
(2) Scheduled or maintenance outages
(3) Overload outages

Forced outages or failures are either permanent forced outages or transient forced outages. Permanent forced outages require repair or replacement of the failed component before it can be restored to service; transient forced outages imply no permanent damage to the component, thus permitting its restoration to service by a simple re-closing or re-fusing operation. Additionally, component failures can be categorized by physical mode or type of failure. This type of failure categorization is important for circuit breakers and other switching devices where the following failure modes are possible:

(1) Faulted, must be cleared by backup devices
(2) Fails to trip when required
(3) Trips falsely
(4) Fails to re-close when required

Each will produce a varying impact on system performance.

The primary result of the FMEA as far as quantitative reliability evaluation is concerned is the list of minimal cut-sets it produces. A minimal cut-set is defined to be a set of components which, if removed from the system, results in loss of continuity to the load point being investigated and does not contain as a subset any set of components that is itself a cut-set of the system. In the present context, the components in a cut-set are just those components whose overlapping outage results in an interruption according to the interruption definition adopted.

An important nonquantitative benefit of FMEA is the thorough and systematic thought process and investigation it requires. Often weak points in system design are identified before any quantitative reliability indexes are computed. Thus, the FMEA is a useful reliability design tool even in the absence of the data needed for quantitative evaluation.

12.5.3 Computation of Quantitative Reliability Indexes. Computation of reliability indexes can proceed once the minimal cut-sets of the system have been found. The first step is to compute the frequency, expected duration, and expected down-time per year of each minimal cut-set. Note that expected down-time per year is the product of the frequency expressed in terms of events per year and the expected duration. If the expected duration is expressed in years, the expected down-time will have the units of years per year and can be regarded as the relative proportion of time or probability the system is down due to the minimal cut-set in question. More commonly, expected duration is expressed in hours and the expected down-time has the number of hours per year.

Approximate expressions for frequency and expected duration of the most commonly considered interruption events associated with first-, second-, and third-order minimal cut-sets are given in Table 35. Note that expressions for the calculation of interruption frequencies and durations are for forced outages (failures) only. A detailed treatment of expressions for the calculation of interruption frequency and duration considering forced outages as well as maintenance outages, switching after faults to restore service, and incomplete redundancy of parallel facilities is given in IEEE Std 493-1980 [1].

Table 35
Frequency and Expected Duration Expressions for
Interruptions Associated with Forced Outages Only

First Order Minimum Cut-Set
$$f_{\text{cs}} = \lambda_i$$
$$r_{\text{cs}} = r_i$$

Second-Order Minimal Cut-Set
$$f_{\text{cs}} = \lambda_i \lambda_j (r_i + r_j)$$
$$r_{\text{cs}} = r_i r_j / (r_i + r_j)$$

Third-Order Minimal Cut-Set
$$f_{\text{cs}} = \lambda_i \lambda_j \lambda_k (r_i r_j + r_i r_k + r_j r_k)$$
$$r_{\text{cs}} = r_i r_j r_k / (r_i r_j + r_i r_k + r_j r_k)$$

Symbols: f_{cs} = Frequency of cut-set event
r_{cs} = Expected duration of cut-set event
λ_i = Forced outage rate of ith component
r_i = Expected repair or replacement time of ith component

NOTES: (1) The time units of r and λ in expressions for f_{cs} must be the same. Once frequencies and expected durations have been computed for each minimal cut-set, system reliability indexes at the load point in question are given by:

f_s = Interruption frequency
$$= \sum_{\substack{\text{min} \\ \text{cut-sets}}} f_{\text{cs}_i}$$

r_s = Expected interruption duration
$$= \sum_{\substack{\text{min} \\ \text{cut-sets}}} f_{\text{cs}_i} r_{\text{cs}_i} / f_s$$

$f_s r_s$ = Total interruption time per time period

(2) These are approximate formulas and should only be used when every $\lambda_i r_i$, $\lambda_j r_j$, $\lambda_k r_k$ (dimensionless) is less than 0.01.

12.6 Reference

[1] IEEE Std 493-1980, IEEE Recommended Practice for the Design of Reliable Industrial and Commercial Power Systems (ANSI).

Chapter 13
Cable Ampacity Studies

13.1 Introduction. The cables that network a power system together form the backbone of the system. It is only logical, therefore, that any complete analysis of a power system should include an analysis of its cable ampacities. This analysis is complicated since the ampacity of a conductor varies with the actual conditions of use. Ampacity is defined as "the current in amperes a conductor can carry continuously under the conditions of use (conditions of the surrounding medium in which the cables are installed) without exceeding its temperature rating." Therefore, a cable ampacity study is the calculation of the temperature rise of the conductors in a cable system under steady-state conditions. The purpose of this chapter is to acquaint the reader with the use of computer software systems in the solution of cable ampacity problems with emphasis on underground installations.

The ampacity of a conductor depends on a number of factors. Prominent among these factors and of much concern to the designers of electrical distribution systems are the following:

(1) Ambient temperature
(2) Thermal characteristics of the surrounding medium
(3) Heat generated by the conductor due to its own losses
(4) Heat generated by adjacent conductors

To account for the various items that affect ampacities of cables, the 1975 National Electrical Code (NEC) Handbook for the first time accepted the Neher-McGrath method (see Reference [3][21]) of determining the ampacities of conductors. Since then, the NEC has added new ampacity tables to account for some limited conditions of use. As an alternative to the ampacity tables, section 310-15 (b) of ANSI/NFPA 70-1990 [1][22] permits, under engineering supervision, the use of an ampacity equation for determining ampacities. A discussion of this evolution and the origin of NEC Tables 310-16 through 310-19 is provided in bibliographic reference [B1]. This equation is based on the Neher-McGrath method, which is the basis for the calculating procedures discussed in this chapter.

---

[21] The numbers in brackets correspond to those in the References at the end of this chapter.

[22] ANSI/NFPA publications can be obtained from the Sales Department, American National Standards Institute, 1430 Broadway, New York, NY 10018, or from Publication Sales, National Fire Protection Association, Batterymarch Park, Quincy, MA 02269.

In subsequent paragraphs, various items that affect cable ampacities are discussed and quantified with the help of ampacity adjustment factor tables and actual computer runs. The computer program from which the ampacity adjustment factors were generated is based on the Neher-McGrath method of calculation and has been corroborated by a second, independently developed computer program of like kind. Under some specific and limited conditions, the ampacity adjustment tables were compared and verified with the NEC ampacity tables, including Appendix B of ANSI/NFPA 70-1990 [1]. Note that the tables provided here generally cover broader conditions of use with greater resolution than the NEC tables.

Since the ampacity adjustment tables have been developed for some specific conditions, they cannot be applied for all cases. In general, these tables can be used to size the cables in the initial stages of a design and to closely approximate ampacities. These preliminary cable sizes can then be used as the basis for a more rigorous computer analysis to determine actual conductor temperatures and to finalize the design.

13.2 Heat Flow Analysis. When designing a power distribution system, the cable ampacity is of primary concern. Once the size and location of electrical loads are determined, an adequate distribution system must be designed. The total number of required circuits, their sizes, and the method of routing are significant elements in the design problem. But in addition, accurate cable sizing becomes especially critical to ensure that the cables are adequate to carry the required load without being subjected to temperatures that exceed their temperature ratings.

As an electrical current flows through a cable, it generates heat. The type of cable and how it is connected and installed determines how many components of heat generation are present, e.g., I^2R losses, sheath losses, etc. The heat flows from these sources through a series of thermal resistances to the surrounding environment. The operating temperature that the cable ultimately reaches is directly related to the amount of heat generated and the net effective value of the thermal resistance through which it flows.

A detailed discussion of all the heat transfer complexities involved is beyond the scope of this section. However, the heat transfer process will be covered briefly in order to establish a basic background from which the discussion to follow can proceed.

The calculation of the temperature rise of cable systems involves the application of a series of thermal equivalents of Ohm's and Kirchoff's laws to a relatively simple thermal circuit, as is illustrated in Fig 155. This circuit includes a number of parallel paths with heat entering at several points. Under conditions of equilibrium, the conductor temperature will be determined by the temperature differential created across a series of thermal resistances as the heat flows to the ambient temperature ($T_c' = T_a' + \Delta T$).

To understand the basic calculation procedure used in cable ampacity programs, consider the fundamental equation for the ampacity of a cable in an underground duct.

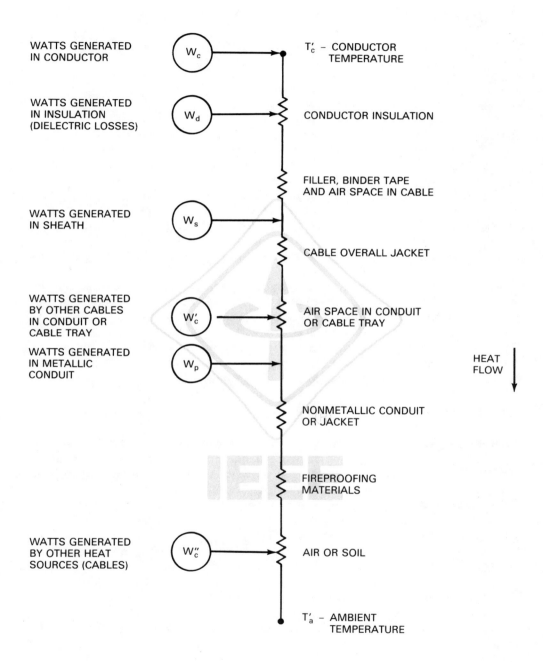

**Fig 155
A Generalized Model for Heat Flow from Heat Sources in a Cable System
to Ambient Temperature Through a Series of Thermal Resistances**

$$I = \left[\frac{T_c' - (T_a' + \Delta T_d + \Delta T_{int})}{(R_{ac})(R_{ca}')} \right]^{1/2} \text{kA} \tag{Eq 205}$$

This equation follows the Neher-McGrath method where:

T_c' = Allowable conductor temperature (°C).

T_a' = Soil ambient temperature (°C).

ΔT_d = Temperature rise of conductor due to dielectric heating (°C).

ΔT_{int} = Temperature rise of conductor due to interference heating from cables in other ducts (°C). (Note that since the temperature rise due to another conductor depends on the current through it, simultaneous solutions of ampacity equations are required.)

R_{ac} = Electrical alternating current resistance of conductor including skin, proximity, and temperature effects (micro-ohms/ft).

R_{ca}' = Effective total thermal resistance from conductor to ambient soil adjusted to include effects of load factor, shield/sheath losses, metallic conduit losses, and the effect of multiple conductors in the same duct (thermal-ohms/ft, °C ft/W).

Note that all effects that produce a conductor temperature rise except the conductor loss ($I^2 R_{ac}$) have been treated as adjustments to the basic thermal system. Fundamentally, the difference between two temperatures (e.g., $T_c' - T_a'$) divided by a separating thermal resistance equals the heat flow in watts (or W/ft of conductor). The similarity of the procedure used in the cable ampacity program to that used with the traditional approach is apparent if both sides of the ampacity equation are squared and then multiplied by R_{ac}. The result is as follows:

$$I^2 R_{ac} = \frac{T_c' - (T_a' + \Delta T_d + \Delta T_{int})}{R_{ca}'} \text{W/ft} \tag{Eq 206}$$

Although it is not necessary to understand these heat transfer concepts in order to use cable ampacity programs, such knowledge may be helpful for understanding how physical parameters affect ampacity. Observation of the ampacity equation shows how lower ampacities are inherent with the following:

- Lower conductor operating temperatures
- Higher soil ambient temperatures
- Smaller conductors (higher R_{ac})
- Higher thermal resistivities of earth, concrete, insulation, duct, etc. (higher R_{ca}')
- Deeper burial depths (higher R_{ca}')
- Closer cable spacing (higher ΔT_{int})
- Cables located in inner, rather than outer, ducts (higher ΔT_{int})

Other factors that decrease ampacity and whose relationship to the ampacity equation is not readily apparent include the following:

- Higher load factor (higher R_{ca}')
- Higher voltage (higher ΔT_d)
- Higher insulation SIC and power factor (higher ΔT_d)
- Lower shield/sheath electrical resistance (higher R_{ca}')

13.3 Application of Computer Program. The calculations used in cable ampacity programs are normally based on the Neher-McGrath method. In computing cable ampacities in duct banks, only power cables need to be considered since control cables, carrying very little current, contribute very little to the overall temperature rise. Cable ampacity programs deal only with the temperature limited, current carrying capacity of cables. Voltage drop, future load growth, and short-circuit capability are also important factors that should be considered when selecting cables.

The calculation of cable ampacity in underground installations is a very complicated procedure requiring the evaluation of a multitude of subtle effects. In order to make the calculations possible for a wide variety of cases, certain assumptions are made. Most of the assumptions are developed in the Neher-McGrath paper [3] and are widely accepted. Some programs may make other assumptions that should be understood.

The basic steps in applying cable ampacity programs are listed below. It is important to follow methodical procedures in order to obtain good results with minimum effort.

(1) The first step in designing an underground cable installation is to establish which circuits are to be routed through the duct bank. Consideration should be given to present circuits as well as to circuits that may be added in the future. Only power cables need to be included as current carrying conductors in analysis; but space allowances must be made for spare ducts or for control and instrumentation circuits.

(2) The duct bank should be designed with consideration given to the circuits contained, the space available for the bank, cable separation criteria, and factors that affect ampacity. For example, cables buried deeply or surrounded by other power cables often have greatly reduced ampacity. It should be decided if ducts will be directly buried or encased in concrete. The size(s) and type(s) of duct to be used should be determined. Finally, a sketch of the duct bank should be prepared with burial depth and spacing of ducts clearly shown. Physical data on the duct installation should be compiled, including thermal resistivity of the soil, ambient temperature of the soil, and thermal resistivity of the concrete. Note that soil thermal resistivity and temperature at some locations (e.g., desert) may be much greater than the typical values often used.

(3) Complete data on all power cables used in the installation must be assembled. Some data may be taken from standard tables; but certain data should be based on manufacturer's specifications. Conductor size, conductor material, operating voltage, type of shield or sheath, temperature rating, insulation type, and jacket type are especially important.

(4) An initial cable placement layout should be designed, based on anticipated loads and load factors. Circuits with high currents and load factors (ratio of average to peak load over a given load cycle) should be placed in outside ducts near the top of the bank to eliminate the need for larger conductors due to unnecessarily reduced ampacity. Frequently, a good compromise between best use of duct space and highest ampacity is achieved by installing

each three-phase circuit in a separate duct. However, nonshielded single-conductor cables may have a higher ampacity with each phase conductor in a separate nonmetallic duct. If the load factor cannot be evaluated readily, a conservative value of 1.0 may be entered, which implies that the circuit always operates at peak load.

(5) The manual method presented in this chapter can be used to initially size the cables based on the ambient temperature, soil thermal resistivity, and grouping of the cables.

(6) Once the initial design is established and all necessary data have been collected, the user should enter the program data interactively or prepare an input data file for a batch program. Normally, the data should be prepared for standard ampacity calculation, using the worst-case conditions. If actual load currents are known, these may be entered to find the temperatures of cables within each duct. Temperature calculations are especially useful if some circuits are lightly loaded, while others carry heavy loads that push ampacity limits. If the lightly loaded circuits were to operate at rated temperature, as the ampacity calculation assumes, the load capability of the heavily loaded circuits would be reduced. Temperature calculations may also be used as a rough indicator of the reserve capacity of each duct.

(7) After a program is run, the user should carefully analyze the results to verify that design currents are less than ampacities (if an ampacity calculation is performed), or that actual temperatures are less than rated temperatures (if a temperature calculation is performed). If the initial design is shown to be inadequate, various corrective measures should be considered. These include increasing conductor sizes, modifying cable locations, and changing the physical design of the bank. The effects of various parameters may be analyzed by repeating these steps until a satisfactory overall design is achieved.

(8) The results of such an analysis should be documented and permanently archived for use in properly controlling and/or analyzing future changes in duct bank usage (i.e., installation of cables in spare ducts).

13.4 Ampacity Adjustment Factors. The ampacity values stated (specified) by the cable manufacturer and/or other authoritative sources, such as ANSI/NFPA 70-1990 [1] and Reference [2] are usually based on some very specific conditions relative to the cable's immediate surrounding environment. The following are examples of some specific conditions:

- Installation under an isolated condition
- Installation of groups of three or six circuits
- Soil thermal resistivity (RHO) of 90 °C-cm/W
- Ambient temperature of 20 °C or 40 °C

In practice, the surrounding medium or environment in which cables are to be installed rarely matches those conditions under which the stated ampacities apply. The differences can be thought of as an intermediate medium (requiring adjustment factors for conditions of use) inserted between the base conditions (an environment at which the base ampacity is specified by the manufacturer or other authoritative sources) and the actual conditions of use. This process is presented pictorially in Fig 156. It illustrates that the nature of the practical problem is to adjust the specified (base) ampacities of the cables by an adjustment factor to account for the effects of the various intermediate elements or conditions of use.

A simple manual method of determining cable ampacities is presented here to illustrate the concept of cable derating and to present the different factors that have a direct effect on the operating temperatures of the conductors.

This method is based on the concept of an adjustment (derating) factor applied against a base ampacity to provide the allowable cable ampacity.

$$I' = F I \qquad \text{(Eq 207)}$$

where:

I' = Allowable ampacity under the actual installation conditions.
F = Overall cable ampacity adjustment factor.
I = Base ampacity, i.e., the ampacity specified by the manufacturers or other authoritative sources, such as the ICEA. For example, the ampacity of a cable that is installed in an underground conduit under isolated conditions with an ambient temperature of 20 °C and soil thermal resistivity RHO of 90 °C-cm/W.

The overall cable adjustment factor is a correction factor that takes into account the differences in the cable's actual installation and operating conditions from the

Fig 156
Simplified Illustration of the Heat Transfer Model Used to Determine
the Cable Ampacity (3-1/C Cables Shown)

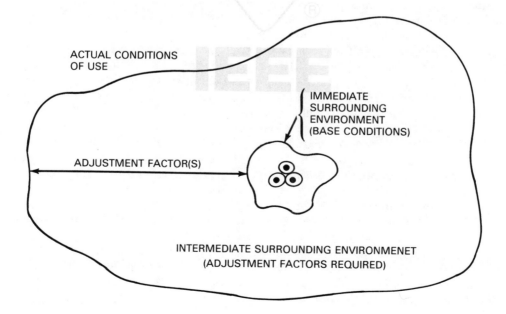

329

base conditions. This factor establishes the maximum load capability that results in an actual cable life equal to or greater than that expected when operated at the base ampacity under the specified conditions.

The overall ampacity adjustment factor is composed of several components as indicated in Eq 208.

$$F = F_t \, F_{th} \, F_g \qquad\qquad\qquad\qquad\qquad \text{(Eq 208)}$$

where:

F_t = Adjustment factor to account for the differences in the ambient and conductor temperatures from the base case.

F_{th} = Adjustment factor to account for the difference in the soil thermal resistivity from the RHO of 90 °C-cm/W at which the base ampacities are specified.

F_g = Adjustment factor to account for cable grouping.

To obtain the values of the adjustment factors F_{th} and F_g, an elaborate computer program was developed based on the Neher-McGrath method and was used to calculate the conductor temperatures for various arrangements. The program takes into account each adjustment factor in Eq 208 which together account for the more significant effects indicated in Fig 155 for underground installations. Thousands of computer runs were made to determine the adjustment factor tables. These tables were then verified with ANSI/NFPA 70-1990 [1], Reference [2], and bibliographic reference [B16]. Reference [4] and bibliographic references [B1] and [B13] report the results of similar efforts for ampacity adjustment factors based on the Neher-McGrath method.

The various adjustment factors in Eq 208 are largely, but not completely, independent from each other. Although the computer program can simulate any complex configuration, for the sake of clarity, the ampacity adjustment tables reported here are based on some simplifying assumptions as listed below.

(1) Cables for some voltage ratings and sizes are combined for the F_{th} tables. For some applications where RHO is considerably high (more than 180 °C-cm/W) and a mixed group of cables are installed, the interdependencies of the adjustment factors for different cable sizes may not be negligible and up to a 4% error in the overall conductor temperatures may be expected.

(2) The effect of the temperature rise due to the insulation dielectric losses is neglected for the temperature adjustment factor, F_t. This temperature rise for rubber and polyethylene insulated cables rated 15 kV and below (sizes 1000 kcmil and below) is less than 2 °C. However, this effect can be included in F_t by adding the temperature rise due to the dielectric losses to the ambient temperatures T_a and T'_a.

(3) The often negligible effects of any applicable sheath, shield, and metallic conduit losses depicted in Fig 155 are ignored.

In the final design case where accuracy and precision are required, the above mentioned assumptions cannot be disregarded and the ampacities obtained from the manual method can be used as an initial approximation for computer simulation of the actual design conditions.

13.4.1 F_t (Ambient and Conductor Temperature Adjustment Factor). This adjustment factor is used to determine the cable ampacity when the operating ambient temperature and/or the maximum allowable conductor temperature differ from the original temperatures at which the cable base ampacity is specified. The expression for calculating the effect of changes in the conductor and ambient temperatures on the base ampacity is given by F_t in Eqs 209 and 210 for copper and aluminum conductors, respectively.

$$F_t = \left[\frac{T_c' - T_a'}{T_c - T_a} \times \frac{234.5 + T_c}{234.5 + T_c'} \right]^{1/2} \text{ Copper} \qquad \text{(Eq 209)}$$

$$F_t = \left[\frac{T_c' - T_a'}{T_c - T_a} \times \frac{228.1 + T_c}{228.1 + T_c'} \right]^{1/2} \text{ Aluminum} \qquad \text{(Eq 210)}$$

where:

T_c = Conductor rated temperature in °C at which the base ampacity is specified.
T_c' = Maximum allowable conductor operating temperature in °C.
T_a = Ambient temperature in °C at which the base ampacity is specified.
T_a' = Actual (maximum) soil ambient temperature in °C.

The maximum operating ambient temperature is usually difficult to obtain and has to be estimated based on historical meteorological data. For application in underground cables, T_a' is the maximum soil temperature at the depth of installation at peak summertime. In general, seasonal temperature variations of the soil follow a roughly sinusoidal cycle with soil temperature peaking during the summer months. The effect of seasonal variation in soil temperature decreases with depth until the depths of 20 to 30 feet are reached, at which the soil temperature remains fairly constant.

Certain characteristics of the soil (texture, density, and moisture content) and soil pavement (asphalt, cement, etc.) have a noticeable effect on the soil temperature profile. For maximum accuracy, it is important to obtain T_a via a field test rather than using an approximate value based on the maximum atmospheric temperature. For cable installation in air, T_a is the maximum air temperature at peak summertime. Special attention should be given for cable applications in the shade or under direct sunlight.

Adjustment factors for typical copper conductor temperatures (T_c = 90 °C and 75 °C) and ambient temperatures (T_a = 20 °C for U/G installation and 40 °C for A/G installation) at which the base ampacities are specified, are calculated from Eq 209 and tabulated in Tables 36 through 39.

13.4.2 F_{th} (Thermal Resistivity Adjustment Factor). Soil thermal resistivity (RHO) indicates the resistance to heat dissipation of the soil in °C - cm/W. Tables 40 through 42 indicate the adjustment factors required when the actual soil thermal resistivity is different from the RHO of 90 °C-cm/W at which the base ampacities are specified. These tables are calculated based on an assumption that the soil has a uniform and constant thermal resistivity.

Table 36

F_t — Adjustment Factor for Various Copper Conductors
and Ambient Temperatures When T_c = 75 °C and T_a = 40 °C

T'_c in °C	T'_a in °C					
	30	35	40	45	50	55
60	0.95	0.87	0.77	0.67	0.55	0.39
75	1.13	1.07	1.00	0.93	0.85	0.76
90	1.28	1.22	1.17	1.11	1.04	0.98
110	1.43	1.34	1.34	1.29	1.24	1.19

Table 37

F_t — Adjustment Factor for Various Copper Conductors
and Ambient Temperatures When T_c = 90 °C and T_a = 40 °C

T'_c in °C	T'_a in °C					
	30	35	40	45	50	55
75	0.97	0.92	0.86	0.79	0.72	0.65
85	1.06	1.01	0.96	0.90	0.84	0.78
90	1.10	1.05	1.00	0.95	0.89	0.84
110	1.23	1.19	1.15	1.11	1.06	1.02
130	1.33	1.30	1.27	1.23	1.19	1.16

Table 38

F_t — Adjustment Factor for Various Copper Conductors
and Ambient Temperatures When T_c = 75 °C and T_a = 20 °C

T'_c in °C	T'_a in °C					
	10	15	20	25	30	35
60	0.98	0.93	0.87	0.82	0.76	0.69
75	1.09	1.04	1.00	0.95	0.90	0.85
90	1.18	1.14	1.10	1.06	1.02	0.98
110	1.29	1.25	1.21	1.18	1.14	1.11

Table 39

F_t — Adjustment Factor for Various Copper Conductors
and Ambient Temperatures When T_c = 90 °C and T_a = 20 °C

T'_c in °C	T'_a in °C					
	10	15	20	25	30	35
75	0.99	0.95	0.91	0.87	0.82	0.77
85	1.04	1.02	0.97	0.93	0.89	0.85
90	1.07	1.04	1.00	0.96	0.93	0.89
110	1.16	1.13	1.10	1.06	1.02	0.98
130	1.24	1.21	1.18	1.16	1.13	1.10

Table 40
F_{th} —Thermal Resistivity Adjustment Factor for 0–1000 V Cables in Duct Banks with Base Ampacity Given at a RHO of 90 °C-cm/W

Cable Size	Number of CKT	RHO (°C-cm/W)							
		60	90	120	140	160	180	200	250
#12	1	1.03	1.0	0.97	0.96	0.94	0.93	0.92	0.90
to	3	1.06	1.0	0.95	0.92	0.89	0.87	0.85	0.82
#1	6	1.09	1.0	0.93	0.89	0.85	0.82	0.79	0.75
	9+	1.11	1.0	0.92	0.87	0.83	0.79	0.76	0.71
1/0	1	1.04	1.0	0.97	0.95	0.93	0.91	0.89	0.86
to	3	1.07	1.0	0.94	0.90	0.87	0.85	0.83	0.80
4/0	6	1.10	1.0	0.92	0.87	0.84	0.81	0.78	0.74
	9+	1.12	1.0	0.91	0.85	0.81	0.78	0.75	0.70
250	1	1.05	1.0	0.96	0.94	0.92	0.90	0.88	0.85
to	3	1.08	1.0	0.93	0.89	0.86	0.83	0.81	0.77
1000	6	1.11	1.0	0.91	0.86	0.83	0.80	0.77	0.72
	9+	1.13	1.0	0.90	0.84	0.80	0.77	0.74	0.69

Table 41
F_{th} —Thermal Resistivity Adjustment Factor for 1001–35 000 V Cables in Duct Banks with Base Ampacity Given at a RHO of 90 °C-cm/W

Cable Size	Number of CKT	RHO (°C-cm/W)							
		60	90	120	140	160	180	200	250
#12	1	1.03	1.0	0.97	0.95	0.93	0.91	0.90	0.88
to	3	1.07	1.0	0.94	0.90	0.87	0.84	0.81	0.77
#1	6	1.09	1.0	0.92	0.87	0.84	0.80	0.77	0.72
	9+	1.10	1.0	0.91	0.85	0.81	0.77	0.74	0.69
1/0	1	1.04	1.0	0.96	0.94	0.92	0.90	0.88	0.85
to	3	1.08	1.0	0.93	0.89	0.86	0.83	0.80	0.75
4/0	6	1.10	1.0	0.91	0.86	0.82	0.79	0.77	0.71
	9+	1.11	1.0	0.90	0.84	0.80	0.76	0.73	0.68
250	1	1.05	1.0	0.95	0.92	0.90	0.88	0.86	0.84
to	3	1.09	1.0	0.92	0.88	0.85	0.82	0.79	0.74
1000	6	1.11	1.0	0.91	0.85	0.81	0.78	0.75	0.70
	9+	1.12	1.0	0.90	0.84	0.79	0.75	0.72	0.67

Table 42

F_{th} — Thermal Resistivity Adjustment Factor for Cables Directly Buried with Base Ampacity Given at a RHO of 90 °C-cm/W

Cable Size	Number of CKT	RHO (°C-cm/W)							
		60	90	120	140	160	180	200	250
#12	1	1.10	1.0	0.91	0.86	0.82	0.79	0.77	0.74
to	2	1.13	1.0	0.90	0.85	0.81	0.77	0.74	0.70
#1	3+	1.14	1.0	0.89	0.84	0.79	0.75	0.72	0.67
1/0	1	1.13	1.0	0.91	0.86	0.81	0.78	0.75	0.71
to	2	1.14	1.0	0.90	0.85	0.80	0.76	0.73	0.69
4/0	3+	1.15	1.0	0.89	0.84	0.78	0.74	0.71	0.67
250	1	1.14	1.0	0.90	0.85	0.81	0.78	0.75	0.71
to	2	1.15	1.0	0.89	0.84	0.80	0.76	0.73	0.69
1000	3+	1.16	1.0	0.88	0.83	0.78	0.74	0.71	0.67

Typical values of thermal resistivity for various materials are listed below. (See Reference [1].)

Material Type	(°C-cm/W)
Solid paper insulation	700
Varnished cambric	600
Polyvinyl chloride (PVC)	650
Paper	550
Neoprene	519
Rubber, jute, textiles	500
Fiber duct	480
Polyethylene (PE)	450
Transite duct	200
Somastic	100
Concrete	55 – 85
Average soil	90
Very dry soil (rocky or sandy)	120
Damp soil (coastal areas, high water table)	60
EPR	400
Crosslinked polyethylene	370

The thermal resistivity of the soil depends on a number of factors, such as soil texture, moisture content, density, and structural arrangement of the soil grains. In general, higher density or moisture content of the soil results in a better heat dissipating ability and lower thermal resistivity. There is a tremendous variation in the soil thermal resistivities ranging from a RHO of less than 40 to more than 300 °C-cm/W. Based on these facts, it is apparent that direct testing of the soil is essential. Furthermore, it is important that this test be conducted after a prolonged dry spell at a peak summer temperature when the soil moisture content is minimal.

The result of such a field test usually indicates a wide range of soil thermal resistance for a given depth over a test site. For the purpose of cable ampacity deratings, the maximum value of the thermal resistivities for a given cable route should be used.

The effect of soil dryout, which is caused by the continuous loading of the cables, can be taken into account by considering a RHO higher than the actual value obtained from the soil test. Use of dense sandy soil as backfill can lower the effective overall thermal resistivity and can offset the soil dryout effect. Dryout curves of RHO versus moisture content can be obtained to help select an appropriate value.

In cases where the soil thermal resistivity is very high and corrective backfill with low thermal resistivity is used, Tables 40 through 42 are inaccurate and may not produce cable ampacity values that are acceptable even on an approximate basis.

13.4.3 F_g (Grouping Adjustment Factor). Grouped cables will operate at a higher temperature than isolated cables. The increase in the operating temperature is due to the presence of the other cables in the group, which act as heat sources. Therefore, the amount of interference temperature rise from other cables in the group depends on the separation of the cables and the surrounding media.

In this section, adjustment factors for cables installed with maintained separation in underground duct banks and for directly buried cables are given in Tables 43 through 46. For cable separations other than those considered in these tables, one can use one's own judgment for estimating the value of F_g or use a computer program directly without an initial approximation for the grouping effect. In general, increasing the horizontal and vertical spacing between the cables would decrease the temperature interference between them and, therefore, increase the value of F_g.

Based on the computer studies for duct bank installations, it was found that the size and voltage rating of the cables make a noticeable difference in the value of F_g. Therefore, the adjustment factors for cable groupings are tabulated as functions of cable sizes and voltage ratings. For applications where a mixed group of cables are installed in a duct bank, the value of F_g will be different for each cable size. In this case, it is recommended that cable ampacities be determined as the location of the cables is progressively changed from the worst (hottest) conduit locations and the best (coolest) conduit locations to establish the most economical arrangement.

Note that no grouping adjustment factor is given for cables installed in air or in conduits in air. Refer to ANSI/NFPA 70-1990 [1] and Reference [2] for the allowable ampacities of cable installed in conduits in air.

13.5 Example. To illustrate the use of the method described in this chapter, a 3×5 duct bank system (3 rows, 5 columns) is considered. The duct bank contains 350 kcmil and 500 kcmil (15 kV, 3/C) copper cables. Ducts are 5 in. diameter (trade size) of PVC, and are separated by 7.5 in. (center-to-center spacing) as shown in Fig 157. The soil thermal resistivity (RHO) is 120 °C-cm/W, and the maximum soil ambient temperature is 30 °C.

The objective of this example is to determine the maximum ampacities of the cables under the specified conditions of use, i.e., to limit the conductor temperature of the hottest location to 75 °C (an NEC requirement for wet locations). To achieve

Table 43
F_g — Grouping Adjustment Factor for 0 – 5000 V 3/C, or Triplexed Cables in Duct Banks (No Spare Ducts, 5 in. Nonmetallic Conduits with 7.5 in. Center-to-Center Spacing)

Cable Size	No. of Rows	Number of Columns														
		1	2	3	4	5	6	7	8	9	10	11	12	13	14	15
#8	1	1.00	.942	.885	.835	.795	.768	.745	.727	.710	.698	.688	.679	.671	.664	.658
	2	.930	.840	.772	.723	.687	.660	.638	.620	.604	.592	.582	.572	.564	.557	.550
	3	.870	.772	.694	.632	.596	.569	.548	.532	.519	.508	.498	.490	.482	.476	.470
	4	.820	.710	.629	.571	.536	.509	.490	.472	.458	.446	.436	.428	.420	.412	.405
#6	1	1.00	.930	.874	.826	.790	.760	.737	.718	.702	.690	.680	.671	.663	.656	.650
	2	.920	.813	.747	.700	.665	.638	.615	.598	.583	.572	.561	.552	.544	.537	.530
	3	.860	.747	.679	.625	.588	.560	.540	.525	.510	.498	.490	.481	.473	.467	.460
	4	.810	.700	.620	.565	.531	.503	.484	.467	.452	.440	.431	.422	.415	.408	.400
#4	1	1.00	.925	.871	.817	.781	.750	.726	.707	.691	.678	.668	.659	.651	.646	.640
	2	.920	.809	.742	.693	.659	.632	.610	.593	.579	.567	.555	.547	.539	.530	.525
	3	.850	.742	.668	.615	.578	.551	.531	.514	.500	.489	.480	.471	.464	.458	.450
	4	.805	.690	.610	.560	.524	.497	.477	.460	.447	.435	.425	.418	.410	.401	.395
#2	1	1.00	.918	.858	.808	.770	.741	.720	.701	.688	.677	.667	.658	.650	.641	.635
	2	.920	.800	.723	.680	.648	.623	.602	.586	.572	.560	.549	.540	.530	.522	.514
	3	.840	.723	.657	.608	.568	.540	.520	.504	.490	.479	.470	.461	.454	.447	.440
	4	.800	.685	.608	.553	.518	.490	.471	.453	.440	.429	.420	.411	.402	.395	.390
#1	1	1.00	.918	.849	.799	.753	.721	.699	.682	.669	.659	.650	.643	.639	.632	.630
	2	.920	.795	.702	.650	.613	.583	.563	.546	.530	.520	.510	.502	.494	.488	.482
	3	.830	.702	.618	.562	.525	.500	.480	.464	.450	.440	.430	.421	.413	.406	.400
	4	.740	.634	.551	.497	.465	.440	.421	.405	.392	.383	.374	.366	.359	.352	.348
1/0	1	1.00	.910	.842	.791	.745	.716	.694	.678	.665	.655	.646	.639	.635	.628	.626
	2	.915	.790	.700	.642	.604	.575	.555	.537	.523	.511	.503	.494	.486	.480	.475
	3	.817	.700	.610	.554	.520	.494	.474	.457	.444	.432	.424	.415	.408	.400	.394
	4	.735	.629	.546	.492	.460	.435	.417	.402	.391	.381	.371	.363	.355	.349	.343
2/0	1	1.00	.910	.842	.791	.745	.716	.694	.678	.665	.655	.646	.639	.635	.628	.626
	2	.915	.790	.700	.642	.604	.575	.555	.537	.523	.511	.503	.494	.486	.480	.475
	3	.817	.700	.610	.554	.520	.494	.474	.457	.444	.432	.424	.415	.408	.400	.394
	4	.735	.629	.546	.492	.460	.435	.417	.402	.391	.381	.371	.363	.355	.349	.343
3/0	1	1.00	.910	.842	.791	.745	.716	.694	.678	.665	.655	.646	.639	.635	.628	.626
	2	.915	.790	.700	.642	.604	.575	.555	.537	.523	.511	.503	.494	.486	.480	.475
	3	.817	.700	.610	.554	.520	.494	.474	.457	.444	.432	.424	.415	.408	.400	.394
	4	.735	.629	.546	.492	.460	.435	.417	.402	.391	.381	.371	.363	.355	.349	.343
4/0	1	1.00	.908	.830	.780	.737	.709	.690	.673	.660	.650	.642	.635	.628	.623	.619
	2	.910	.770	.684	.635	.599	.570	.550	.532	.518	.506	.498	.489	.481	.475	.470
	3	.810	.684	.602	.548	.515	.489	.469	.452	.440	.429	.420	.411	.403	.397	.391
	4	.730	.624	.541	.487	.456	.431	.414	.399	.388	.378	.368	.360	.352	.346	.341
250	1	1.00	.905	.830	.777	.725	.692	.668	.646	.628	.615	.603	.597	.590	.583	.580
	2	.890	.770	.675	.609	.570	.542	.519	.500	.485	.474	.466	.458	.450	.445	.440
	3	.780	.675	.579	.518	.480	.454	.434	.420	.408	.398	.390	.383	.378	.373	.370
	4	.694	.588	.512	.460	.422	.397	.379	.364	.352	.345	.338	.331	.327	.323	.320

Table 43 *(Continued)*

Cable Size	No. of Rows	Number of Columns														
		1	2	3	4	5	6	7	8	9	10	11	12	13	14	15
350	1	1.00	.905	.830	.770	.720	.688	.661	.640	.622	.608	.597	.590	.583	.578	.573
	2	.887	.749	.664	.609	.570	.540	.518	.499	.484	.474	.465	.458	.450	.445	.440
	3	.775	.664	.575	.515	.479	.453	.433	.419	.406	.397	.389	.382	.377	.372	.369
	4	.690	.587	.511	.457	.421	.395	.377	.362	.351	.343	.336	.330	.325	.321	.318
500	1	1.00	.897	.815	.762	.708	.678	.652	.630	.613	.599	.588	.581	.575	.570	.565
	2	.882	.745	.656	.608	.569	.539	.516	.498	.483	.473	.463	.457	.450	.444	.439
	3	.770	.656	.570	.514	.478	.452	.432	.417	.404	.395	.388	.381	.375	.370	.367
	4	.685	.585	.510	.454	.420	.393	.374	.360	.349	.340	.333	.328	.323	.319	.315
750	1	1.00	.890	.802	.747	.700	.670	.640	.622	.605	.590	.580	.572	.566	.560	.555
	2	.870	.725	.641	.591	.552	.522	.500	.484	.469	.457	.448	.440	.434	.430	.425
	3	.760	.641	.560	.507	.470	.445	.425	.410	.398	.389	.380	.374	.369	.363	.360
	4	.680	.579	.501	.448	.413	.389	.371	.357	.346	.337	.330	.323	.318	.314	.310
1000	1	1.00	.885	.795	.740	.695	.665	.639	.618	.600	.585	.574	.567	.561	.555	.551
	2	.858	.716	.632	.582	.544	.513	.493	.474	.460	.448	.439	.431	.425	.420	.415
	3	.748	.632	.551	.499	.464	.439	.419	.403	.392	.383	.375	.369	.363	.358	.355
	4	.676	.574	.497	.444	.409	.385	.367	.353	.342	.333	.326	.319	.315	.311	.308

Table 44
F_g — Grouping Adjustment Factor for 5001–35 000 V 3/C, or Triplexed Cables in Duct Banks (No Spare Ducts, 5 in. Nonmetallic Conduits with 7.5 in. Center-to-Center Spacing)

Cable Size	No. of Rows	Number of Columns														
		1	2	3	4	5	6	7	8	9	10	11	12	13	14	15
#6	1	1.00	.920	.854	.803	.758	.726	.699	.678	.660	.646	.635	.628	.620	.615	.610
	2	.920	.800	.714	.660	.620	.590	.570	.552	.540	.530	.521	.515	.509	.503	.500
	3	.840	.714	.625	.569	.530	.501	.484	.470	.459	.450	.442	.436	.429	.423	.420
	4	.770	.642	.560	.506	.469	.441	.422	.406	.394	.385	.378	.371	.367	.362	.358
#4	1	1.00	.920	.852	.800	.755	.722	.695	.673	.655	.642	.630	.623	.615	.610	.605
	2	.920	.795	.714	.660	.620	.590	.570	.552	.540	.530	.521	.515	.434	.430	.425
	3	.835	.709	.615	.561	.521	.493	.474	.459	.488	.439	.430	.424	.420	.416	.412
	4	.760	.630	.548	.498	.460	.430	.410	.395	.382	.374	.367	.361	.356	.352	.350
#2	1	1.00	.910	.836	.784	.748	.714	.688	.665	.649	.635	.625	.616	.609	.602	.598
	2	.920	.782	.689	.639	.599	.570	.548	.531	.518	.508	.500	.494	.489	.484	.480
	3	.820	.689	.600	.544	.505	.479	.460	.445	.433	.424	.417	.410	.405	.400	.395
	4	.746	.622	.539	.484	.445	.415	.396	.382	.370	.361	.353	.348	.342	.338	.334
#1	1	1.00	.905	.827	.777	.731	.697	.670	.645	.626	.610	.598	.588	.579	.571	.565
	2	.920	.771	.681	.629	.590	.560	.538	.519	.502	.491	.480	.471	.462	.455	.450
	3	.816	.681	.588	.532	.497	.469	.448	.432	.418	.407	.397	.389	.382	.376	.370
	4	.785	.605	.524	.471	.435	.410	.390	.376	.364	.353	.347	.340	.333	.328	.323

Table 44 *(Continued)*

Cable Size	No. of Rows	1	2	3	4	5	6	7	8	9	10	11	12	13	14	15
								Number of Columns								
1/0	1	1.00	.904	.825	.775	.729	.695	.668	.643	.624	.609	.597	.587	.578	.570	.564
	2	.912	.765	.671	.619	.580	.549	.527	.509	.494	.481	.471	.462	.453	.446	.440
	3	.811	.671	.581	.525	.488	.460	.440	.423	.409	.398	.387	.379	.372	.365	.359
	4	.730	.604	.518	.464	.431	.406	.385	.372	.359	.349	.341	.335	.329	.324	.320
2/0	1	1.00	.904	.823	.773	.728	.694	.668	.643	.624	.609	.580	.597	.587	.578	.570
	2	.903	.761	.667	.612	.573	.542	.520	.500	.488	.475	.463	.455	.448	.441	.434
	3	.800	.667	.578	.520	.482	.454	.433	.418	.402	.391	.382	.374	.367	.360	.353
	4	.722	.597	.511	.460	.425	.400	.380	.365	.353	.343	.335	.329	.322	.317	.312
3/0	1	1.00	.898	.814	.765	.722	.690	.661	.637	.618	.602	.590	.580	.571	.563	.556
	2	.898	.752	.664	.609	.570	.539	.451	.498	.483	.471	.461	.451	.443	.437	.429
	3	.802	.664	.572	.514	.479	.451	.430	.414	.399	.388	.379	.371	.364	.357	.350
	4	.720	.593	.508	.456	.421	.396	.377	.362	.350	.340	.332	.327	.320	.314	.310
4/0	1	1.00	.894	.811	.762	.717	.682	.653	.631	.612	.597	.585	.574	.566	.558	.550
	2	.896	.743	.656	.603	.565	.536	.513	.496	.480	.468	.459	.449	.441	.434	.427
	3	.795	.656	.564	.513	.474	.447	.427	.411	.397	.386	.377	.369	.362	.355	.349
	4	.711	.584	.502	.450	.417	.392	.374	.359	.348	.338	.329	.324	.317	.311	.307
250	1	1.00	.892	.811	.762	.715	.679	.645	.620	.600	.583	.572	.564	.557	.552	.550
	2	.885	.741	.654	.594	.552	.523	.500	.482	.469	.457	.447	.438	.430	.422	.416
	3	.785	.654	.559	.498	.459	.429	.408	.388	.373	.361	.351	.342	.335	.328	.321
	4	.701	.580	.500	.448	.414	.385	.365	.348	.332	.321	.311	.302	.295	.288	.281
350	1	1.00	.890	.807	.754	.700	.661	.634	.609	.589	.572	.561	.552	.548	.542	.540
	2	.872	.733	.641	.580	.538	.510	.488	.470	.455	.443	.432	.423	.415	.408	.400
	3	.772	.641	.550	.492	.451	.420	.396	.377	.362	.350	.340	.331	.323	.316	.310
	4	.681	.572	.491	.440	.402	.375	.354	.337	.322	.311	.300	.292	.285	.278	.271
500	1	1.00	.885	.801	.745	.692	.650	.620	.593	.573	.559	.548	.539	.533	.529	.526
	2	.862	.728	.634	.572	.531	.502	.480	.462	.447	.435	.425	.415	.407	.400	.391
	3	.765	.634	.542	.483	.446	.415	.391	.373	.358	.346	.335	.327	.319	.311	.305
	4	.676	.574	.497	.444	.409	.385	.367	.353	.342	.333	.326	.319	.315	.311	.308
750	1	1.00	.879	.790	.780	.682	.647	.615	.589	.570	.556	.545	.536	.530	.524	.520
	2	.850	.710	.622	.560	.520	.490	.469	.450	.436	.424	.412	.402	.394	.388	.381
	3	.755	.622	.530	.479	.441	.410	.387	.368	.352	.341	.331	.322	.314	.307	.300
	4	.671	.560	.480	.430	.392	.366	.345	.328	.314	.302	.292	.284	.277	.270	.263
1000	1	1.00	.873	.786	.730	.680	.642	.609	.582	.562	.548	.537	.528	.521	.516	.512
	2	.844	.705	.614	.554	.514	.485	.463	.445	.430	.418	.406	.397	.390	.383	.376
	3	.745	.614	.523	.472	.434	.403	.381	.363	.348	.337	.327	.318	.309	.301	.294
	4	.663	.552	.473	.422	.385	.359	.338	.321	.307	.295	.285	.278	.270	.263	.256

this, the base ampacities of the cables are found first. These ampacities are then derated using the adjustment factors. The computer program is then used to verify the derated ampacities by calculating the actual conductor temperatures.

The depth of the duct bank is set at 30 inches for this example. For average values of soil thermal resistivity, the depth can be varied by approximately ±10% without drastically affecting the resulting ampacities. However, larger variations in

Table 45
F_g — Grouping Adjustment Factor for Directly Buried 3/C, or Triplexed Cables (7.5 in. Horizontal and 10 in. Center-to-Center Vertical Spacing)

Number of Layers	Number of Horizontal Cables						
	1	2	3	4	6	9	12
1	1.0	0.82	0.70	0.63	0.56	0.51	0.49
2	0.81	0.62	0.53	0.48	0.41	—	—

Table 46
F_g — Grouping Adjustment Factor for Directly Buried 1/C Cables (7.5 in. Horizontal and 10 in. Center-to-Center Vertical Spacing)

Number of Layers	Number of Horizontal Cables			
	3	6	9	12
1	1.0	0.79	0.71	0.68
2	0.73	0.58	—	—

Fig 157
3 × 5 Duct Bank Arrangement

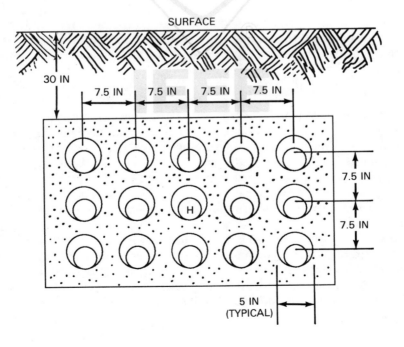

the bank depth, or larger soil thermal resistivities may significantly effect ampacities.

13.5.1 Base Ampacities. From the NEC ampacity tables, the base ampacities of 15 kV three-conductor cables under an isolated condition and based on a conductor temperature of 90 °C, ambient soil temperature of 20 °C, and thermal resistivity (RHO) of 90 °C-cm/W are:

I = 375 amperes (350 kcmil)
I = 450 amperes (500 kcmil)

13.5.2 Manual Method. The required ampacity adjustment factors for the ambient and conductor temperatures, thermal resistivity, and grouping are as follows:

F_t = 0.82   For adjustment in the ambient temperature from 20 °C to 30 °C and conductor temperature from 90 °C to 75 °C (see Table 39).

F_{th} = 0.90   For adjustment in the thermal resistivity from a RHO of 90 to 120 °C-cm/W (see Table 41).

F_g = 0.479 For grouping adjustment of 15 kV, 3/C 350 kcmil cables installed in a 3 × 5 duct bank (see Table 43).

F_g = 0.478 For grouping adjustment of 15 kV, 3/C 500 kcmil cables installed in a 3 × 5 duct bank (see Table 43).

The overall cable adjustment factors are:

F = 0.82 × 0.90 × 0.479 = 0.354 (350 kcmil cables)
F = 0.82 × 0.90 × 0.478 = 0.353 (500 kcmil cables)

The maximum allowable ampacity of each cable size is the multiplication product of the cable base ampacity by the overall adjustment factor. This ampacity adjustment would limit the temperature of the hottest conductor to 75 °C when all of the cables in the duct bank are loaded at 100% of their derated ampacities.

I' = 375 × 0.354 = 133 amperes (350 kcmil cables)
I' = 450 × 0.353 = 159 amperes (500 kcmil cables)

13.5.3 Computer Method. As the last step, a computer program is run to simulate the actual conductor temperature using the ampacities determined by the manual method. The computer program used here is the same program that was used to generate the ampacity adjustment factors. The output report of the program is shown below where the first page indicates all input parameters and the second page indicates conduit locations and conductor temperatures.

The objective for this design was to find the cable ampacities that would limit the conductor temperature to 75 °C. The results of the computer study indicate that the hottest conductor is located in the middle row (2) and middle column (3) with a temperature of 74.3 °C. The ampacities obtained from the manual method for this simplified example case exactly agree with the ampacities obtained by the computer calculations. In more general cases, however, where the assumptions listed in 13.4 do not apply, computer calculation would be necessary to establish final ampacities.

CABLE AMPACITY DERATING

Project: Example
Location: Irvine, California
Contract: 1234567
Engineer: F.S.

Page: 1
Date: 09-01-1989
Study: SC-100

Cable Ampacity Derating Example — 3 × 5 Duct Bank Application

Cable Size	No. of Cond.	Volt kV	Type	DC Res. μohm/ft	O.D. (in)	Insul. Ther. R ohm-ft	Dielect. Losses (W/ft)	Yc	Ys
500	3	15	CU	21.60	2.590	1.430	0.056	0.018	0.000
350	3	15	CU	30.80	2.290	1.564	0.048	0.009	0.000

Instal-lation	Conduit Type	# of Rows	# of Cols.	Ref. Depth (in)	Height (in)	Width (in)	RHO Soil	RHO Fill	Ambient Temp. °C
Duct Bank	PVC	3	5	30.0	27.0	42.0	120.0	90.0	30.0

Row	Col.	Horiz. Dist. (in)	Vert. Dist. (in)	Load Current (A)	Cable No.	C/C	Size	kV	Type	Conduit (in) Size	Thickness
1	1	6.00	6.00	159.0	1	3	500	15	CU	5.040	0.260
2	1	6.00	13.50	159.0	1	3	500	15	CU	5.040	0.260
3	1	6.00	21.00	159.0	1	3	500	15	CU	5.040	0.260
1	2	13.50	6.00	159.0	1	3	500	15	CU	5.040	0.260
2	2	13.50	13.50	159.0	1	3	500	15	CU	5.040	0.260
3	2	13.50	21.00	159.0	1	3	500	15	CU	5.040	0.260
1	3	21.00	6.00	133.0	1	3	350	15	CU	5.040	0.260
2	3	21.00	13.50	133.0	1	3	350	15	CU	5.040	0.260
3	3	21.00	21.00	133.0	1	3	350	15	CU	5.040	0.260
1	4	28.50	6.00	33.0	1	3	350	15	CU	5.040	0.260
2	4	28.50	13.50	133.0	1	3	350	15	CU	5.040	0.260
3	4	28.50	21.00	133.0	1	3	350	15	CU	5.040	0.260
1	5	37.00	6.00	133.0	1	3	350	15	CU	5.040	0.260
2	5	37.00	13.00	133.0	1	3	350	15	CU	5.040	0.260
3	5	37.00	21.00	133.0	1	3	350	15	CU	5.040	0.260

13.6 Conclusion. Analytical derating of cable ampacity is a complex and tedious process. A manual method was developed in this chapter that uses adjustment factors to simplify cable derating for some very specific conditions of use and produce close approximations to actual ampacities. The results from the manual method can then be entered as the initial ampacities for input into a cable ampacity computer program. The speed of the computer allows it to use a more complex

CABLE AMPACITY DERATING

Project: Example			Page:	2	
Location: Irvine, California			Date:	09-01-1989	
Contract: 1234567			Study:	SC-100	
Engineer: F.S.					

Cable Ampacity Derating Example — 3 × 5 Duct Bank Application

	Columns	1	2	3	4	5
Row 1	Cable:	500	500	350	350	350
	Amp. :	159.0	159.0	133.0	133.0	133.0
	Temp. :	66.8	69.7	70.9	69.9	66.6
Row 2	Cable:	500	500	350	350	350
	Amp. :	159.0	159.0	133.0	133.0	133.0
	Temp. :	69.7	73.0	74.3	73.1	69.3
Row 3	Cable:	500	500	350	350	350
	Amp. :	159.0	159.0	133.0	133.0	133.0
	Temp. :	69.3	72.3	73.5	72.4	69.0

model, which considers factors specific to a particular installation and can iteratively adjust the conductor resistances as a function of temperature. The following is a list of factors that are specific for the cable system:

- Conduit type
- Conduit wall thickness
- Conduit inside diameter
- Asymmetrical spacing of cables or conduits
- Conductor load currents and load cycles
- Height, width, and depth of duct bank
- Thermal resistivity of backfill and/or duct bank
- Thermal resistance of cable insulation
- Dielectric losses of cable insulation
- AC/DC ratio of conductor resistance

The results from the computer program should be compared with the initial ampacities found by the manual process to determine whether corrective measures, i.e., changes in cable sizes, duct rearrangement, etc., are required. Many computer programs alternatively calculate cable temperatures for a given ampere loading or cable ampacities at a given temperature. Some recently developed computer programs perform the entire process to size the cables automatically. To find an optimal design, the cable ampacity computer program simulates many different cable arrangements and loading conditions, including future load expansion requirements. This optimization is important in the initial stages of cable system design since changes to cable systems are costly, especially for underground installations. Additionally, the downtime required to correct a faulty cable design may be very long.

13.7 References

[1] ANSI/NFPA 70-1990, National Electrical Code Handbook.

[2] ICEA Power Cable Ampacities, IEEE publication no. S-135 (ICEA publication no. P-46–426), New York, 1978.

[3] NEHER, J. H. and McGRATH, M. H. "The calculations of the temperature rise and load capability of cable systems," *AIEE Transactions on Power Applications Systems*, vol. 76, pt. III, Oct. 1957, pp. 752–772.

[4] SHOKOOH, F. and KNUTSON, H. M. "Ampacity Derating of Underground Cables," IEEE paper no. CH2581-7/88, presented at the I&CPS conference in Baltimore, Maryland in May 1988.

13.8 Bibliography

[B1] KNUTSON, H. M. and MILES, B. B. "Cable derating parameters and their effects," IEEE paper no. PCIC-77-5, 1977.

[B2] *National Electrical Code Committee Report*, 1980 Annual Meeting.

[B3] *National Electrical Code Committee Report*, 1986 Annual Meeting.

[B4] *National Electrical Code Committee Report*, 1989 Annual Meeting.

[B5] *National Electrical Code Technical Committee Documentation*, 1980 Annual Meeting.

[B6] *National Electrical Code Technical Committee Documentation*, 1983 Annual Meeting.

[B7] *National Electrical Code Technical Committee Documentation*, 1986 Annual Meeting.

[B8] *National Electrical Code Technical Committee Documentation*, 1989 Annual Meeting.

[B9] *National Electrical Code Technical Committee Report*, 1983 Annual Meeting.

[B10] NEMA Subcommittee Final Report, *Determination of Maximum Permissible Current Carrying Capacity of Code Insulated Wires and Cable for Building Purposes, Part IV.*[23]

[B11] ROSCH, S. J. "The Current-Carrying Capacity of Rubber Insulated Conductors," paper presented at Jan. 27, 1938 midwinter AIEE meeting in New York City.

[B12] SCHURIG, O. R. and FRICK, G. W. "Heating and Current-Carrying Capacity of Bare Conductors for Outdoor Service," *General Electric Review*, vol. 33, Schenectady, NY, 1930, p. 141.

---

[23] NEMA publications are available from the National Electrical Manufacturers Association, 2101 L Street, NW, Washington, DC 20037.

[B13] SHOKOOH, F. and KNUTSON, H. M. "A simple approach to cable ampacity rating," IEEE paper no. PCIC-83-16, presented at the PCIC conference in Denver, Colorado in Sept. 1983.

[B14] SIMMONS, D. M. *Calculation of the Electrical Problems of Underground Cables*, The Electric Journal, East Pittsburgh, PA, May-Nov. 1932.

[B15] Study NBSIR 78-1477, Department of Energy, Washington, DC.

[B16] *Underground Systems Reference Book*, EEI publication no. 55-16, Edison Electric Institute, New York City, NY, 1957.

[B17] ZIPSE, D. W. "Ampacity Tables — Demystifying the Myths," presented at the PCIC conference in Sept. 1988.

Chapter 14
Ground Mat Studies

14.1 Introduction. A ground mat study has one primary purpose: to determine if a ground mat design will limit the neutral-to-ground voltages normally present during ground faults to values that the average person can tolerate. Equipment protection or system operation is rarely an objective of a ground mat study. Historically, only utilities and unusually large industrial plants have been concerned with this type of study. However, the trend of power systems toward ever-increasing short-circuit capability has made safe ground mat design a criterion for all sizes of substations. This chapter will briefly review the theoretical background behind ground mat studies and discuss its application in the design of a ground mat by computer program.

14.2 Justification for Ground Mat Studies. Virtually every exposed metallic object in an industrial facility is connected to ground, either deliberately or by accident. Under normal operating conditions, these conductors will be at the same potential as the surrounding earth. However, during ground faults, the absolute potential of the grounding system will rise (often to thousands of volts) along with any structural steel tied to the grounding system. Because any metal is a relatively good conductor, the steelwork everywhere will be at essentially the same voltage. Most soils are poor conductors, however, and the flow of fault current through the earth will create definite and sometimes deadly potential gradients. Ground mat studies calculate the voltage difference between the grounding grid and points at the earth's surface and evaluate the shock hazard involved. Moreover, a computerized ground mat analysis of the type described herein allows the designer to specifically identify unsafe areas within a proposed mat and to optimize the mat design while verifying that the design is safe throughout the area in question.

14.3 Modeling the Human Body. To properly understand the analytical techniques involved in a ground mat study, it is necessary to understand the electrical characteristics of the most important part of the circuit: the human body. A normal healthy person can feel a current of about 1 milliampere. (Tests have long ago established that electric shock effects are the result of current and not voltage.) Currents of approximately 10–25 milliamperes can cause lack of muscular control. In most men, 100 milliamperes will cause ventricular fibrillation. Higher currents can stop the heart completely or cause severe electrical burns.

For practical reasons, most ground mat studies use the threshold of ventricular fibrillation, rather than muscular paralysis or other physiological factors, as their design criterion. Ventricular fibrillation is a condition in which the heart beats in an abnormal and ineffective manner, with fatal results. Accordingly, most ground mats are designed to limit body currents to values below this threshold. Tests on animals with body and heart weights comparable to those of a man have determined that 99.5% of all healthy men can tolerate a current through the heart region defined by

$$I_b = \frac{0.116}{\sqrt{T}} \qquad \qquad \text{(Eq 211)}$$

where

I_b = Maximum body current in amperes
T = Duration of current in seconds

without going into ventricular fibrillation. This equation is only valid for 60 Hz currents. In practice, most fault currents have a dc offset. This dc component is represented by a correction factor described in a subsequent section in this chapter.

Tests indicate that the heart requires about 5 minutes to return to normal after experiencing a severe electrical shock. This implies that two or more closely spaced shocks (such as those that would occur in systems with automatic reclosing) would tend to have a cumulative effect. Present industry practice considers two closely spaced shocks (T_1 and T_2) to be equivalent to a single shock (T_3) whose duration is the sum of the intervals of the individual shocks ($T_1 + T_2 = T_3$).

Although there are many possible ways that a person may be shocked, industry practice is to evaluate shock hazards for two common, standard conditions. Figures 158 and 159 show these situations and their equivalent resistance diagrams. Figure 158 shows a touch contact with current flowing from the operator's hand to

**Fig 158
Touch Potential**

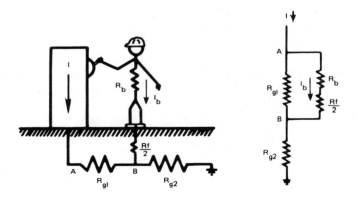

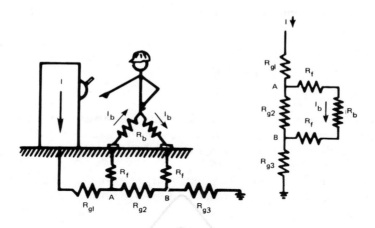

**Fig 159
Step Potential**

his feet. Figure 159 shows a step contact where current flows from one foot to the other. In each case, the body current I_b is driven by the potential difference between points A and B. Exposure to touch potential normally poses a greater danger than exposure to step potential. The step potentials are usually smaller in magnitude, the corresponding body resistance greater, and the permissible body current higher than for touch contacts. (The fibrillation current is the same for both types of contact. In the case of step potentials, however, not all current flowing from one leg to the other will pass through the heart region.) The worst possible touch potential (called "mesh potential") occurs at or near the center of a grid mesh. Accordingly, industry practice has made the mesh potential the standard criterion for determining safe ground mat design. In most cases, controlling mesh potentials will bring step potentials well within safe limits. Step potentials can, however, reach dangerous levels at points immediately outside the grid.

Since the body of an individual who is exposed to an electrical shock forms a shunt branch in an electrical circuit, the resistance of this branch must be determined to calculate the corresponding body current. Generally, the hand and foot contact resistances are considered to be negligible. However, the resistance of the soil directly underneath the foot is usually significant. Treating the foot as a circular plate electrode gives an approximate resistance of $3 \rho_s$, where ρ_s is the soil resistivity. The body itself has a total measured resistance of about 2300 ohms hand to hand or 1100 ohms hand to foot. In the interest of simplicity and conservatism, IEEE Std 80-1986 (ANSI) [1][24] recommends the use of 1000 ohms as a reasonable

---

[24]The numbers in brackets correspond to those in the References at the end of this chapter. IEEE publications are available from the Institute of Electrical and Electronics Engineers, IEEE Service Center, 445 Hoes Lane, Piscataway, NJ 08855-1331.

approximation for body resistance in both models. This yields a total branch resistance

$$R = 1000 \ \Omega + 6 \ \rho_s \qquad \text{(Eq 212)}$$

for foot-to-foot resistance, and

$$R = 1000 \ \Omega + 1.5 \ \rho_s \qquad \text{(Eq 213)}$$

for hand-to-foot resistance where ρ_s is the surface resistivity in ohm-meters and R is expressed in ohms. If the station surface has been dressed with crushed rock or some other high resistivity material, the resistivity of the surface layer material should be used in Eqs 212 and 213.

Because potential is easier to calculate and measure than current, the fibrillation threshold given by Eq 211 is normally expressed in terms of voltage. Combining Eqs 211, 212, and 213 gives the maximum tolerable step and touch potentials:

$$E_{\text{step-tolerable}} = \frac{(1000 \ \Omega + 6 \ \rho_s)(0.116)}{\sqrt{T}} \qquad \text{(Eq 214)}$$

$$E_{\text{touch-tolerable}} = \frac{(1000 \ \Omega + 1.5 \ \rho_s)(0.116)}{\sqrt{T}} \qquad \text{(Eq 215)}$$

Because these voltages are dependent on surface resistivity, most industrial facilities have several different values for each to match the various surface materials found in the plant.

Although in each of the cases discussed, body resistance shunts a part of the ground resistance, its actual effect on voltage and current distribution in the overall system is negligible. This becomes obvious when the normal magnitude of the ground fault current (as much as several thousand amperes) is compared to the desired body current (usually no more than several hundred milliamperes).

14.4 Traditional Analysis of the Ground Mat. The voltage rise of any point within the grid depends upon three basic factors: ground resistivity, available fault current, and grid geometry.

14.4.1 Ground Resistivity. Most ground mat studies assume that the ground grid is buried in homogeneous soil. This is a good model for most soils and simplifies the calculations considerably. Also, many nonhomogeneous soils can be modeled by two-layer techniques. Although reasonably straightforward, these methods involve quite a bit of calculation, making computation by hand difficult. Normally, the two-layer model is necessary only for locations where bedrock and other natural soil layers are close enough to the surface to severely affect the distribution of current.

Of far more serious concern are soils that experience drastic and unpredictable changes in resistivity at various points on the surface. These situations present the following problems:

(1) Difficulty of modeling the soil in calculations
(2) Physical difficulties in finding the area boundaries in the field and measuring each area's local resistivity

At present, these cases are normally handled by the inclusion of a safety margin in the value used for soil resistivity.

IEEE Std 80-1986 (ANSI) [1] and bibliographic reference [B17] contain descriptions of simple methods of measuring soil resistivity for homogeneous and two-layer soils. Because soil resistivity varies with moisture content and, to a lesser degree, with temperature, ideally these measurements should be made over a period of time under different weather conditions. If, for some reason, an actual measurement of resistivity is impractical, Table 47 gives approximate values of resistivity for different soil types. These values are only approximations and should be replaced by measured data whenever possible.

14.4.2 Fault Current — Magnitude and Duration. Since shock hazard is a function of both time and current, a strictly rigorous ground mat analysis would require checking every possible combination of time and current. In practice, the worst shock hazard normally occurs at the maximum fault current. Determination of ground fault current and clearing time normally requires a separate system study. The techniques and problems of making fault studies are covered in numerous sources (including this book). Therefore, this section will only cover aspects peculiar to ground grid studies.

After the system impedance and grid resistance have been determined, the maximum ground fault current (assuming a bolted fault) is given as:

$$I = \frac{3\,V}{3R_g + (R_1 + R_2 + R_0) + j\,(X_1'' + X_2 + X_0)} \qquad \text{(Eq 216)}$$

where:

I = Maximum fault current in amperes (note that this is not the same as the current in I_b in Eq 211).
V = Phase-to-neutral voltage in volts.
R_g = Grid resistance to earth in ohms.
R_1 = Positive sequence system resistance in ohms.
R_2 = Negative sequence system resistance in ohms.
R_0 = Zero sequence system resistance in ohms.
X_1'' = Positive sequence subtransient system reactance in ohms.
X_2 = Negative sequence system reactance in ohms.
X_0 = Zero sequence system reactance in ohms.

This current will, in general, be a sinusoidal wave with a dc offset. Since dc current can also cause fibrillation, the current value I must be multiplied by a

Table 47
Representative Values of Soil Resistivities

Type of Ground	Resistivity in Ohm-meters
Wet organic soil	10
Moist soil	10^2
Dry soil	10^3
Bedrock	10^4

correction factor called the decrement factor to account for this effect. Table 48 gives approximate values for this factor. For more accurate results, the exact value for the decrement factor D is given by the equation:

$$D = \sqrt{\frac{1}{T}\left[T + \frac{1}{\omega} \cdot \frac{X}{R}\left(1 - e^{\frac{-2\omega T}{X/R}}\right)\right]}$$

(Eq 217)

where:

T = Duration of fault in seconds.
ω = System frequency in radians per second.
X = Total system reactance in ohms.
R = Total system resistance in ohms.

The current value calculated in Eq 216 must be multiplied by this factor to find the effective fault current. Note that the time T in Eq 217 is the same as that used in Eqs 211, 214, and 215.

A common mistake in calculating ground mat current is to ignore alternate current paths. In most systems, only a portion of the ground fault current will return to the source through the earth. Because of the time and expense involved in running a full scale short-circuit study to accurately account for the division of fault current, the worst-case situation based upon the full short-circuit capability of the fault source is generally used.

To determine the fault duration, it is necessary to analyze the relaying scheme to find the interrupting time for the current calculated by Eq 216. The choice of the clearing time of either the primary protective devices or the backup protection for the fault duration depends upon the individual system. Designers must choose between the two on the basis of the estimated reliability of the primary protection and the desired safety margin. Choice of backup device clearing time is more conservative, but it will result in a more costly ground mat installation. Substitution of this time in Eqs 214 and 215 will fix the maximum allowable step and touch potentials at the appropriate values.

14.4.3 Fault Current—The Role of Grid Resistance. In most power systems, the grid resistance is a significant part of the total ground fault impedance. Accurate calculation of ground fault currents requires an accurate and dependable

Table 48
Decrement Factor for Use in Calculating Electrical
Shock Effect of Asymmetrical AC Currents

Shock and Fault Duration		Decrement Factor
Seconds	Cycles (60 Hz)	
0.008	½	1.65
0.1	6	1.25
0.25	15	1.10
0.5 or more	30 or more	1.0

value for the grid resistance. Equation 218 (taken from IEEE Std 80-1986 (ANSI) [1]) gives a quick and simple formula for the calculation of resistance when a minimum of design work has been completed.

$$R = \frac{\rho}{4r} + \frac{\rho}{L} \qquad \text{(Eq 218)}$$

where:

R = Grid resistance to ground in ohms.
ρ = Soil resistivity in ohm-meters.
L = Total length of grid conductors in meters.
r = Radius of a circle with area equal to that of the grid in meters.

The first term gives the resistance of a circular plate with the same area as the grid. The second term compensates for the grid's departure from the idealized plate model. The more the length of the grid conductors increases, the smaller this term becomes. This equation is surprisingly accurate and is ideal for the initial stages of a study where only the most basic data about the ground mat is available.

By inspecting Eq 218, it also becomes evident that adding grid conductors to a mat to reduce its resistance eventually becomes ineffective. As the conductors are crowded together, their mutual interference increases to the point where new conductors tend only to redistribute fault current around the grid, rather than lower its resistance.

Any computer program that can calculate the grid voltage rise can also calculate the grid resistance (with greater accuracy than the method described immediately above). The grid resistance is simply the total grid voltage rise (relative to a "remote" ground reference) divided by the total fault current. In many cases, such programs perform this calculation automatically. This method can be applied to any grid configuration with any number of conductor elements. However, because the more advanced of these programs calculate grid voltages by solving hundreds of simultaneous equations, the same procedure is usually not practically achievable with hand calculations.

Since grid resistance is viewed as a measure of the grid's ability to disperse ground fault current, many designers are tempted to use resistance as an indicator of relative safety of a ground mesh. In general, however, there is no direct correlation between grid resistance and safety. At high fault currents, dangerous potentials exist within low resistance grids. The only occasion where a low grid resistance can guarantee safety is when the maximum potential rise of the entire grid (that is, grid potential) is less than the allowable touch potentials. In these cases, the ground mat is inherently safe.

14.4.4 Grid Geometry. The physical layout of the grid conductors plays a major role in ground mat analysis. The step and touch potentials depend upon grid burial depth, length and diameter of conductors, spacing between each conductor, distribution of current throughout the grid, and proximity of the fault electrode and the system grounding electrodes to the grid conductors, along with many other factors of lesser importance. A perfectly rigorous analysis of all these variables would require both simultaneous linear and complex differential equations to exactly describe the distribution of current throughout the grid. It is hardly surprising that the quantitative effect of these factors upon touch and step potentials is one of the least discussed aspects of grid analysis. Paradoxically, these factors include most of the elements of grid design that normally can be changed to control grid voltages.

Until recently, IEEE Std 80-1986 (ANSI) [1] provided the only practical method for computing the effects of the grid geometry upon the step and touch potentials (Eqs 219 and 220).

$$E_{mesh} = K_m K_i \rho \frac{I}{L} \tag{Eq 219}$$

$$E_{step} = K_s K_i \rho \frac{I}{L} \tag{Eq 220}$$

where:

E_{mesh} = Worst-case touch potential at the surface above any individual grid area (i.e., "mesh") within the mat.
E_{step} = Worst-case step potential anywhere above the mat.
ρ = Soil resistivity in ohm-meters.
I = Maximum total fault current in amperes (adjusted for the decrement factor).
L = Total length of grid conductors in meters.
K_m = Mesh coefficient.
K_s = Step coefficient.
K_i = Irregularity factor.

Coefficients K_m and K_s are calculated by two reasonably simple equations based upon the number of grid elements, their spacing and diameters, and the burial depth of the grid. These equations are not meant to rigorously model a grid design, but are instead intended to make hand calculation of touch and step potentials feasible. Equations 219 and 220 incorporate an irregularity factor K_i to compensate for the inaccuracies introduced by these simplifying assumptions. Except for applications involving very simple grid configurations, proper selection of a value for K_i is totally dependent upon the experience and judgment of the designer. K_m and K_s can only be calculated for regular grid designs. Most often, a high value is picked for K_i in the interest of conservatism, which usually results in an overdesigned mat. Conversely, there is no way to determine if the value is too low, resulting in an unsafe ground mat design.

The values of E_{mesh} and E_{step} calculated by Eqs 219 and 220 must be compared to the tolerable touch and step potentials, $E_{touch-tolerable}$ and $E_{step-tolerable}$ as determined from Eqs 214 and 215 in order to establish whether or not the design is safe. If, in fact, one of the tolerable voltage limits is exceeded, it is sometimes possible by inspection of the grid to determine mesh locations where additional cross-conductors should be added in order to achieve a safe design. The more general approach, however, is to uniformly increase the number of grid conductors.

Although this traditional hand calculation method for determining step and mesh potentials was considered acceptable in the past, modern ground mat studies normally use one of the new generation of computer programs. There are two types of computer programs available for ground mat studies. One type performs the aforementioned traditional hand calculations for empirically determining step and mesh potentials, but does it faster and more efficiently than possible by hand. The other type of program calculates the step and touch potentials for each individual grid (i.e., "mesh") within the overall ground mat. The results, therefore, allow a more detailed analysis of ground mat design effectiveness, pinpointing any mesh locations where shock hazards may exist. The discussion that follows will concentrate on the latter of the two program types.

14.5 Advanced Grid Modeling. The key to an accurate ground grid analysis is the individual modeling of each single grid element, rather than the en masse treatment used in IEEE Std 80-1986 (ANSI) [1]. For example, Fig 160 shows a single grid element located at some depth h below the earth's surface. The element runs from point (x_1, y_1, z_1) to (x_1, y_2, z_1) and is radiating current to the surrounding earth at the linear current density σ_l (the current per unit length). By integrating σ_l over the length of the grid element, the current flux ξ can be found at any desired point a as follows:

$$\xi = \int_{y_1}^{y_2} \frac{\sigma_l}{4\pi} \frac{dy}{R^2} \, r \qquad \text{(Eq 221)}$$

**Fig 160
Physical Model Used in Calculating Voltage
at Point a Due to a Single Conductor**

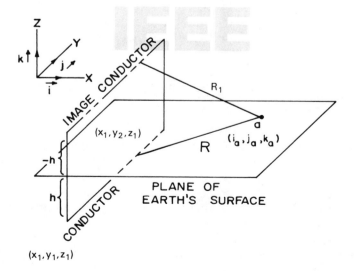

where:

ξ = Current per unit area at any point.

σ_l = Current flowing to ground per unit length of conductor (current density).

$$R = \sqrt{(i-x_1)^2 + (j-y_1)^2 + (k-z_1)^2}$$

$$r = \frac{(i-x_1)\,i + (j-y_1)\,j + (k-z_1)\,k}{\sqrt{(i-x_1)^2 + (j-y_1)^2 + (k-z_1)^2}}$$

NOTE: For the purposes of illustration, Eq 221 shows a special-case expression that is only valid for lines running parallel to the y axis. The more general form is derived in the same manner, but is much more difficult to follow.

Once ξ has been determined, the E field at the same point can be expressed as follows (assuming a homogeneous soil):

$$E = \rho \xi \tag{Eq 222}$$

where ρ is the soil resistivity.

From this, the voltage at point a can be obtained by performing the integration:

$$V_{al} = -\int_{\infty}^{a} E \cdot dl \tag{Eq 223}$$

or

$$V_{al} = -\frac{\rho \sigma_l}{4\pi} \ln \left[j_a - y_2 + \sqrt{(i_a-x_1)^2 + (j_a-y_2)^2 + (k_a-z_1)^2} \right]$$

$$+ \frac{\rho \sigma_l}{4\pi} \ln \left[j_a - y_1 + \sqrt{(i_a-x_1)^2 + (j_a-y_1)^2 + (k_a-z_1)^2} \right] \tag{Eq 224}$$

where V_{al} is the absolute potential at any point a due to line ℓ. This process must be repeated for every element in the grid.

NOTE: This process is complicated somewhat by the presence of the "current density" factor σ_l in the equations. Although Eqs 211 and 212 treat σ_l as a constant, in actuality it varies continuously along the length of each grid element, as well as from element to element. In practice, the variation of σ_l along the length of an element has little effect upon the calculated voltages, expecially when calculating mesh potentials. The variation between elements is very significant, however, and must be obtained by solving a set of simultaneous equations. These can be written by using Eq 224 to calculate the voltage at known points (the surface of each grid element, for example). When the variation of current density along an element is important, it can be approximated by modeling the element as several segments, each with its own value of σ_l.

Finally, the individual contribution of each grid element can be summed to determine the total voltage at point a.

The advantages of this analytical method are immediately apparent. This technique automatically accounts for the finite length of each element, a particularly important consideration when finding the potential at points near the end of an element. It can handle grid designs with large degrees of asymmetry with no sacrifice in accuracy. Furthermore, since point a can be located anywhere and any number of points can be examined, detailed analysis of the grid design is possible.

The grid layout also determines which points should be checked for touch and step potentials. Touch potentials are normally calculated at the mesh centers, at control stations (where operators may be present), at the entrances to the facility, and at the corners of the grid. Step potentials are rarely a problem inside the grid. However, they may be a concern in the areas just outside the grid. The worst step potentials usually occur along a diagonal line at the corners of a grid (see Fig 166).

14.6 Benchmark Problems. Every analytical procedure (including computer programs) needs to be checked against a benchmark problem. Figure 161 shows six different grid layouts where the mesh potentials were measured on small scale models. These designs are the best available benchmark problems for ground mat analysis programs. The calculated voltages (expressed as a percentage of the grid voltage) are shown in the center of each mesh. For easy reference, the actual measured voltages are also shown in parentheses. For this analysis, a program calculated the absolute voltage at the center of each mesh. In an actual ground mat study, the calculated point voltages would be subtracted from the calculated grid potential to determine the touch potential hazard. Figure 161 (especially grids E and F) illustrates the accuracy of the program. Errors are typically on the order of 5%, and never worse than 10%.

The most impressive feature of this type of program, however, is its ability to calculate voltages at any point of interest within or around the mat's geometric boundaries. By repeated use of the program throughout the design process, a ground mat layout can be fine-tuned to achieve the desired protection without the need to overdesign any section of the mat.

14.7 Input/Output Techniques. The increased use of personal computers in the workplace has led to higher expectations for all software. Ground mat analysis programs are no exception. Since the grid layout is so important to this class of programs, they are especially well suited to graphical input/output methods. First generation programs typically required the user to input the end coordinates of each grid element, and simply printed the calculated voltages at user-designated coordinates. Figure 167 shows a typical output report from an early ground mat analysis computer program (in this particular case, grid E of Fig 161). Although the basic calculations are correct, interpretation is difficult and checking the data entry is laborious and time consuming.

Most modern ground mat analysis programs support some form of graphical output. Many allow the user to draw the grid design and then calculate the endpoints internally. With proper preparation, others are capable of reading the grid design directly from CADD drawing files. Some programs can select the points to calculate automatically or plot equipotential lines. Ground mat programs can also do material take-offs as well as material and labor cost estimates. The latest programs not only calculate raw voltages, they compute the mesh and touch potentials and compare them to the limits (also automatically calculated). Typically, they give the user several different views of the same data. New programs can also manage several different surface materials on the same grid, showing the true shock hazard at every point simultaneously. They can also store intermediate calculations for

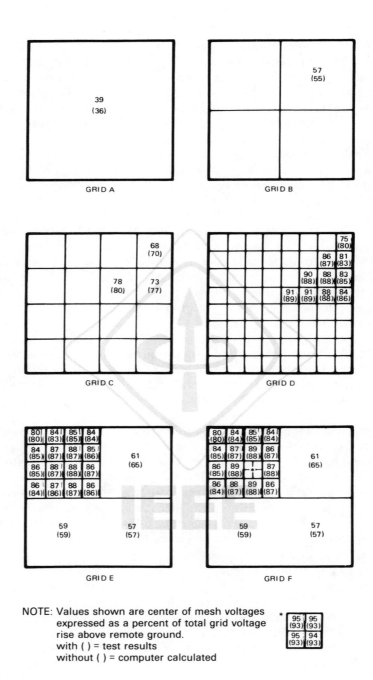

Fig 161
Experimental Grids Showing Various (Mesh) Arrangements

later use. Advanced programs can edit their input data and ignore unimportant data. Figure 168 shows a graphical output from a typical ground mat analysis program of recent vintage. The conductor arrangement shown affords a ground mat design that has a 5% minimum safety margin.

14.8 Sample Problem. Figures 162 through 165 demonstrate the use of a ground mat analysis program on a typical ground mat design taken from IEEE Std 80-1986, Appendix B [1]. Figure 162 shows the grid layout and all pertinent data. Figure 163 gives the results of a computer analysis of the grid clearly indicating that grid corners are unsafe. Figures 164 and 165 show the results of modified grid designs. Note that in Fig 165 the amount of additional grid conductor required to safely control mesh potentials in the grid has been minimized and that the use of the computer program has permitted the location of this conductor to be optimally determined (that is, the conductor has been added only where required). This is the great advantage of a computer ground mat analysis.

Fig 162
Typical Ground Mat Design Showing
All Pertinent Soil and System Data

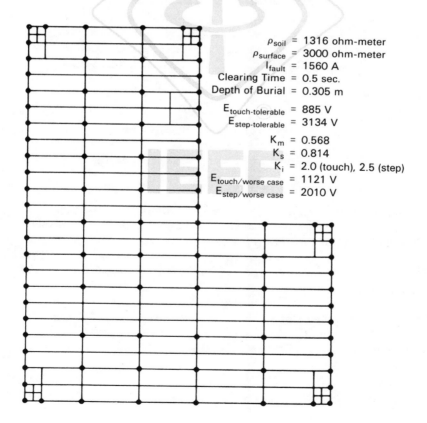

ρ_{soil} = 1316 ohm-meter
$\rho_{surface}$ = 3000 ohm-meter
I_{fault} = 1560 A
Clearing Time = 0.5 sec.
Depth of Burial = 0.305 m

$E_{touch\text{-}tolerable}$ = 885 V
$E_{step\text{-}tolerable}$ = 3134 V

K_m = 0.568
K_s = 0.814
K_i = 2.0 (touch), 2.5 (step)
$E_{touch/worse\ case}$ = 1121 V
$E_{step/worse\ case}$ = 2010 V

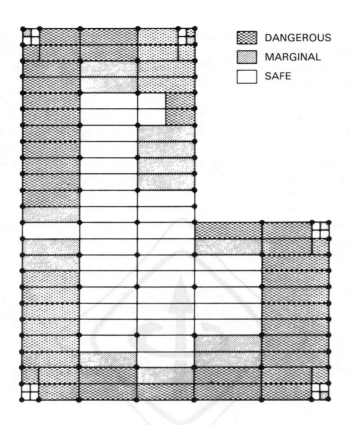

Fig 163
Typical Ground Mat Design Showing
Meshes with Hazardous Potentials
as Identified by Computer Analysis

14.9 Conclusion. The adaptation of classical analytical techniques and calculating procedures to the digital computer has made ground mat analysis much more precise, reliable, and useful. Many ground mat analysis programs can all but eliminate unnecessary grid overdesign and, properly applied, all ground mat analysis programs can detect unsafe conditions that might otherwise go undiscovered until made apparent by serious mishap.

A ground mat study requires, as a minimum, the following data:
  (1) Soil resistivity, both at the surface and at the level of the grid
  (2) Resistivity of any special soil surface dressing material
  (3) Estimated duration of a ground fault
  (4) System frequency
  (5) System X/R ratio
  (6) Maximum symmetrical ground fault current, both future and present

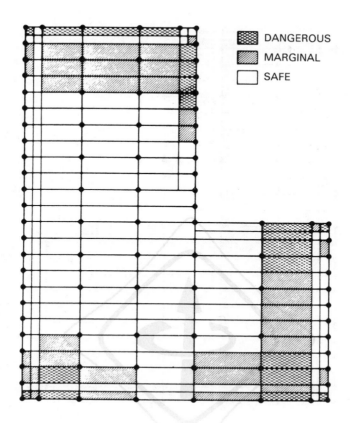

Fig 164
Typical Ground Mat Design, First Refinement
Showing Meshes with Hazardous Touch Potentials

(7) Grid layout showing the precise location of every conductor
(8) Coordinates where the potential rise must be calculated
Consideration of all this information will lead to a reliable, accurate, and useful study.

Two basic types of ground mat programs have been discussed. When considering the purchase or use of such programs, it first must be determined what level of analytical accuracy is required and, accordingly, which type of calculating method is desired; the traditional empirical method discussed in 14.4 of this chapter, or the detailed mesh-by-mesh method described in 14.5 through 14.8. Care should then be taken to select a program that is appropriate for the application.

Although ground mat analysis programs provide an invaluable design tool, they are by no means infallible. If at all possible, a followup investigation should be made of each grid after it has been installed. This should include a measurement of grid

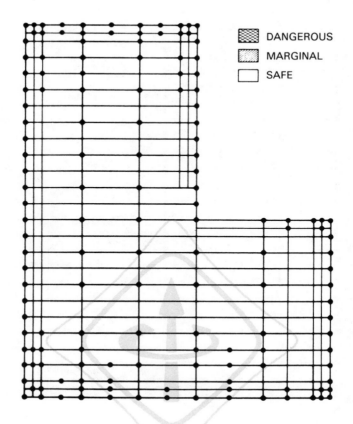

Fig 165
Typical Ground Mat Design, Final Refinement
with No Hazardous Touch Potentials

resistance at the very least, and preferably the measurement of the ac mesh potential at several locations within the grid. If these measured values differ appreciably from the calculated ones, the results of the grid study should be rechecked and supplemental rods or buried conductors provided as required to establish safe conditions (see IEEE Std 80-1986 (ANSI) [1]).

14.10 Reference

[1] IEEE Std 80-1986, IEEE Guide for Safety in AC Substation Grounding (ANSI).

14.11 Bibliography

[B1] DALZIEL, C. F. "Dangerous Electric Currents," *AIEE Transactions*, vol. 65, pp. 579–585 and 1123–1124.

[B2] DALZIEL, C. F. "Threshold 60-cycle Fibrillating Currents," *AIEE Transactions*, vol. 79, pp. 667–673, 1960.

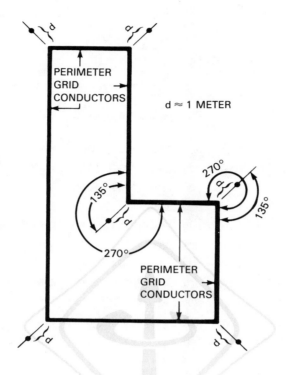

**Fig 166
Typical Ground Mat Layout Showing
Possible Locations of Critical Step and
Touch Potentials Near Grid Corners**

[B3] DALZIEL, C. F. "A Study of the Hazards of Impulse Currents," *AIEE Transactions*, vol. 72, pp. 1032–1043, 1953.

[B4] DALZIEL, C. F. and LEE, W. R. "Re-evaluation of Lethal Electric Currents," *IEEE Transactions on Industry General Applications*, vol. IGA-4, pp. 467–476, Sept./Oct. 1968.

[B5] DAWALIBI, F. and MUKHEDKAR, D. "Multi-Step Analysis of Interconnected Grounding Electrodes," *IEEE Transactions on Power Applications Systems*, vol. PAS-95, pp. 113–119, Jan./Feb. 1976.

[B6] DAWALIBI, F. and MUKHEDKAR, D. "Optimum Design of a Substation Grounding in a Two-Layer Earth Structure," *IEEE Transactions on Power Applications Systems*, vol. PAS-94, no. 2, pp. 252–272, Mar./Apr. 1975.

[B7] FERRIS, L. P., KING, B. G., SPENCE, P. W., and WILLIAMS, H. B. "Effect of Electrical Shock on the Heart," *AIEE Transactions*, vol. 55, pp. 498–515 and 1263, May 1936.

GROUND GRID VOLTAGE STUDY   EXAMPLE PROBLEM
FIGURE 161, GRID E

———————————————— RESISTIVITY = 1000.0 ————————————

FAULT CURRENT (AMPERES)		500.		1000.		4000.			
GRID COORDINATES (MILLIMETERS)		VOLTAGE A = ABSOLUTE POTENTIAL B = TOUCH POTENTIAL							
		A	B	A	B	A	B	A	B	A	B
(30.0,	30.0)	1372.	951.	2743.	1902.	10973.	7610.				
(90.0,	30.0)	1307.	1016.	2613.	2032.	10453.	8129.				
(90.0,	90.0)	1418.	905.	2836.	1809.	11344,	7238.				
(7.5,	67.5)	1984.	339.	3968.	678.	15871.	2711.				
(7.5,	82.5)	1975.	348.	3950.	696.	15798.	2784.				
(7.5,	97.5)	1942.	381.	3883.	762.	15534.	3048.				
(7.5,	112.5)	1851.	472.	3701.	944.	14805.	3777.				
(22.5,	67.5)	2015.	307.	4031.	615.	16123.	2459.				
(22.5,	82.5)	2023.	300.	4046.	599.	16184.	2398.				
(22.5,	97.5)	2001.	322.	4001.	644.	16005.	2577.				
(22.5,	112.5)	1936.	387.	3871.	774.	15484.	3098.				
(37.5,	67.5)	2019.	304.	4037.	608.	16150.	2432.				
(37.5,	82.5)	2032.	291.	4064.	582.	16254.	2328.				
(37.5,	97.5)	2016.	307.	4032.	614.	16128.	2454.				
(37.5,	112.5)	1963.	360.	3926.	720.	15702.	2880.				
(52.5,	67.5)	1981.	342.	3961.	684.	15846.	2736.				
(52.5,	82.5)	1975.	348.	3950.	696.	15800.	2782.				
(52.5,	97.5)	1960.	363.	3920.	726.	15679.	2903.				
(52.5,	112.5)	1931.	392.	3862.	784.	15447.	3136.				
GRID POTENTIAL		2323.		4646.		18582.					

NOTE: GRID IS 120×120 MILLIMETERS, ORIGIN (0,0)
IS LOCATED AT LOWER LEFT CORNER.

**Fig 167
Sample Output Report—Computer Calculated Ground Grid Potentials**

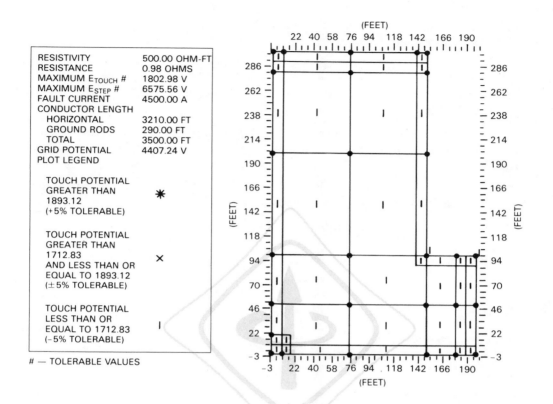

RESISTIVITY   500.00 OHM-FT
RESISTANCE   0.98 OHMS
MAXIMUM E_{TOUCH} #   1802.98 V
MAXIMUM E_{STEP} #   6575.56 V
FAULT CURRENT   4500.00 A
CONDUCTOR LENGTH
  HORIZONTAL   3210.00 FT
  GROUND RODS   290.00 FT
  TOTAL   3500.00 FT
GRID POTENTIAL   4407.24 V
PLOT LEGEND

  TOUCH POTENTIAL
  GREATER THAN   ✳
  1893.12
  (+5% TOLERABLE)

  TOUCH POTENTIAL
  GREATER THAN
  1712.83   ✕
  AND LESS THAN OR
  EQUAL TO 1893.12
  (±5% TOLERABLE)

  TOUCH POTENTIAL
  LESS THAN OR
  EQUAL TO 1712.83   I
  (−5% TOLERABLE)

— TOLERABLE VALUES

Fig 168
Sample Computer Generated Graphical Output Report for a Ground Mat
Similar to Fig 161 But with Different Soil and Ground Fault Conditions

[B8] GEDDES, L. A. and BAKER, L. E. "Response to Passage of Electric Current Through the Body," *J. Association Advancement of Medical Instruction*, vol. 2, pp. 13–18, Feb. 1971.

[B9] GROSS, E. T. B., CHITNIS, B. V., and STRATTON, L. J. "Grounding Grids for High-Voltage Stations," *AIEE Transactions*, vol. 72, pp. 799–810, 1953.

[B10] GROSS, E. T. B. and HOLLITCH, R. F. "Grounding Grids for High-Voltage Stations — III, Resistance of Rectangular Grids," *AIEE Transactions*, vol. 75, pt. III, pp. 926–935, 1953.

[B11] GROSS, E. T. B. and WISE, R. B, "Grounding Grids for High-Voltage Stations — II," *AIEE Transactions*, vol. 74, pt. III, pp. 801–809, 1955.

[B12] KOCH, W. "Grounding Methods for High-Voltage Stations with Grounded Neutrals," *Electrotechnische Zeit*, vol. 71, no. 4, pp. 89–91, 1950.

[B13] LAURENT, P. "General Fundamentals of Electrical Grounding Techniques," *Bulletin de la Societe Francaise des Electriciens*, vol. I, series 7, pp. 368–402, Jul. 1951.

[B14] NIEMANN, J. "Changeover from High-Tension Grounding Installation to Operation with a Grounded Star Point," *Electrotechnische Zeit*, vol. 73, no. 10, pp. 333–337, May 15, 1952.

[B15] RUDENBERG, R. "Grounding Principals and Practice I—Fundamental Considerations on Ground Currents," *Electrical Engineering*, vol. 64, pp. 1–13, Jan. 1945.

[B16] SCHWARTZ, S. J. "Analytical Expression for Resistance of Grounding System," *AIEE Transactions*, vol. 73, pp. 1011–1016, 1954.

[B17] SUNDE, E. D. *Earth Conduction Effects in Transmission Systems*, New York: Van Nostrand, 1949.

Chapter 15
Coordination Studies

15.1 Introduction. An important step in the design of any power distribution system is the time-current coordination of all overcurrent protective devices required for the protection of the equipment. When a short circuit or an abnormal power flow occurs for a sustained period of time, the protective devices should react to isolate the problem with a minimum of disruption to the balance of the system.

The reader should be aware that this chapter addresses only one aspect of system protection. For most large, medium voltage, and high voltage systems, overcurrent protection acts only as backup for primary protection and, as such, is not necessarily a complete study of system protection.

The operation of protective devices can be estimated by a graphic representation of the time-current characteristics curves (TCC's) of these devices. By plotting these characteristics on a common graph, the relationship of the characteristics among the devices is immediately apparent. Any potential trouble spots, such as overlapping of curves or unnecessarily long time intervals between devices, are revealed. By indicating on the current scale the maximum and minimum value of short-circuit currents (three-phase and line-to-ground) that can occur at various points in the circuit, the operation of circuit protective devices can be estimated for various fault conditions. An accompanying single-line diagram can indicate the components and define their location in the circuit.

The time-honored method of plotting these curves, as illustrated in Fig 169, is to superimpose a transparent sheet of graph paper over the manufacturer's published curves on an illuminated drafting table. When the time scales have been carefully matched and the current scales adjusted, the curve is traced onto the graph paper using a French curve or flexible spline. The process is then repeated for the remaining protective devices. Damage curves for equipment, such as motors, transformers, and cables, are also plotted to assess the level of protection provided for the equipment and to provide a graphical representation of the protection achieved. This process can be both tedious and time consuming.

Since the selection of device parameters follows, for the most part, well-defined rules, computer programs are available for the task of producing these curves. Some programs will select the settings necessary to achieve a well-coordinated system as well as plotting curves and related data.

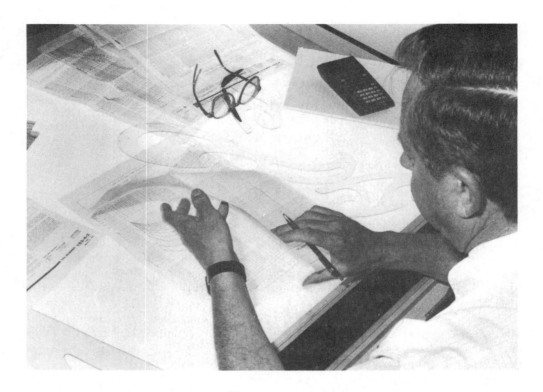

Fig 169
Manual Method of Producing Time-Current Curves
by Using a Light Table

The application of the computer and computer software to time-current coordination studies is a viable alternative to the manual approach. With the availability of dependable hardware for digitizing, plotting, computing, and communicating with a computer, time-current coordination studies using these tools are a practical alternative that is now possible. Device settings and ratings may be calculated and tabulated. The TCC drawings may be displayed on graphic monitors, and the output reports may be routed to graphical printers and plotters. The popular K&E form 48-5258 may be used as a plot background, when desired.

15.2 Basics of Coordination. Whether the coordination is done manually or by computer, it is necessary for the engineer to "describe" the system. The information needed to perform a coordination study is a single-line diagram showing:

- Protective device manufacturer and type
- Protective device ratings
- Trip settings and available ranges
- Short-circuit current at each system bus (three-phase and line-to-ground)
- Full load current of all loads

- Voltage level at each bus
- Transformer kVA, impedance, and connections (delta-wye, etc.)
- CT and PT ratios
- Cable size, conductor material, and insulation
- All sources and ties

Figure 170 shows a section of the single-line diagram presented in Chapter 1 with protective devices and other information added using the normal conventions employed for coordination studies. Figure 171 is a copy of the figure in IEEE Std 141-1986, Recommended Practice for Electric Power Distribution for Industrial Plants (ANSI),[25] page 257, showing a coordination study with a manually drawn plot. Figure 172 uses the same single-line diagram and shows the study as done with a computer program.

Information is needed regarding the time-current characteristics of the devices in the circuit. Traditionally this information is in the form of manufacturer's TCC curves on 11 inch by 17 inch, 4½ by 5 cycle log-log paper. When coordination is performed on a computer, this information is stored in a device data file or "library."

The coordination process starts with the farthest downstream device. This downstream device is sketched with a characteristic and setting that allows full load current to flow and prevents tripping for transient and normal load conditions. The device immediately upstream is next, with its characteristics and settings selected to satisfy the specified current requirements and to coordinate with the downstream device. This procedure is followed for each device either by use of a light table and shifting and sketching or by giving the information to the computer and allowing it to show the coordination. When a transformer is encountered, the impedance, connection, and rating are needed to properly select settings and ratings of upstream devices.

The engineer, when working on a coordination study either manually or by computer, will encounter situations where there are devices of a specific size or setting that cannot be changed or when other constraints make it impossible to obtain perfect coordination. In a situation such as this, the design engineer must make a compromise judgment, based on his or her training and experience. The ability to try various settings to determine the best coordination must be a feature of any computer program.

Once the overcurrent coordination study is made, a ground-fault current coordination study should be performed using separate plots because of different fault current levels. The results should be compared with the phase overcurrent protection to verify the coordination.

15.3 Computer Programs for Coordination. Several types of computer programs for the coordination of circuit protective devices are available to the design engineer. For some time to come, this area is expected to be a rapidly changing technology. For purposes of illustration, two types of programs will be introduced. One

---

[25] IEEE publications are available from the Institute of Electrical and Electronics Engineers, IEEE Service Center, 445 Hoes Lane, Piscataway, NJ 08855-1331.

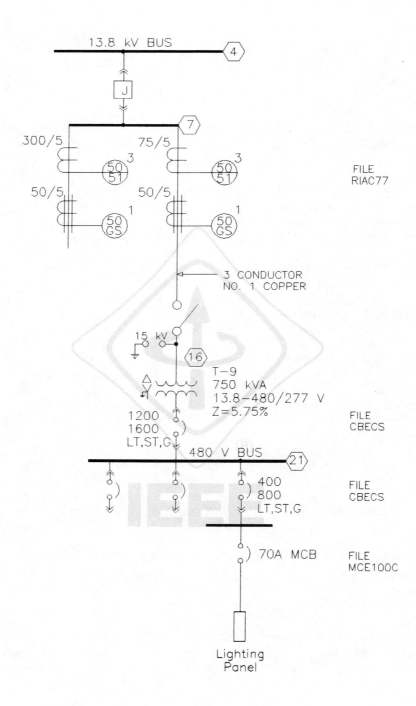

Fig 170
Single-Line Diagram Showing Notations Relative to Coordination

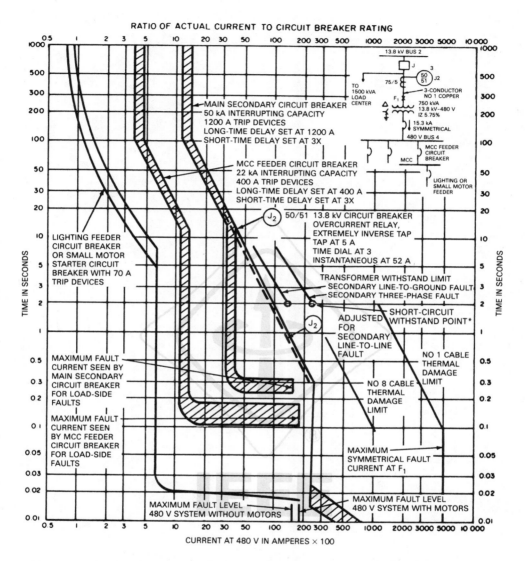

*As defined in ANSI C57.12.00-1973, which has been superseded by IEEE C57.12.00-1987.

Fig 171
Manually Produced Time-Current Curve

type of program is designed to select circuit breaker trip settings, overcurrent relay tap and time dial settings, and fuse ratings and to plot the TCC curves on the standard 4½ by 5 cycle format. The second type stores the data, makes it accessible to the user, but does not perform the coordination. It produces the plot based on the settings chosen by the user.

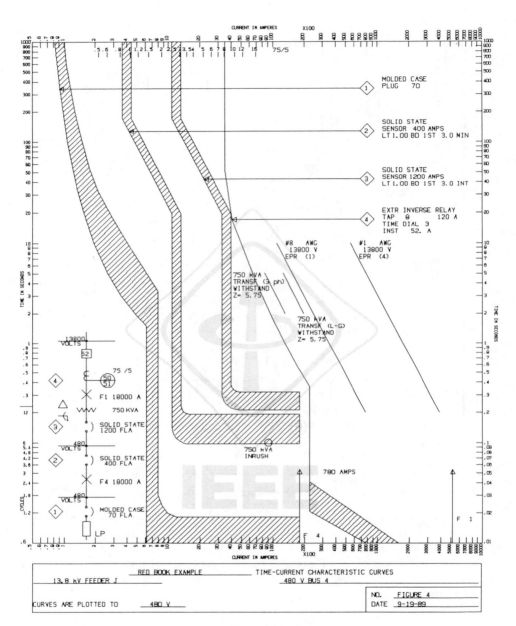

Fig 172
Computer-Produced Time-Current Curve
Plotted on a Printer

15.3.1 Coordination Programs. The first type of program to be discussed makes the selections of settings and ratings unless these items are input by the user. In such a program, the time-current characteristic data for various manufacturers' fuses, relays, and circuit breakers must be stored. The program chooses the ratings and settings to satisfy the stated input conditions. The engineer should accept the responsibility of reviewing the selections and making the final determinations in any coordination study.

Such programs may include the necessary logic to calculate and plot transformer inrush and withstand as well as cable damage and motor starting curves.

If this is the case, the program will select the setting of the first device to ensure that it will not trip on motor starting and will provide the commands to plot the curve. From that point on, the program will perform the coordination based upon the type of device used and the parameters built into the program for clearance with downstream devices using accepted coordination criteria for separation between curves.

Programs that perform the coordination are usually structured so that any device parameter may be specified by the user. If the settings are input, the program may not perform the usual checks for that device, since the logic of the program may be based on the assumption that the user is aware of what he or she is doing. With this method, the engineer may add one device at a time and view it before continuing, or may input the entire set of data before viewing it on a graphics screen.

With this type of program, information on the type of device selected may be available for use in developing and plotting the single-line diagram.

15.3.2 TCC Plotting Programs. The second type of coordination program is the type that provides a library for drawing coordination curves and other features usually shown on a TCC curve. The same type of plot can be obtained as in the first coordination type by having the user enter the settings and ratings and construct the curves using the library files. All decisions associated with the selection of device settings are performed by the engineer.

15.4 Common Structure for Computer Programs. The primary task of the protective device coordination computer program is to permit the engineer to produce coordination studies that are similar to the studies that would be created using manual techniques.

A well-structured software program will contain features to model various types of protective devices and equipment damage characteristics and to store these characteristics in a device library. The software should be able to call these devices from the library and to faithfully reproduce the manufacturer's curves on the graphic output device. In addition, the software should generate documentation of the studies including output reports indicating the device settings and single-line diagrams indicating the elements of the power system described on the TCC drawing.

15.4.1 Project Data Base Files. A data base is a method by which the computer program stores the information contained in the study on some type of permanent magnetic media, usually a magnetic disk. Ideally, the data base will be structured to hold all of the information on all of the devices located in the power system. Coordination studies usually consist of a number of different TCC drawings. A single device, say a main breaker, may be shown on more than one drawing. Keeping the characteristics of the devices in specified groups will eliminate duplicate data entries for displaying the same device on a second or third drawing. This data base structure will eliminate the possibility of plotting the same device with different settings on different drawings.

15.4.2 Interactive Data Entry. The selection of the devices is normally an interactive procedure whereby the software prompts the user for the designation and description of the devices to be used.

It is common for the program to display information on all the devices in the library when requested by the user.

Regardless of whether the program is the type that performs the coordination, the type that gives the engineer the choice of entering his or her own settings, or the type that does both, rapid feedback of the information to the engineer via the computer monitor is important. Viewing the screen should provide accurate information so that a plot is not necessary until the study shows what the engineer wants it to show.

It is convenient when the software provides the opportunity for the engineer to modify the data, by changing plotting voltage, current setting, or any device parameter, without leaving the program. Features to zoom into areas of the display and directly measure time intervals between device operating characteristics may be available.

Figure 173 shows an engineer zooming a plot on the screen to determine which area needs to be checked further.

15.4.3 User-Defined Device Libraries. Good software will provide for a user-defined device library. Protective devices, when not available in the program library, are modeled by the engineer using some type of digitizing procedure, mathematical modeling, or a combination of the two. Since output devices cannot draw curves, curves are simulated by the software as very short straight lines.

For some types of protective devices, mathematical modeling is preferred. For example, the solid-state trip device for a low voltage circuit breaker may have literally thousands of possible settings. It is not practical to store each of these as a separate curve in the data base. Mathematical modeling can accomplish this task with a minimum of data.

15.4.4 Single-Line Diagram Generator. To assist in documentation for the study, the software may include some means of generating a single-line diagram. One method of accomplishing this is to have the software automatically generate the single-line diagram as the devices are entered. With other programs, the single-line diagram may be developed using computer-aided design and drafting (CADD) software. The CADD software may be included with the coordination software. Some programs produce graphic commands that may be used with popular CADD systems, so that the single-line diagram may be added or enhanced with the software residing on the computer.

15.4.5 Graphics Monitor. Most software has provisions to view the coordination study on a graphics monitor before requesting a plot on the printer or plotter.

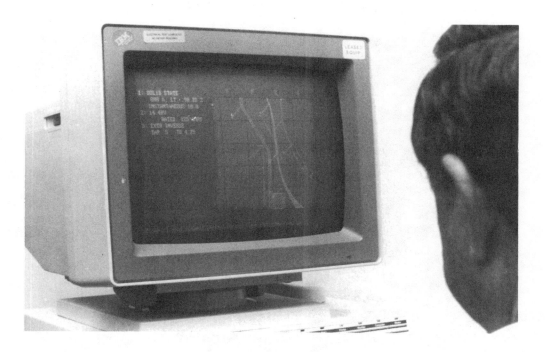

Fig 173
Engineer Shown Viewing a CRT Monitor While Using
the Computer Method

Some method should be provided with the screen plot so that the engineer can determine if the results of the coordination are satisfactory and can accurately establish the clearance between devices. An opportunity to modify, add, or delete devices should be available. Figure 174 is an example of a screen plot.

15.4.6 Plotter/Printer Graphical Interface. In today's computer environment, the engineer is able to select from a wide range of graphic display plotters and printers. Programs are normally written to use plot drivers so that the software will be able to support a wide range of hardware. Plot drivers are programs that translate the data from the internal computer representation to a form usable by the plotting hardware.

Software is used to provide the output format of the data that the graphics software sends to the plot driver. The use of this method permits a single program to support most types of hardware without modification of the basic software package. Such programs permit the engineer to upgrade hardware without fear that the software will become obsolete.

15.4.7 Graphical Output Reports. The computer program should be capable of producing the TCC drawings in a wide range of formats on a wide range of graphical output devices. As a minimum, the software should be able to produce the drawings on a 4½ by 5 cycle K&E form 48-5258 preprinted log-log paper. Figure 175 shows such output.

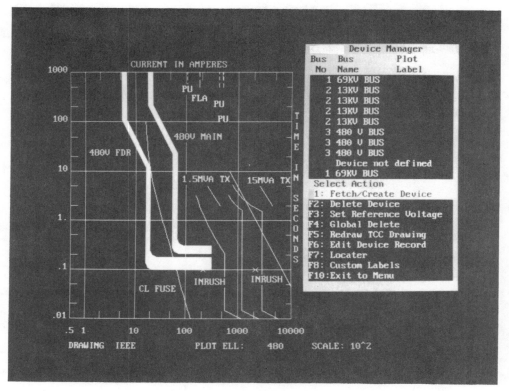

Fig 174
Example of Screen Plot

Most computer programs support graphical output to both printers and plotters. Features are provided for the engineer to specify if the software is required to generate a background grid. Pen colors and fill patterns may also be an option available to the user. Figure 172 is a plot produced by a printer.

15.4.8 Device Setting Report Generator. The software should contain provisions for reporting the device settings and ratings of all devices. These reports may be sent to a printer when desired.

15.5 How to Make Use of Coordination Software. There are various ways that an engineer can make use of existing coordination programs, depending on the type of equipment that is used for computing and plotting, the frequency of program to use, and the money invested in the program. The choices are use of a mainframe computer in-house, use of a personal computer in-house, use of a mainframe computer on time share, and use of a program through a consultant or manufacturer. A description of equipment needs, cost, and advantages or disadvantages follows.

15.5.1 In-House Mainframe Computer. Most large companies have mainframe computers that are used for accounting and business functions. They may permit

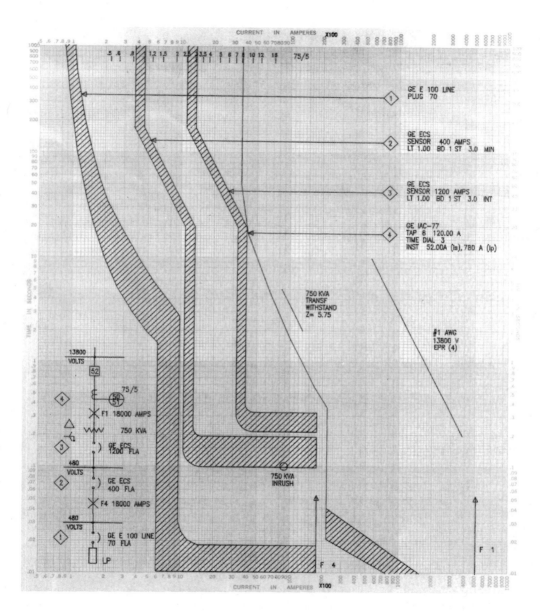

Fig 175
Example of Plot on K&E 48-5258 Form

access to the engineering department for a limited amount of scientific or technical computing. Programs for a mainframe computer are usually more expensive than for a personal computer but usually provide more flexibility since memory in the mainframe is not limited. Once the program is purchased, unless there are maintenance fees, there is no outside expense.

Many engineers in large companies complain of the lack of communication between computer departments and technical departments. Business programs are given preference, and often computer personnel are not familiar with technical programs and do not encourage use of programs. Also, computer departments often assess expensive data processing charges for use of mainframe programs. Since electrical equipment manufacturers are constantly producing devices with new characteristics and methods of setting, the program must be updated to make use of the changing protective device market. If there is good communication between the departments and if there is a large amount of coordination being done, the mainframe version may be acceptable provided, of course, the costs are also acceptable.

15.5.2 Personal Computer. Many engineering departments have purchased personal computers, which are single-user, stand-alone computers that sit on (or under) the desk of the engineer. This type of computer provides engineering departments better control over and flexibility in their computing chores. Time-current coordination programs are available for these computers. Display graphics is an important feature common to this type of computer and, with low-cost graphic printers and desktop plotters, results of the studies may be quickly generated.

Early personal computers suffered from a lack of random access memory and a limited amount of permanent data storage capacity. There have been very rapid technical enhancements in hardware and computer operating systems so that engineers have very powerful personal computers with large memory and high speed readily available.

The personal computer offers all firms, large and small, the opportunity to use advanced software. The personal computer offers engineers the opportunity to examine many design alternatives in no more time than a single manual analysis would have taken.

An additional use of the personal computer is as a storage device for coordination curves. Storing the 11 inch by 17 inch forms often is done haphazardly and, when one is needed, it is hard to find. When the curves are done by the computer, the data can be copied to a disk. This disk can then be stored and, when a copy of the curve is needed, it can be printed again. Some companies are using the computer to "re-plot" old curves and store the information to be available when needed.

15.5.3 Time Share. Time share is a method in which the engineer accesses a remote computer via a communication device. The owner of the remote computer charges for the amount of time the engineer is connected to the computer and a fee for the software being used. The use of coordination programs on time share is preferred by some companies. Instead of an initial investment, they can pay as they use.

When studies are not made on a regular basis, time share is an inexpensive way to make use of a coordination program. The disadvantage is speed of response by the computer. The speed of response is limited by the communication rate. At 1200 baud (120 characters per second), the response can be agonizingly slow; but higher speed modems are available. With any telephone communication, there is the possibility of noise on the line; although this problem is not nearly as prevalent as in past years.

15.5.4 Consulting Service. Some manufacturers will provide coordination studies (usually for a fee) for customers using their equipment. Often, their device libraries are limited, particularly on existing devices of another manufacturer.

Consulting companies who have purchased or written a coordination program are another source for having the studies done. Consulting engineers develop experience with these programs. They can produce the time-current curves and it would be less expensive than for the company to add manpower or learn how to use a time share service to produce the curves. There is a learning curve with any new program and, if the knowledge will not be used on a continuing basis, it is often better to let someone else do it. Also, consultant's experience in performing coordination studies can provide important "know-how" and expertise that may be required, especially when complex power systems and circuit arrangements are involved.

15.6 Equipment Needs. Equipment needs depend on the way the program is accessed. For time share users, a simple terminal and modem are required. The plotter and printer requirements may be provided by the service bureau. For central mainframe systems, the requirements are normally the same. For personal computer users, a local printer and plotter are required.

Before purchasing equipment, the supplier of the software should be contacted for assistance. The required hardware configuration may vary with the software being purchased.

15.7 Conclusion. With the increased popularity of the computer and the availability of it in most engineering facilities, using it has become a way of freeing the design engineer from the tedious task of manually drawn coordination curves, thereby, permitting him or her to design. With the use of simplified input of interactive software, even those engineers who have an inherent fear of computers can become confident. The engineer is still needed to make those judgment decisions and to establish the criteria that should not be left to a computer. The engineer should not be required to struggle with the mundane tasks of manually drawing curves and tabulating results.

15.8 References

[1] IEEE Std 242-1986, IEEE Recommended Practice for Protection and Coordination of Industrial and Commercial Power Systems (ANSI).

[2] BLACKBURN, J. LEWIS "Protective Relaying Principles and Applications," Marcel Dekker, Inc., New York, 1987.

[3] "Industrial Power System Data Book," Section .54, General Electric Company, Schenectady, NY, 1955.

[4] KENNEDY, R. A. and CURTIS, L. E. "Overcurrent Protective Device Coordination by Computer," *IEEE Industry Applications Society Conference Record*, Annual Meeting 1981, p. 464.

[5] LANGHANS, JOHN D. and RONAT, A. ELIZABETH "Protective Devices Coordination via Computer Graphics," *IEEE Transactions on Industry Applications*, pp. 404–412, May/Jun. 1980.

[6] "Selection Guide for Transformer-Primary Fuses in Medium-Voltage Industrial, Commercial, and Institutional Power Systems," *Data Bulletin 240-110*, S&C Electric Company, 1981.

[7] STAGG, G. W. and EL-ABIAD, A. H. "Computer Methods in Power System Analysis," McGraw-Hill Inc., New York, 1968.

Index